IEE HISTORY OF TECHNOLOGY SERIES 31

Series Editors: Dr B. Bowers
Dr C. Hempstead

Electric Railways
1880 – 1990

Other volumes in this series:

Volume 1   **Measuring instruments—tools of knowledge** P. H. Sydenham
Volume 2   **Early radio wave detectors** V. J. Phillips
Volume 3   **A history of electric light and power** B. Bowers
Volume 4   **The history of electric wires and cables** R. M. Black
Volume 5   **An early history of electricity supply** J. D. Poulter
Volume 6   **Technical history of the beginnings of radar** S. S. Swords
Volume 7   **British television—the formative years** R. W. Burns
Volume 8   **Hertz and the Maxwellians** J. G. O'Hara and D. W. Pricha
Volume 9   **Vintage telephones of the world** P. J. Povey and R. A. J. Earl
Volume 10 **The GEC Research Laboratories 1919–84** R. J. Clayton and J. Algar
Volume 11 **Metres to microwaves** E. B. Callick
Volume 12 **A history of the world semiconductor industry** P. R. Morris
Volume 13 **Wireless: the crucial decade, 1924–1934** G. Bussey
Volume 14 **A scientist's war—diary of** Sir Clifford Paterson R. Clayton and J. Algar (Editors)
Volume 15 **Electrical technology in mining: the dawn of a new age** A. V. Jones and R. Tarkenter
Volume 16 **'Curiosity perfectly satisfyed': Faraday's travels in Europe 1813–1815** B. Bowers and L. Symons (Editors)
Volume 17 **Michael Faraday's 'Chemical notes, hints, suggestions and objects of pursuit' of 1822** R. D. Tweney and D. Gooding (Editors)
Volume 18 **Lord Kelvin: his influence on electrical measurements and units** P. Tunbridge
Volume 19 **History of international broadcasting** J. Wood
Volume 20 **The very early history of radio, from Faraday to Marconi** G. Garratt
Volume 21 **Exhibiting electricity** K. G. Beauchamp
Volume 22 **Television: an international history of the formative years** R. W. Burns
Volume 23 **History of international broadcasting, Volume 2** J. Wood
Volume 24 **The life and times of** A. D. Blumlein R. W. Burns
Volume 25 **History of electric light and power, 2nd edition** B. Bowers
Volume 26 **A history of telegraphy** K. G. Beauchamp
Volume 27 **Restoring Baird's image** D. F. McLean
Volume 28 **John Logie Baird, television pioneer** R. W. Burns
Volume 29 **Sir Charles Wheatstone FRS** B. Bowers
Volume 30 **Radio Man: the remarkable rise and fall of C.O. Stanley** M. Frankland

Electric Railways
1880 – 1990

Michael C. Duffy

The Institution of Electrical Engineers

Published by: The Institution of Electrical Engineers, London,
United Kingdom

© 2003: The Institution of Electrical Engineers

This publication is copyright under the Berne Convention and the
Universal Copyright Convention. All rights reserved. Apart from any
fair dealing for the purposes of research or private study, or criticism or
review, as permitted under the Copyright, Designs and Patents Act, 1988,
this publication may be reproduced, stored or transmitted, in any forms or
by any means, only with the prior permission in writing of the publishers,
or in the case of reprographic reproduction in accordance with the terms
of licences issued by the Copyright Licensing Agency. Inquiries concerning
reproduction outside those terms should be sent to the publishers at the
undermentioned address:

The Institution of Electrical Engineers,
Michael Faraday House,
Six Hills Way, Stevenage,
Herts. SG1 2AY, United Kingdom

While the author and the publishers believe that the information and guidance
given in this work are correct, all parties must rely upon their own skill and
judgment when making use of them. Neither the author nor the publishers
assume any liability to anyone for any loss or damage caused by any error or
omission in the work, whether such error or omission is the result of
negligence or any other cause. Any and all such liability is disclaimed.

The moral right of the author to be identified as author of this work has
been asserted by him in accordance with the Copyright, Designs
and Patents Act 1988.

British Library Cataloguing in Publication Data

Duffy, M.C. (Michael C.)
Electric railways 1880 -1990. - (IEE history of technology series; no. 31)
1. Railroads - Electrification 2. Railroads - Electrification - History 3. Railroads -
Great Britain - Electrification 4. Railroads - Great Britain - Electrification -
History 5. Electric railroads 6. Electric railroads - Great Britain
7. Electric railroads - History 8. Electric railroads - Great Britain - History
I. Title II. Institution of Electrical Engineers
621.3'3

ISBN 0 85296 805 1

Typeset by RefineCatch Ltd, Bungay, Suffolk
Printed in England by MPG Books Ltd, Bodmin, Cornwall

This book is dedicated to the memory of
J. GRAEME BRUCE, O.B.E. (1913–2001)
of London Transport. Engineer and historian.

Contents

Foreword		xi
Introduction		xv
Acknowledgments		xix
1	**Telegraphs, track circuits and signals before 1890**	1
	1.1 Introduction	1
	1.2 Evolution of the basic signalling and communications system	3
	1.3 First use of the track circuit	8
2	**Electric railways and American practice**	11
	2.1 The pioneer period	11
	2.2 Electrical engineering and railway systems	16
3	**The electrification of street railways**	23
	3.1 The work of F. J. Sprague	23
	3.2 The extension of the Sprague system	28
4	**Heilmann, Ward Leonard and the electric railway**	35
	4.1 Introduction	35
	4.2 J. J. Heilmann and electric railway traction	35
	4.3 The 'Fusée Electrique' of 1893	38
	4.4 Heilmann and the Ward Leonard system	45
5	**Electrification of British rapid-transit lines**	51
	5.1 Electric railways and the power supply	51
	5.2 Alternating current power stations and transmission	53
	5.3 The electrification of the Mersey Railway	55
	5.4 Economic benefits of electrifying the Mersey Railway	58
	5.5 The prospects for electric traction in Britain after the electrification of the Mersey Railway	61
	5.6 Electrification of the London Underground	64
	5.7 The influence of American electrical engineering	65

viii Contents

6	**Electrification 1900–20**	73
	6.1 The Great Eastern Railway and the Liverpool Street Station experiment	73
	6.2 Electric rapid-transit railways and general railway electrification	77
	6.3 Locomotive working	81
	6.4 LVDC lines and main line traffic	85
	6.5 The electric railway and internal-combustion engined traction	92
7	**Track circuits, describers and electrical signalling 1890–1920**	95
	7.1 Introduction	95
	7.2 Pioneer systems	96
	7.3 Automatic signalling and related innovations	98
	7.4 Main line signalling developments	106
	7.5 Wireless telegraphy in train operations	109
	7.6 Telephony and railway communications	112
8	**Evolution of the electric railway 1920–40**	115
	8.1 Introduction	115
	8.2 The three-phase railway electrification system	115
	8.3 Development of the single-phase traction system	121
	8.4 The emergence of the single-phase archetype	126
	8.5 Converter locomotives	130
	8.6 Motor-converter locomotives	132
	8.7 Phase-splitting locomotives	133
	8.8 The Kando system	137
	8.9 The high-voltage DC railway	139
	8.10 The LVDC railway and the interurban network	144
	8.11 The Presidents' Conference Car	146
9	**Railway electrification and the thermal-electric locomotive**	149
	9.1 The thermal-electric locomotive and general railway electrification	149
	9.2 The 'Electro-Turbo-Loco' of 1909	151
	9.3 The Ramsay-Armstrong Whitworth locomotive of 1922	155
	9.4 Internal-combustion locomotives and the transmission question	157
10	**Converters, the mercury-arc rectifier and supply to electric railways**	169
	10.1 Introduction	169
	10.2 Rectification of power supply	169
	10.3 Development of the industrial mercury-arc rectifier	174
	10.4 The mercury-arc rectifier in railway traction	177
	10.5 British railway rectifiers	180

		Contents	ix

11	**Signalling, communications and control 1920–40**	185
	11.1 Introduction	185
	11.2 Signalling and interlocking	188
	11.3 Speed signalling	193
	11.4 Powered operation and interlocking	195
	11.5 American practice 1920–40	202
	11.6 British railway communications and control 1920–40	203
	11.7 Train describer-recorders	204

12	**Railway electrification 1920–40**	207
	12.1 Introduction	207
	12.2 Electric traction and the efficient use of fuel	207
	12.3 The stagnation in British railway engineering	213
	12.4 Lomonossoff and state ownership of railways	217
	12.5 Railway electrification in Britain 1920–40	219

13	**Electro-Motive, General Motors and oil-electric traction**	225
	13.1 Introduction	225
	13.2 Development of the Electro-Motive (General Motors) diesel locomotive	230
	13.3 General Electric and the 'Steamotive' project	236
	13.4 Conclusion	240

14	**The mercury-arc rectifier locomotive**	245
	14.1 Introduction	245
	14.2 Westinghouse, General Electric and the Pennsylvania Railroad rectifier locomotives	249
	14.3 The advent of the solid-state rectifier	255
	14.4 The mercury rectifier in France and Britain	255
	14.5 The static rectifier and post-war modernisation	256
	14.6 The first British mercury-arc rectifier locomotive	258

15	**Railway electrification in Britain 1920–60**	261
	15.1 Introduction	261
	15.2 The Weir Report	263
	15.3 The acceptance of the HVAC standard in Britain	268
	15.4 The impact of the new standard	275
	15.5 Traction policy 1945–60	277

16	**Signalling, communications and control 1940–70**	283
	16.1 Introduction	283
	16.2 Centralising signal boxes 1945–70	285
	16.3 Automatic train describers and Program Machines	289
	16.4 The automatic railway	293

x *Contents*

	16.5	Automatic warning, control and driving systems on main line railways	297
	16.6	General trends 1940–70	300

17	**Main line direct current traction in Britain**	305
	17.1 The electrical industry and the railways after 1945	305
	17.2 LVDC electric locomotive development after 1945	309
	17.3 The high-voltage direct current system	315

18	**Main line alternating current traction in Britain**	321
	18.1 Advantages and disadvantages of the standard systems	321
	18.2 Gaining AC expertise	322
	18.3 The first production classes: AL1–AL5	325
	18.4 The Class 87 and its significance	336

19	**Solid-state electronics, motor control and the locomotive**	339
	19.1 The advent of solid-state control systems for locomotives	339
	19.2 The Brown-Boveri system	341
	19.3 The Deutsches Bundesbahn Class 120	344
	19.4 Solid-state electronics and locomotive drive systems	347
	19.5 The modular system	351

20	**Solid-state electronics in signalling, communications and control**	355
	20.1 Centralised traffic control and interlocking 1965–85	355
	20.2 The computer in signalling and control before 1985	360
	20.3 Reversible working	363
	20.4 The locomotive-mounted computer	364
	20.5 Radio communications, signalling and control	368
	20.6 Large-scale integration of systems	370
	20.7 The moving block system	377
	20.8 Automatic control of main line trains	381

21	**The electric railway 1965–95**	385
	21.1 The need for increased performance 1965–95	385
	21.2 The origins of very high speed train operations	388
	21.3 The development of the SNCF train à grande vitesse	392
	21.4 The British Rail Advanced Passenger Train project	399
	21.5 Developments subsequent to the TGV	407
	21.6 Current developments in main line electric traction	410

Bibliography	419
Index	441

Foreword

To trace, as historians, the beginnings, early development, growth and setbacks in the applications of electricity to railways on a step by step basis may seem to us philosophising about the past. Of course, this requires many quotations of facts, reflections and analysis of trends for the future and of time consuming and costly setbacks.

From first reading, Dr Duffy's book is a most important work on this subject. Its approach is new because, perhaps for the first time, it recognises and emphasises the interconnection between the many components that are covered within the broad description of electric traction. It focuses on aspects of the subject which require the most careful consideration by all those having specialist knowledge – aspects which often receive little attention from students or even those contributing to, or sharing, engineering enthusiasm for such research. Frequently the various aspects of the subject are considered separately, i.e. signalling, communications and train control, from power supply, distribution and current collection – even if each of these has something, such as automation, in common with the others.

Of course a professor of electric traction, previously involved in research and development in the same subjects in the industrial world, should be well equipped to know the subject that he teaches. But the limited time available for teaching on current techniques and technology and those in the course of development for the future restricts the opportunities to outline the linkage between the past and the present. In addition to old and proven technology, we know of innumerable examples from the past as essential bases for relevant and possible future developments. This underlines the essential desirability that history and philosophy of the subject should be introduced into the education of young and not so young engineers, through a new concept of sufficiently reliable in-depth knowledge. This should be based on comparisons between sources, also between thought processes and philosophies and not merely from the personal views shared by many of the writers on, and exponents of, electric traction as expressed on especially controversial cases.

Such cases of subjects arousing strongly partisan arguments, sometimes of great bitterness, are fairly dealt with in Dr Duffy's book, although these arguments cannot now be considered as being unbiased or disinterested. Thus,

Chapter 15 deals with the difficult evolution of electrification policy in Britain after 1920, whilst Chapters 17 and 18 clearly describe and document the application of low voltage DC and high voltage AC systems in Britain. To quote what Phillip Dawson said on the subject in 1909, "I have in the past, and shall always, fully realise the many and just claims, as regards merit, of the continuous current system . . . I am, at the same time, a thorough believer in the many advantages of the single phase system, which is applicable in cases where the continuous current system is either inapplicable or undesirable." Of course, such general principles remain of value today, as they did almost a century ago.

One important factor running through Dr Duffy's historical chronology is the very important part played in many developments and successes (as well as in some disappointments) encountered by the large electrical manufacturing companies, such as: General Electric, Westinghouse, Siemens, AEG, Brown-Boveri, Oerlikon and their affiliates. The fact is that manufacturers rather than operators or railway companies, were, and still are, often the pioneers and innovators in applying electricity to railway use. An early example of this, before the invention of the dynamo, was in the application of the electric telegraph to signalling and track or trains circuits on railways.

It may be noted that in these early times and until the First World War, all new discoveries in technology, testing and railway applications were well known and described in detail in many languages and technical publications in Europe and America. Later, as seems evident from the bibliography to Dr Duffy's book, the largest number of references were in the English language, concentrating on papers and technical books covering British and North American technology and to a lesser extent on published material in continental Europe (the latter mainly in France and Germany). These USA and UK developments are generally less well known or regarded by continental European engineers, apart of course from the outside influence of diesel-electric 'classical' models, still to be observed today throughout Europe. This book also stresses the major influence of USA and UK signalling, communications and control technology incorporating solid state electronics, which are not only, as is sometimes thought, the prerogative of the Japanese.

The trend towards more centralised and partially automated supervisory operation is also discussed. In Europe the latest combinations of function achieved in control centres bringing together all operational aspects are dealt with. One recent example being the railway control centre for the Channel Tunnel at Folkestone, which regulates all operations over 37 miles of double track over which pass up to 420 trains per day, with running speeds ranging from 62 to 100 miles per hour, with a headway of three minutes between trains. In Chapter 20 Dr Duffy postulates the probable integrated control centres of the future, which will be capable of regulating operation over a complete railway network. This is surely an indication of things to come.

Being an optimist, I am sure that there still remains much scope for new development in the field of railways (quite apart from the basic superiority of the electric railway) on two major issues: environment and energy usage. In the first case, the unobtrusive quietness, existing or economically obtainable with electric traction and, in the second case, its versatility, reliability and relatively low cost. The essential 'ethical, philosophical, social and economic problems' cannot be solved in guided land transport 'using obsolete thought-systems carried over from the past' (I quote the best authors!). But, as previously mentioned, and as stressed repeatedly in Dr Duffy's book, a broadly based combined systems approach is the only one historically leading to an acceptable and effective future for the electric railway.

Y. Machefert – Tassin
C. Eng., F. I. Mech. E.

(Professeur de la Traction Electrique, ESTP (Public Works Technical University), Paris.

Introduction

Heavy-duty electrical engineering developed between 1880 and 1900. Many modern engineering systems and methods of organisation and management grew with it. Heavy-duty electrical systems were built to generate, transmit and utilise power on the industrial scale. Methods were devised for analysing electrical networks, and equipment was introduced for measuring electrical power accurately. Electrical engineers set new standards for monitoring industrial systems, and they improved mechanical as well as electrical engineering. This pioneer era of Gramme, Edison, Westinghouse, Siemens, Hopkinson, Tesla and others saw the growth of great industrial combines – like General Electric and AEG. Electrical engineering was dominated by electric lighting and electric traction on heavy-duty rapid-transit railways. Electric traction was used in rapid-transit service where steam traction could not cope, and its transfer from light-duty street-car tramways to heavy-duty railways was a crucial step in the growth of large-scale electrical industrial systems.

In the 1920s the United States led the world in the construction of long-distance transmission lines, usually supplied from hydroelectric stations. During this period, the thermal efficiency of a network fed from coal-fired stations was low. This restricted electrification until new power stations were built, using high-pressure, high-capacity boilers, turbine-alternators and vacuum condensers, which greatly increased economy. Innovators prepared for the electrification of town and country, and demonstrated electric prototypes for commercial, industrial and domestic use, to replace mechanical engines. The availability of high-capacity arc-rectifiers and inverters after 1930, and a marked improvement in the thermal efficiency of the coal-fired power plant, liberated electrical supply from dependence on hydroelectric power, and enabled the grid to extend. In the United States in the 1930s, and in Europe in the 1950s, most small towns and villages and many farmsteads had access to the grid supply. After 1960 there was a steady transformation of all components of electrified systems by solid-state electronics.

Electric railway traction was important during the first stage of general electrification. In the 1890s, electric railways represented electrical engineering in an advanced state. Rapid-transit systems were electrified because steam traction could not cope with rising demands. The electrification of general-purpose railways was a separate issue, demanded for operational reasons, and was often reliant on state support to overcome the economic objections to main line electrification. Electrification of main lines was restricted to limited sections, usually supplied with power from hydrostations, and before the mid-1920s engineering and operational considerations determined whether or not a railway was electrified. After that date, political and economic issues played a deciding role. The main agent of progress was the electric rapid-transit railway, which is still a major source of innovations for transforming railway engineering. The techniques which transformed the general-purpose main line system into the new railway of the present era, with performance far in excess of any steam railway or earlier electric system, originated in rapid-transit railways. On electric rapid-transit railways between 1890 and 1910, the traction, signalling and control elements of the railway system were integrated with a higher degree of unity than on the steam railways. In 1890, the steam railway provided the industrial world with a fine example of a large integrated mechanical system, particularly in the signalling and switching mechanisms. In the pioneer electric railways – City & South London Railway; Great Northern & City Railway; Post Office Railway – the degree of integration, and automatic operation, was greater than on any steam-worked line. Today, the automatic electric railway, in which the traction, communications and control components function as one, is a reality – thus actualising the vision of far-sighted signalling engineers in the 1920s.

The major theme of this book is the progressive integration of the traction, signalling, control and communication elements of the railway into one system. Charting the 'integration of the machine-ensemble' has given structure to the work, which is not an exhaustive catalogue of every engineering development for all the main kinds of electric railway. Tassin, Nouvion and Woimant have done this in their 'L'Histoire de la Traction Electrique', a work this author greatly respects. The objective of the present book is to tell the story of electrification and the union of the traction, signalling, power supply, communication and control systems. Much simplification was necessary and was effected by identifying those innovations (chiefly in traction and signalling) which in each era set the pace of engineering development. No attempt was made to record all variants of a particular system. Once an important development was illustrated by one case study, equivalent ones were left to footnotes. Much detail, for example relating to motor control, solid-state rectification, electric point motor design or motor generators, was omitted to keep the length of the book within the limit set by the publisher.

The author hopes he has succeeded in identifying the main course by which electric railways evolved. The Heilmann thermal-electric locomotive is identified as being important in the history of electric and thermal-electric traction, and the diesel-electric locomotive is presented as an agent in general electrification. American engineers always regarded the diesel as a means for achieving 'cheap electrification', and without its availability on non-electrified lines, general electrification would have been restricted, and the total transformation of the railways could not have been effected. The inclusion of the diesel-electric locomotive in the history of electric traction is essential.

The material is presented to show the drawing together of the major components into closely integrated unity. The narrative passes backwards and forwards between traction, signalling and power supply, with each topic given its own chapter or subsection. Key systems (such as R Dell's equipment for the automatic control of London Transport trains) and key projects (such as the Heilmann locomotives, or British Railways Advanced Passenger train) are described in relative detail. Otherwise, priority is given to general matters. Concentration is on examples representative of best practice in each period, and – regrettably – it has not been possible to include studies of experimental and unusual systems which failed. For reasons of space, the author has been compelled to leave out studies of electric rack railways, mine railways and – most serious omission of all – street tramways and light rail.

The history of electric traction begins with a chapter on signalling and control before 1890. Though electric locomotion began on the street tramways and mining railways of the USA and Germany, the electrical element of the signalling, control and communications system dates from the early years of the steam railway. In this period, 1840–90, electricity was an auxiliary to a basically mechanical system. It was essential for the telegraph, without which efficient railway operations would scarcely have been possible. With the advent of electric signalling, track circuiting and telephony, and with other developments following after 1900, the signalling component of the 'machine-ensemble' became as important as the traction system and the power supply. Unfortunately, many accounts of electric traction concentrate on the rolling stock, and say little about signalling and the power supply. This book argues that the present-day electric railway cannot be understood without affording each of these equal status. Within the limitations of space, the author has described the coming together of these diverse components from their 19th-century origins to their present-day unification into an integrated whole.

Acknowledgments

The author must begin by acknowledging the influence of his friend and colleague O. M. Sanne, engineer on Norwegian State Railways, and the Oslo Metro system, whose enthusiasm for electric traction and its history aroused a similar interest in the author. The visits to railway and tramway workshops and installations in the Oslo district did much to cultivate the author's deepening interest in electric traction and led to this book being written.

The author thanks the Institution of Electrical Engineers for the opportunity to publish this volume, in particular those who provided advice and help from first draught, through revisions to the final text: B. Bowers, C. Hempstead, D. Levy and W. Hiles. He is grateful for the information provided by colleagues through the IEE History of Electrical Engineering Group (S7) and its lectures and meetings: B. Bowers, J. Bridge, O. B. Charlton, C. Hempstead, K. Thrower and K. A. Yeomans. He is grateful to K. A. Yeomans for the information supplied concerning Ward Leonard. Valuable help was received by colleagues working within London Transport, through meetings, papers, visits and conversations, often in conjunction with the Locomotive & Carriage Institution. He is indebted to N. Agnew, and to the late J. Graeme Bruce OBE, for insights gained into the workings of electric rapid transit railways. The London Transport Museum (Covent Garden) and the National Railway Museum (York) proved vital sources of information and illustrations. H. Robertson of London Transport Museum Photographic Library was very helpful.

The University of Sunderland provided support for the research activities vital to the writing of this book, and sponsored the series of symposia "Evolution of Modern Traction" where much was learned which found a place in the text. The author is grateful to the former history of science and technology department at University of Manchester Institute of Science & Technology, where he studied the principles of historical analysis and researched the history of engineering. His deepest gratitude is expressed to the late Prof. Cardwell, and to Mr J. O. Marsh, of UMIST. The author is likewise grateful to the historians of engineering he has met through the agency of the International Committee for the History of Technology (ICOHTEC) and the Newcomen Society of London, which have allowed him to put his ideas about the history of railway engineering before a critical audience. The constructive criticism received from the members of these societies has proved invaluable. He is indebted to G. W. Carpenter for information on all aspects of traction: steam, diesel and electric. Many members of the Newcomen Society have made essential direct and indirect contributions to the author's understanding of the history of engineering which he is glad to acknowledge: M. R. Bailey, A. Buchanan, C. Billows, E. F. Clark, C. Ellam, A. Smith, D. Smith and N. Smith. Similarly his knowledge has increased through contact with railwaymen who are active in many railway-related professional bodies including the Locomotive & Carriage Institution (N. Agnew, J. Lunn), the Permanent Way Institution, and the Institution of Railway Signal Engineers (E. J. Eden, R. J. Post and A. C. Howker). Many railway-related societies have provided information, advice and assistance: the Electric Railway Society; London Underground Railway Society; Signalling Record Society; Southern Electric Group; and Stephenson Locomotive Society.

Many individuals in addition to those acknowledged above have earned the author's gratitude. He thanks E. J. Eden for data on signalling and interlocking; A. J. Heywood for information concerning G. V. Lomonossoff and developments in Russia; J. Sandiford for comments on economics and post-war traction; A. J. Sprought for his descriptions of the Post Office Railway and other technologies; T. H. Tayler for his papers about the Southern Railway and Southern Region LVDC locomotives; W. Turner for his historical work on railway power supply; J. Talbot for his accounts of the evolution of railway signalling; R. L. Vickers for data on the early diesel locomotive, Durtnall, and low voltage direct current railways; R. Whittaker for notes on Wellington and G. Woodward for information concerning the Liverpool Overhead Railway.

The author further acknowledges his debt to the work "Histoire de la Traction Electrique" by Y. M. Machefert-Tassin, F. Nouvion and J. Woimant which has become the standard work on the subject, and to the authoritative studies of signalling and its history by G. M. Kichenside and A. Williams. The works of the late J. G. Bruce have been a reliable guide in all matters relating to traction on London Underground, and in private correspondence Mr Bruce offered additional, detailed information. The seminars and papers of the Institute of Railway Studies, York, shaped the author's general ideas concerning railway engineering, research and development and he thanks Prof. C. Divall for the deeper insights which have come through the activities of this institute.

Thanks are due to the Newcomen Society for permission to base sections of the book related to 19th century power supply, and the development of asynchronous traction motors on papers first read by the author at the London meetings of the society, and published in its Transactions. Thanks are also due to the Institution of Electrical Engineers for allowing sections of the book to use material first presented by the author at the annual History of Electrical Engineering summer meetings.

The photographs used throughout this book were kindly provided by the information officers, publicity executives, and librarians of the following establishments: the author is grateful for the contribution they have made towards the successful completing of this work. Adtranz; Amtrak (USA); British Rail; Brown – Boveri; Burlington Northern & Sante Fe RR; Daimler-Benz; Deutsches Bundesbahn; Electro-Motive Division of General Motors; GEC-Alsthom (now Alstom); General Electric (USA); Krauss-Maffei; London Transport; National Railway Museum (York); Norfolk Southern RR; Oerlikon; Railtrack; SNCF (French National Railways); Union Pacific-Missouri Pacific RR.; Westinghouse (USA). Photographs were also kindly provided by S. Dovgvillo, A. J. Heywood and R. L. Vickers. Valuable help in selecting the pictures, and in checking the original typescript was provided by the author's wife, Mrs C. A. Duffy. Finally, the author expresses his gratitude to Y. M. Machefert-Tassin for writing a generous Foreword.

M. C. Duffy

Chapter 1
Telegraphs, track circuits and signals before 1890

1.1 Introduction

In the 1840s, main line railways required methods of monitoring train positions, and signalling movements, in which information was exchanged between control points faster than the trains moved. By 1850, train speeds were high enough for stopping distances to exceed sighting distances in some locations and weather conditions. In the 1850s, busy terminal stations, carriage sidings, junctions and crossings needed a method for controlling the state of the system before it changed too much. The deciding factors were train speed; frequency of services; intersection of train paths by other trains; and the limited ability to control an area from one point. It became necessary to anticipate changes in the system by monitoring activities in distant sectors. Much of the history of signalling, communications and control concerns apparatus for monitoring and controlling larger sectors, accurately and reliably, from fewer centres by fewer personnel. It is also the history of integrating the traction system (the machine-ensemble of rolling stock and permanent way) with the apparatus of control. A high degree of integration was found on steam railways in the 19th century using both electrical and mechanical equipment, but the most important feature of the system was the calibre of the personnel who worked it. In the 19th century, nations like Britain, Germany, France and the USA drew on the skills of a plentiful workforce which was low paid and not effectively organised to demand better conditions until well into the 20th century. Despite labour unrest, there were enough diligent, intelligent, dutiful, sober and conscientious workers to fill the vital posts of train crew, shed staff, station staff, signalmen, shuntsmen and permanent way foremen. Without them, the vast apparatus of the combined traction and signalling system, which depended on individuals remaining alert and performing hundreds of mechanical actions according to strict rules, could never have worked. As it was, it worked well until rising labour costs, and the loss of high-calibre workers to better-paid jobs more worthy of their skills, encouraged the introduction of automatic and semi-automatic devices. Powered signalling and centralised control, with signals worked from boxes 50 miles distant, are examples of engineering innovations in 1970s Britain that owed much to the social changes of the 1960s. In the 19th century, the designers of mechanical and electrical signals, switches or telegraphs could assume that there were well-trained, obedient, dutiful men willing to work them for little pay.

This book concentrates on electric traction, and the stress is on signalling developed for electric rather than steam railways, but there was considerable exchange of practice between the two. Many electrical techniques applied to track circuits, signalling and communications originated in the 19th-century steam railway, and techniques were suggested – such as automatic signalling – that were not successfully applied until the 20th century. The electric railways of the 1890s were a better environment in which to perfect these techniques than the steam railways, and the degree of integration of traction system and signalling network achieved in electric railways in later years could never have been approached with steam traction. Electric traction offers repeatedly uniform performance, and automatic operation, which greatly assists integration with automatic or semi-automatic signalling and control systems. The ability of modern diesel traction to be likewise integrated owes much to the electronic control systems, and the electric transmissions now standard on most units above moderate power.

On main line railways there have been two general philosophies for controlling trains, with considerable diversity within the practice of each. The one commonly used in Britain, and found abroad on densely used routes with junctions and complex intersections, relies on lineside signals, controlled from a box, to pass information to train crew. Each box serves as a centre for gathering information about adjacent sectors and other parts of the system, and for passing this on to other boxes or trains. This system was developed with the use of the electric telegraph, the telephone and the radio. The alternative was best suited to sections of railway where trains were infrequent, intersections were few, and uninterrupted lengths of route were common. It used train orders, despatched to the crew from an information centre, often in the station. Orders could be handed to crew in the form of written messages, and for effective operation despatching needed the telegraph. This system was widely used in the USA following the adoption of the Morse telegraph, and was later upgraded by telephony, wireless and radio communications. In the USA, Minot on the Erie Railroad is credited with the first use in 1851 of the telegraph and the train order (Armstrong, 1978), and 'T & T O' became standard on a national network where, to this day, half the route miles have no signals. It enjoys wide use still, though busy routes near junctions, stations, terminals, crossings and goods yards, and heavily used sections of line, were usually controlled by lineside signals. Lineside signals increase line capacity and efficiency of working, but they are expensive and need not be installed on many US lines, where 'T & T O', supported by radio communications, is adequate. For most British routes, traffic was so dense, and operations so complex, that signalling was essential.

In the days before telegraphy was accepted by railways, approximately down to the 1850s, operations were governed by the time interval system, with trains sent out from stations over a sector of route, guided by the working timetable. Following trains were sent out after a time interval thought safe. Dangers arose with trains running over single track, relying on loops at stations for passing – a common state of affairs in the Americas and in colonies where double track was rare. If an expected train failed to arrive at a station on schedule, a period was allowed for it to show up. If it did not, a train held in the station, and due to run in the opposite direction, was then permitted to leave, proceeding down the single track. The danger is obvious, despite slow speeds and the repeated blowing of the whistle to warn any oncoming, late-running train. Whistles and headlights were inadequate warning in fog, rain, snow and darkness, especially on curving track, and head-on collisions did occur – but how else could a railway operate in an era when the trains themselves were the fastest information carriers about the system which they determined? If the system was to continue working when a train failed to arrive at a passing loop, the risk had to be taken of sending a train down a single track to the next station. Before the mid-1850s, night operations were rare, and pilot engines were sent ahead of passenger trains when ignorance governed movements, but accidents happened and the public became alarmed. Railway operation in the USA in the period 1830 to 1860 affords examples of working a system in which the fastest signal speed was no greater than the speed that determined the system change. In this period, trains grew heavier and faster, and the stopping distance began to outrun the sighting distance from the engine cab. There was a marked increase in major railway disasters in the 1850s, and advances in heavy-duty railway transport required a system for monitoring the movement of trains using signals that travelled far faster than the fastest train, and which transmitted messages much further than did lamps or whistles. The electric telegraph met these requirements.

During the development of signalling and communications, innovations were suggested decades before reliable, economic apparatus based on them was available; and within a complex system like the railway, modern equipment could be found working close by apparatus from much earlier times. It was difficult to fit the new into the old without causing disruption of the total system and demanding too-expensive changes. The most extensive, modern signalling and communications systems were found on new lines because they were not trammelled by established practice. Modern electric signalling, communications and control were introduced, perfected and integrated into an evolving whole in new electric railways or on steam railways that were electrified.

1.2 Evolution of the basic signalling and communications system

An outline history of signalling is provided by Kichenside and Williams (1980, 1998), Nock (1962, 1969) and Westinghouse (1956). The earliest signals were lamps, flags and hand signals, derived from military and civilian semaphore, which gave the railways the structure of the fixed signals themselves. Signalling by lamp, flag and hand is still used in emergencies. Originally, signalling was the lot of the railway policeman, who combined the duties of patrolman, signalman and shuntsman, and who lived close by his work, sometimes in a cottage beside the tunnel portal he guarded. In the 1840s, the use of fixed signals in the form of arms, boards, balls and lights on posts became general and the railway police worked the levers for signals and points, which were not centralised, but were located next to the individual signal or switch. Signals were of a wide variety, with no standardisation between companies, and groupings of small companies into larger ones resulted in many designs being found within one railway. A very early use of the semaphore in railway service was a three-position lower quadrant signal on the London & Croydon line in 1841, of the type employed to telegraph messages overland and between shore and ship. This type of signal, still in use on secondary railways, was adopted by other companies and became the most common form on the steam railway (Wilson, 1900; Dutton, 1928 and Lewis, 1932).

The three-position signal became standard in the USA, but though used in Britain in the 19th century and early 20th century, it gave way to the two-position signal once the distinction between 'home', 'distant', 'starter' and 'advanced starter' signals was established. In 1876, the caution signal position of 45 degrees to horizontal, with vertical signalling clear, was abandoned, and two-position signals became the norm in Britain. By the mid-1850s, the levers working signals and points controlling stations and junctions were grouped into centralised frames or control centres, sometimes provided with shelters, but shunting yards with points worked by non-centralised hand levers survived to the end of steam traction (Westinghouse, 1956; Signalling Study Group, 1986).

In the 1850s, there was fruitful experimentation with locking bars and levers taken from locksmiths' technology, to prevent the contradictory setting of signals and switches. In 1846, partial interlocking of signals and switches was carried out by Charles Gregory at Bricklayers Arms, and in 1856 John Saxby – one of the greatest pioneers of signalling – introduced a reliable method for fully interlocking signals and points. In 1860, levers in a centralised frame controlled interlocked points and signals at Kentish Town. These innovations heralded a closer integration of the traction system (machine-ensemble) with the signalling and control system. Interlocking machinery was large and expensive. Each frame was designed for a particular location, and required a protective shelter of wood or brick, underneath the signal lever platform, which was sometimes exposed. This led to the signal box in a form that is still seen on secondary lines. In the 1850s, the value of centralised levers; interlocked points and signals; and communication between signal boxes, stations and control centres by electric telegraph was recognised. Costs, disputes over which equipment to use and arguments over patents and royalties delayed implementation, even on first-class railways. In 1850, the time interval system was usual, and trains were sent out from stations with allocated periods between them. The electric telegraph was understood, but was thought too expensive and demanding of operator skill to be adopted universally. The Cooke & Wheatstone telegraph, perhaps the best known on British railways, was based on a patent of 1837, tried in 1838 between

Paddington and West Drayton, and later extended to Slough on the Great Western Railway. This Great Western telegraph, installed by I. K. Brunel from April 1838 onwards, used wires laid in an underground tube. It began work in July 1839, but suffered damage, and was replaced by Cooke in 1842, and extended to Slough using wires resting on pole-supported insulators, the most common form of telegraph line. This proved an economic form, especially over company-owned rights of way. Two-wire operation with a two-needle telegraph needed skilled operators, but these were available from an expanding pool of skilled, diligent artisans. After a slow start, the system became successful and railway telegraphs became the largest private communications networks at home and abroad, offering a public service until 1870. As early as 1837, a five-needle telegraph was installed in Euston Station for signalling to the Camden engine sheds and winding engines, which pulled trains up the steep gradient out of the terminus, but this was replaced by a pneumatic pipe and whistle system. The simple electric telegraph was used on the four-mile-long rope-worked Blackwall Railway in 1840 to signal to the engine house from stations along the route, and simple block instruments were used in the 1840s to control traffic in tunnels. Electric telegraphs were used by several industries and government departments, and innovations made in one sector found ready use elsewhere. In 1847, Bain's perforated tape telegraph was introduced and the many later improvements aimed at speeding transmission, such as printing telegraphs, were used in railway signalling, control and administration.

The electric telegraph was recognised as having application outside the railways. Its military, social, political and strategic importance to central government were obvious, when used in conjunction with railways to control troop movements and gather information. Cooke was an energetic projector of electric telegraph lines and proposed one for the Admirality, which operated a system of mechanical telegraphs from London to points on the coast. These have given the name to many hills – such as Telegraph Hill in Guildford. Cooke persuaded the South Western Railway to participate and allow the telegraph to run beside 88 miles of their route. The London to Dover telegraph of the South Eastern Railway was another important link in a developing system which the government and civilians could use. In 1871, the 1868 and 1869 telegraph acts came into effect and the system was nationalised. Until then, the public telegraph service was provided by the railways, which had 11,000 miles of line by 1851. A great pioneer was C. E. P. della Diana Spagnoletti, a pupil of Alexander Bain who was appointed telegraph superintendent of the Great Western Railway (GWR) in 1855, and who served as telegraph superintendent on the Metropolitan Railway when it opened in 1863. He patented the disk block telegraph, which was used throughout the GWR, and his many inventions included a portable apparatus transportable by train which could be used to establish communications at any point on the route. The several kinds of electric telegraph, the signal box with its interlocking frame, and the block system using lineside signals formed a system that controlled the steam railway for most of its existence. It absorbed the improved telegraphs; telephony; radio communication; and automatic, powered and colour-light signalling. Its history is one of steady improvement

6 Electric railways 1880–1990

Figure 1.1 Amersham Signal Cabin (Metropolitan Line, London Transport) photographed in 1960 with Spagnoletti Lock & Block apparatus arranged above levers for working signals and points.
Source: Colin Tait/London Transport Museum.

and gradual transformation, down to the end of steam traction, and before the widespread use of electronics effected a radical and extensive revolution (H. R. Wilson, 1900, 1908; RGI, 1984).

The telegraph was used in the 1840s to signal through long tunnels between policemen at each portal. The basic methods for communicating between signal boxes and control centres existed in the 1860s, being the simple electric telegraph which sent 'state of the line' messages in simple code by bell or needle displacement, and the 'speaking telegraph' which exchanged messages. Many components of practical installations had to be greatly improved, and seemingly simple, reliable devices failed in service and attracted much attention from innovators. Late 19th-century texts (von Urbanitzky, 1896; Wormell and Walmsley, 1896) devote much space to bells and audible signals, suggesting that it was difficult to get a reliable bell loud enough to be heard in signal boxes. The telegraph signal could not power the bells or whistles, which were driven by strong clockwork or compressed air, and it simply controlled the audible signal. Systems were tried on trains for communicating between engine, guard and passengers. Some were entirely electric, supplied from cells, and others used electric signals to control air, steam or clockwork alarms.

The block system of signalling a route was used in 1851 on the South Eastern Railway between London and Dover, but the expense of installing electrical devices and training staff delayed its general use. By 1870, semaphore signals were normal, usually as the slotted post type, working in the lower quadrant. There were variations, like the Great Northern Railway 'somersault' signal, which was pivoted to danger, however weighted with ice and snow, if the control linkage failed. In the 1870s, the Board of Trade Railway Inspection Department could only recommend action, not enforce it on companies, and many resisted advice – particularly in the matter of continuous brakes. After important tests of types of brake at Newark in 1875, the Board of Trade recommended that standard working include interlocking of points and signals; the block system of working; and automatic continuous brakes which would halt all sections of a parted train if couplings failed. The Board of Trade got compulsory powers, supported by public outcry, when accidents occurred because its recommendations were not heeded. They were mandatory after 1889 – a considerable period after their value was proven. In 1876, green became the colour light for 'clear' rather than white. Engineers long anticipated the features of the late 19th-century mechanical signalling system which was to serve the steam railway down to the 1970s, when widespread application of electronics and electromechanical devices removed the surviving components from main lines. They survive on secondary lines to this day.

The late 19th-century system incorporated the 'lock and block' principle of using the passage of trains to limit the action of signalmen, or swing bridge controllers. One early system was the Sykes Lock and Block, first installed in 1875 between Shepherd's Lane and Canterbury Road Junction. This system, and others invented by Hodgson, Spagnoletti and Tyer spread rapidly. The first systems used treadles, held firm by the train wheel flanges to prevent the points

8   *Electric railways 1880–1990*

being changed with a train passing over them – a further integration of the machine-ensemble with the control apparatus. Lock and block was developed using mechanical and electromechanical locking to interlock section signals with single-line working staffs, and to interlock section signal setting with condition of the block instrument reading 'Line clear' or 'Train on line'. Following a series of catastrophes in the USA in the 1850s and 1860s, when trains plunged into rivers from open drawbridges, interlocking systems were installed in such locations. By the 1880s, despite the uneven rate with which safety techniques spread through the hundred-odd private companies, the British railway was perhaps the safest in the world. Automatic control of signals by trains, and the first successful use of powered signals and points, were made from the 1880s onwards, such as the automatic semaphore signalling used on the Liverpool Overhead Railway in 1893 (H. R. Wilson, 1900, 1908; S. T. Dutton, 1928).

First attempts were made much earlier. Light signals were used in the 1840s, as in the 'lighthouse' on the Croydon and Greenwich lines, and coloured glasses were fitted to semaphores for night signalling, though colour codes were not standardised until 1876. Automatic signalling by mechanical contact was tried in the 1850s, and tests with track circuits were made in the 1860s. Electric communication cords were tried in the 1860s and 1870s. In the USA, on 21 May 1877, telephone communications were attempted in railway service at Altoona, Pennsylvania. Before 1890, there were efforts to improve train working in fog by mounting an electrically controlled steam whistle or bell on locomotives, interconnected with the signals through a ground contact. These were proposed as part of an automatic signalling system, making use of relays, by which the trains worked the signals themselves – such as Putnam's system for line signalling. Preece, the distinguished British Post Office engineer, and Walker invented electrical signalling mechanisms of all kinds, but most of the systems were not then reliable. Their promise was never lost to view, and the experiments with electrical signalling cultivated an expertise, and an interest on the part of established makers of signals and telegraphs, which proved valuable in the 1890s when such methods of working became desirable and then essential on the first electric rapid-transit railways.

1.3 First use of the track circuit

The railway track circuit originated in the 19th century. An American report of 1910, quoted by Ruffell (1996) assesses the importance of the track circuit, which in the USA was already controlling automatic signalling over some nine thousand miles of railway.

Perhaps no single invention in the history of the development of railway transportation has contributed more towards safety and dispatch in that field than the track circuit. By this invention, simple in itself, the foundation was obtained for the development of practically every one of the intricate systems of railway block signalling in use today wherein the train is under all conditions continuously active in maintaining its own protection.

The successful use of the track circuit began that ceaseless, increased integration of the traction system with the signalling, control and communications systems, which today is found in the transmission-based signalling and communications networks used to control the German ICE trains. The idea of using track circuits to monitor train position, and to work signals, is as old as the Wheatstone telegraph and Morse apparatus, though the patents of the 1840s were not translated into practical devices and were seldom put to serious trial. Ruffell gives 1848 as the date of one of the first English patents for using the running rails as conductors for electric currents intended to work signals, and he mentions a patent of 1853, granted to Dugmore and Milward, for communicating between trains via the rails, using electrical signals. More serious work was done in the 1860s. Bull's patent of 1860 showed how insulated sections of rail could be used to indicate the progress of a train, and during the 1860s, W. R. Sykes carried out experiments with track circuits on the London, Chatham & Dover Railway, and at Crystal Palace. Considerable British initiative was shown in the 1860s, but it was American engineers who developed practical systems and set up industries for making and marketing them. Dr W. Robinson developed automatic railway signalling in the late 1860s, and in 1870 he installed a test section at Kinjua, Pennsylvania, on the Philadelphia and Erie Railroad. Dr Robinson's contribution was the invention of the closed rail track circuit in place of the open rail track circuit used in the tests of the 1860s.

In 1872, the closed rail track circuit worked automatic signalling at Kinjua. Robinson introduced rail bonding, and experimented with material for insulating rail joints. These were tried at Kinjua, and at Irvineton, also in Pennsylvania. In 1877, Robinson is credited with installing the first automatic signals, controlled by more than one track circuit, in the Tehauntepec tunnel in California, using relays derived from telegraph practice. In 1878, he founded the Union Electric Signal Company of Boston, Massachusetts, serving as its president. George Westinghouse bought the company in 1880 and combined it with his own interests to form the Union Switch & Signal Company, which played a vital role in establishing automatic signalling and powered operations throughout the world's railways, starting with the USA. An engineer of the Union Switch & Signal Co., J. B. Struble, invented the AC track circuit (in place of the original DC system) for use on DC railways, where DC return current usually passed through the running rails.

According to Ruffell, work in Britain proceeded more slowly than in the USA. In 1886, W. R. Sykes installed track circuits to protect platform routes at St Pauls Station, London, on the South Eastern & Chatham Railway. The Great Northern Railway experimented with a track circuited section in 1893, and in 1894 used circuits to protect the routes through Gas Works Tunnel, just outside Kings Cross Station, London. In 1900, the Lancashire & Yorkshire Railway installed them at Victoria and Exchange stations, Manchester. These pioneer British systems used track circuits to monitor train position, and helped signalmen working from manual boxes, but the circuits did not control signals automatically. The first use of circuits to work automatic signals on main lines in Britain was in

1901, on the London & South Western Railway at Grately, near Andover, using circuits one mile in length between signals. In 1907, the Great Western Railway installed automatic signals worked by track circuits on a 2.75-mile (4.4-km) length of main line between Pangbourne and Goring. At this time, 1907, there were some 8500 miles (13,600 km) of track in the USA which were automatically signalled, with track circuits. Ruffell (1996) implies that the early British track circuits did not perform reliably, and at a time when a large pool of diligent, disciplined workmen were available to work manual signalling, there was no compelling need to embrace automatic signalling as the Americans did. Complex junctions were beyond the capacity of early automatic systems, and complete automation was limited to simpler track-work, leaving several manual boxes to control major junctions, extensive sidings and termini. In the USA there were longer sections of route, where track layout was simple, and where the early automatic signalling systems were adequate and economically justified.

Chapter 2
Electric railways and American practice

2.1 The pioneer period

The period between 1790 and 1840 witnessed the experiments of Volta, Galvani, Oersted, Ampere, Faraday, Barlow, Jacobi, Henry, Elias, Froment, Davenport, Page, and others. These pioneers produced primitive, low-powered motors, of uncertain reliability, which used batteries to produce motion, by rotation of a shaft, the oscillation of a lever, or the reciprocation of a rod (Cardwell, 1992; Gee, 1991). Major steps towards a successful motor were Barlow's spur wheel motor (1826), Henry's oscillating or rocking motor (1831), and the motors of Jacobi (1834), Elias (1842), and Froment (1845) (Dunsheath, 1962; von Urbanitzky, 1896). Important research into magneto-electric generators and motors was carried out by Pixii (1832), Wheatstone and Cook (1845), William Scoresby and James Prescott Joule (1846). By the 1840s, powered locomotion had been widely demonstrated with models, boats and wheeled vehicles. The idea of electrical motive power gained in credibility when improved engines were produced after 1840.

In the mid-1840s, electromagnetic power was discussed in papers such as 'Experiments and observations of the mechanical powers of electro-magnetism, steam and horses' by Scoresby and Joule in 1846 (Stamp and Stamp, 1975), and in 1847 a model electromagnetic traction system was demonstrated by Moses Farmer. Robert Davidson built a full-sized battery-powered locomotive between 1838 and 1842 and ran it on standard gauge track (Anderson, 1975). A similar trial with a battery locomotive on standard gauge track was carried out in the USA in 1851 by Grafton Page on the Baltimore & Ohio Railroad (Post, 1972). They aroused interest in a new mode of traction, publicised pioneer work, and identified many of the problems which needed solving if the new mode was to be a technical and a commercial success. Experience showed electricity to be unsuitable for traction in the 1840s. Batteries were inefficient, vehicles were heavy and of limited range, and the only available motors were of the unsatisfactory magneto-electric and electro-magnetic type.

In 1851, an employee of Page called Hall proposed sending power to the car through the rails, and he built a demonstration model. He used Grove batteries for the stationary power source, and the moving vehicle was driven by a simple motor of limited potential, but an important principle was established: it was not necessary to carry the power source with the locomotive. If a means of transmitting power without great loss from a source to a locomotive car could be found, and a reliable motor developed, electric traction on a commercial scale would be possible. In the event, this was delayed for 30 years by the imperfect nature of essential components, such as cells, generators, motors, control equipment and driving engines. Whipple (1889) wrote:

Hall, a scientific instrument maker, built a small model of an electric locomotive, and successfully ran it, taking the electricity from the rails. This is to be especially noted since it was a most important step in the right direction. The cells were eliminated from the car and maintained at the station where no harm could possibly come to them. The most important commercial factor to note is the ultimate development of our ability to produce economically and under complete control, large quantities of electricity. This accomplished, we had only to avail ourselves of Thomas Hall's idea – that of transmitting electrical energy to the moving car – and our commercial railway became an immediate success. We should not, however, forget the most vital and fundamental discovery concerning the reversibility of the Gramme dynamo.

In the 1870s, Gramme and other engineers brought generators, motors, transmissions, control systems and transformers to the stage when they formed a working system worthy of commercial support (Guillemin, 1891; von Urbanitzky, 1896). Gramme's work enabled large quantities of direct current to be generated and transmitted to car-mounted motors which were superior to the magneto-electric engines and electro-magnetic motors tried between 1840 and 1870. Early electric motors reproduced the action of Newcomen and Watt beam engines, with the cylinders replaced by solenoids. Others copied the action of the Trevithick engine by linking the moving core of a solenoid to a connecting rod and flywheel (Dunsheath, 1962; Gee, 1991). This was not a promising line of development, and a rotary motor was needed for locomotive purposes. Gramme's 'reversible' motor-dynamo of 1870 was the first practical example. In 1860, Dr Pacinotti produced an armature generating continuous current without rectification when rotated by a prime mover, and he was singled out for especial mention by Whipple:

(Pacinotti's engine) could be used either for a magneto-electric generator or for an electro-magnetic engine. This man deserves a place among the foremost in our history. The machine of Gramme, patented in 1870 was almost identical with Pacinotti. It was, however, in some features so modified and mechanically improved that it became at once the first real practical commercial machine for the generation of direct absolutely continuous current put on the market. Although many other exceedingly ingenious machines had been brought out, yet they all gave intermittent, instantaneous or alternating currents.

The first commercially successful electric railways used direct current, transmitted through rails or wires to a locomotive or carriage. They emerged as a largely American enterprise, with important contributions by Italian, British, German and French engineers between 1840 and 1880, when the 'philosophical machines' of the earlier phase gave way to the components which first demonstrated electric traction with commercial potential. Between 1870 and 1885, crude models of locomotives and cars of the first phase were developed into the locomotives of Siemens (1879), Edison (1880), and Daft (1885), with current supplied from a lineside station via a third rail or the running rails. The basic form of the first commercially successful electric railway was thus created in the 1880s.

The evolution of the electric railway between 1870 and 1890 took it beyond its streetcar origins, to create a new railway which was more than a contemporary railway with electric-motive power substituted for steam traction. On heavy-duty rapid-transit lines, elevated or underground, the electrical system of rolling stock, current distribution and signalling were integrated in a manner not possible on a steam-worked line, and performance was raised to new levels. The transformation on rapid-transit lines brought about by electrification was total and irreversible, and a new railway was created. On the sections of main line which were first electrified, more of the features defining a steam railway were retained, such as signalling, loose-coupled wagons, 'steam' passenger stock, goods sidings, complex working of passenger and goods traffic, and thus the total transformation was less marked. The electric rapid-transit railway was more technically advanced, and remained so for much of the history of electric traction. The electric rapid-transit system was created by European and American contributions. In Britain, American practice dominated between 1880 and 1920, because of the American lead in heavy-duty electric engineering, especially in railway technology and large-scale power generation. German and French contributions were great, but were not as influential in Britain as the American achievements, and it was in Chicago, New York and London where the electric railway emerged following American practice, and driven by an American enthusiasm for general electrification. Histories of American railway electrification are provided by CERA Bulletin 113 (1973), Condit (1977, 1980), Cudahy (1979) and Middleton (1974, 1977).

The new kind of railway, which set an example the world was to follow, was established in Chicago and New York, before it appeared in London, Paris, Berlin, Koln or Vienna. The LVDC system spread to displace steam traction from existing lines, and was used on newly built light-rail systems. Its origins were in the streetcar lines worked by electricity in the 1880s. These were horse-worked lines, cable systems or steam tramways, electrified by installing batteries in the trams, or supplying DC current from a lineside station through overhead wire or a ground-level rail. They were the proving ground for components later used in heavy-duty rapid-transit, and they built up electrical engineering expertise necessary for extending electrification in the 1890s (Lind, 1974, 1979; Reid, 1899; Rohrbeck, 1980; Schramm and Henning, 1978). During the 1880s, electric locomotion evolved rapidly. Sprague worked with Edison on locomotives after

Figure 2.1 The locomotive 'Ampere' built in 1883 by the Daft company, and tested on the Saratoga & Mount McGregor Railroad, Saratoga, N.Y. It drew current from a central conductor rail. It was later operated as a battery locomotive.
Source: Out of copyright.

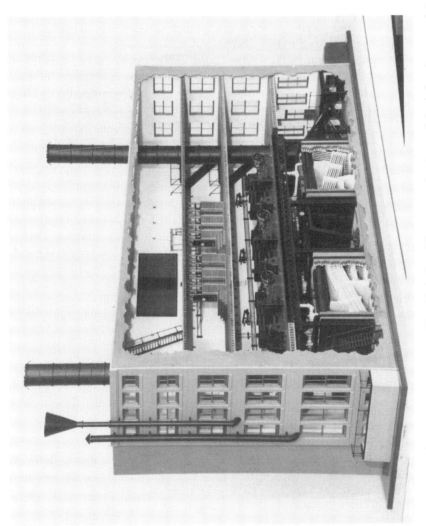

Figure 2.2 Model of Edison's Pearl Street power station in New York City. This non-condensing coal-fired plant began work in September 1882 with Babcock & Wilcox boilers in the basement and Porter-Allen reciprocating engines on the floor above. Fuel economy was too low for general electrification.

Source: Science Museum, South Kensington, London.

1880, Raffard operated a battery-electric tramway in 1881, the Daft company produced the outstanding 'Ampere' in 1883, the Field-Edison 'Judge' appeared in 1883, Thury tested electric traction on rack railways in 1884, and Daft constructed a full-sized electric rack locomotive in 1888 (Condit, 1977; Middleton, 1974; Tassin, Nouvion and Woimant, 1980). By the end of the 1880s, several specialist applications of electric traction had been demonstrated. By 1890, there were motorised cars and locomotives at work on street railways, light railways, elevated lines, mine and quarry systems, contractors' railways, rack railways, and on tunnel railways. Commercial success was first won in the 1880s on the passenger-carrying light railways and encouraged Sprague to pursue heavy-duty electrification in the 1890s. In the 1880s, powers, speeds, and fuel efficiency were too low on electric railways to displace the more powerful, fast and thermally efficient steam locomotive from main line services, heavy-duty suburban work, or rapid-transit duty on underground and elevated lines. However, the expertise gained on the light-duty street tramways in the 1880s was used in the following decade to tackle the difficult job of electrifying the elevated lines in Chicago or the underground railways in London. When electric traction was able to work these heavy-duty rapid-transit lines, it could displace steam traction from all suburban operations where high acceleration and deceleration were needed. By 1900, the heavy-duty electric railway was working in the United States, and was establishing itself in Britain. By 1910, it was transforming the suburban networks round major cities in Europe and the United States. This is described in contemporary texts, review articles and general histories: Ashe and Keiley (1905); Barbillon (1923); Barbillon and Griffith (1903); Blondel and Dubois (1898); Burgh (1911); Calisch (1913); Carter (1922); Dawson (1909, 1923); Gerard (1897); Marechal (1904); Smith (1900); Tassin, Nouvion and Woimant (1980); Whipple (1889); Whyte (1911); Wilson and Lydall (1907, 1908).

2.2 Electrical engineering and railway systems

Railway engineering progressed when electric traction took over from steam locomotion, cable haulage and horse traction. The heavy-duty electric railway depended on supply of current, and before nationwide, standardised supply grids were built, companies had to build their own power stations. Railway electrification between 1890 and 1910 was closely related to availability of electricity supply, and costs were high because systems efficiency was low (Bowers, 1982; Hannah, 1979; Martin, 1902). The supply networks, with a systems efficiency high enough to guarantee low power costs, grew after 1910, and were not always accessible where electrification was wanted. In North America large supply networks preceded their construction in Britain (Quigley, 1925). By 1920, long-distance grids were being built in Canada and the United States to rival the construction of the transcontinental railway lines as an engineering achievement (Hughes, 1983; Talbot, 1920). The American lead was obvious, and Sinclair

Figure 2.3 Early hydro-electric stations at Niagara Falls (USA side) in June 1910 with Schoellkopf station 3A under construction in the background and the 1895 No. 2 station in the foreground. The Niagara installations pioneered large-scale generation and distribution of electrical power.

Source: Niagara Mohawk Inc., Buffalo, N.Y.

warned in the early 1920s, when American engineers were developing the 'superpower' station, that three big stations in Chicago provided more electricity than 77 small stations in London, and New York City generated more electricity, by a sizable margin, than did the entire United Kingdom. Britain was slow to build up expertise in electrical engineering and to develop industries founded upon it. The United States built up a lead which put it first amongst industrial nations from 1890 down to the 1970s, when Japan and Germany challenged its position. Between 1890 and 1930, the USA did what Britain had done between 1770 and 1840, which was to build up expertise of strategic importance in a new technology and associated activities, which other nations could not easily overtake. In the case of the USA and electrical engineering, there was not the near monopoly of expertise which Britain enjoyed with steam engines because Germany, France and Italy were centres of electrical innovation. The strength of the USA lay in visionary thinking, large-scale planning, and the use of rational methods (Kranzberg and Pursell, 1967; Mumford, 1934). In the USA, there were widespread efforts to displace steam power (Langdon, 1901), and electrical equipment manufacturers staged publicity contests intended to win orders for the 'new motive power' (Talbot, 1920). There were comparative tests between large steam-shovels and electric shovels during the cutting of the Welland Canal, in Ontario, and tractive effort trials between steam and electric locomotives. Between 1910 and 1920, it became clear that economic rather than technical considerations were limiting the electrification of activities powered by steam. In the early 20th century, engineers built transmission lines across the wilderness of Canada and the United States and developed a new industrial technology which electrified farms and isolated homes as well as factories and cities. The union of high-efficiency generating stations and grids was to be the single most important act which enabled power systems to spread, and it added commercial and domestic markets to the industrial market. The grid brought electrical goods, such as typewriters, dictaphones, telephones and radios, into commerce. Mechanical inventions dating from the 19th century were replaced by electrical versions, including sewing machines, gramophones, washing machines and water boilers. This first phase of electrification, limited to traction, industrial power and public lighting, left many homes, schools, houses and offices reliant on gas, coal or muscle-power for facilities that were to be electrified during the post-1920s phase (Hennessey, 1972). This general use of light-current electrical apparatus built up American expertise in electrical signalling, control and communications in railways, where the Americans pioneered the advances.

The lack of electric street tramways in Britain reduced opportunities for cultivating expertise in heavy-duty electrical engineering. Britain had textile mills, iron works and coal mines in the 18th century, which needed water-wheels and steam engines, and gave the land the steam-powered industries of the 19th century. The British steam-powered technologies and industries did not lead to an electric revolution, or any systems likely to foster one. Such a system – in the form of light street tramways – did exist in the USA. In Britain in the 19th

Electric railways and American practice 19

Figure 2.4 The generator hall of Edward Dean Adams station No. 1 Niagara Falls, N.Y., in December 1909 with vertical-axis polyphase generators driven by Francis turbines.

Source: Niagara Mohawk Inc., Buffalo, N.Y.

century, there was no large, growing network of electrified street lines to foster electrical engineering on the industrial scale. This developed first in the United States, then in Germany and France. British electric tramways developed remarkably late, granted that tramcars and street railways date from the mid-1860s and that there were many experiments made with different modes of traction. There were no extensive electric street tramways within British 19th-century cities to cultivate electrical expertise (Barker and Robbins, 1963, 1974). Much of the early electric industry (power, light and traction) was hampered by legislation. The big networks were in the United States, and therefore the expertise was there. Electric rapid transit in London was of necessity an American import. Yet the British had tried many modes of public transport. The horse-bus appeared in 1829, and there were several unsuccessful experiments with steam coaches in the 1820s and 1830s. The horse tram was tried in 1861–62, but removed after legal objection, though successful horse-drawn tramcar services began in 1870. Many vehicles were imported from the United States following the introduction of horsedrawn street railways in New York and other cities in the 1850s. Limited success with the steam tramcar was achieved in 1873, and general use of steam trams dates from expansion of the steam-worked network in 1885–91. The internal-combustion engined motor-bus was first tried in 1899. There were experiments with battery buses in the 1890s, but the electric tram proper did not appear in London until 1901. Only the steam bus, 1904, and the trolleybus, post-1918, came later. The steam tram was withdrawn in 1891, the horse bus disappeared in 1914, the horse tram in 1915, the steam bus in 1919, and the electric tram in 1952. The London electric tramway came late and disappeared early, and it never provided a pool of talent based in the capital with which to electrify the Underground. Yet, as the useful chronology compiled by the Electric Railway Society makes clear, there were electric railways in the British Isles in the 1880s, when Sprague, van der Poele, Daft, Houston, Thomson and others were developing similar railways in the United States (Beecroft, Frew, Holmewood, Rayner and Stevenson, 1983; Frew, 1983; Hardy, Frew and Willson, 1981). The Giant's Causeway, Portrush and Bushmills Railway and Tramway dates from 1883 and 1889; Volk's Electric Railway, Brighton, from 1884; Ryde Pier Tramway, 1886; the Bessbrook and Newry Railway from 1885; the Manx Electric Railway from 1893; Blackpool Tramways (overhead, 550 V DC) from 1893, though an earlier installation, using the Hopkinson conduit system, was tried in 1886; a section of the Liverpool Overhead Railway from 1893; and the Snaefell Mountain Railway, 1895. There were several hotel and estate tramways, with electric traction, dating from the mid-1880s. Most of these early systems relied on American or Continental technology, and their total mileage was low compared with the USA (Tassin, Nouvion and Woimant, 1980). A nation can possess a few electric lines yet lack that inter-related network of industries, entrepreneurs, educational facilities, sympathetic politicians, experienced engineers and managers to make these pioneer efforts the start of a global enterprise. When British heavy-duty rapid-transit lines needed to be electrified, American expertise was ready to do all that was required, rapidly and with confidence and competence. Americans took advantage of the lack of an extensive, experienced and organised British electrical industry. They enjoyed the sort of advantage possessed by the British engineers between 1790 and 1840 in nations with no steam-engine expertise. The development of electrical engineering gave the initiative to the Americans and diminished the advantages of possessing expertise in the engineering that had created older industries. In the 1850s, American gun-makers tried without success to break into the British arms market with the Colt revolving pistol. The successful introduction of American electrical engineering in the 1890s marks a change. The United States, with Germany, had seized the initiative in strategic technology. The introduction of Edison power stations at Holborn Viaduct in 1882, and the American electrification of British steam railways in the heart of London in the opening years of the 20th century, are convincing demonstrations of this fact (Bowers, 1982; Parsons, 1939; Vickers, 1986).

Chapter 3

The electrification of street railways

3.1 The work of F. J. Sprague

A. G. Bell, T. A. Edison, G. Stephenson, J. Watt, H. S. Maxim, F. J. Sprague, Z. T. Gramme, W. Siemens, G. Marconi, S. de Ferranti, G. W. Westinghouse and H. Ford are examples of men who gave the world new engineering systems and components. Sprague, more than any other, created the heavy-duty electric railway (Agnew, 1937), as distinguished from the electric street tramway, or the first electrified main lines on steam railways (Middleton, 1967; Sprague, 1888, 1899). The considerable achievements of Siemens, Westinghouse, van de Poele, Brush, Daft, Edison, Crompton, Thomson, Hopkinson, Houston and other pioneers are likewise acknowledged. These pioneers turned the electric street railway of 1880 into the electric heavy-duty railway of 1900. The earliest efforts to electrify tramways were done in ignorance of electrical engineering (Whipple, 1889).

The earliest attempts of commercially equipping and operating a street railway with electricity were made by the Daft Electric Light Co., of New York. In the fall of 1883 they obtained the privilege to make the trial from the company whose tracks join Newark to Bloomfield, N.J. After several heroic efforts to get the track into a proper electrical condition, the experiment was abandoned. The attempt here was to use the two rails for the complete circuit, but it proved, as naturally it would, a dismal failure. The loss of current through leakage was so great that all indications were lost within a distance of 800 or 1000 feet. It was claimed by the electrician that since he used a very low potential, but 80 or 100 volts, this loss would not be large.

In 1883, the Daft Company experimented with a two-ton four-wheeled locomotive, called the 'Ampere', which was fitted with a 30 hp (22.3 kW) motor driving two pairs of wheels through belts and drawing power from a centre conductor rail laid along 15 miles (24 km) of the standard-gauge Saratoga and Mount McGregor Railroad, New York State. Contemporary prints show the Ampere pulling a car of the Pullman type, so that it must have been able to haul at least 24 tons, or possibly even 35. The engine was derailed and wrecked on the return trip, but the test attracted much attention, and is claimed to be the 'first recorded instance of an electric locomotive hauling a standard steam railroad coach over a steam road'. The power was taken from the Saratoga Gossamer Company's engine house close by the track. The test demonstrated the technical (if not the economic) potential of electric railway traction drawing power from lineside generator stations. The 1880s was a period of experiment and application. Whipple dates the first American conduit tramway in 1884, installed by the Bentley-Knight Company in Cleveland, Ohio. Electric locomotive haulage was tried on street railways, light railways and on the heavier-duty passenger systems, but in the end it gave way to traction using cars fitted with motorised bogies, and – after the mid-1890s – fitted with some form of multiple-unit control. These experiments facilitated the introduction of main line electric traction within steam railways, where locomotives were needed for goods working or for hauling passenger trains to a point where steam traction took over. The Daft Company tried third-rail distribution in commercial service on the Union Passenger Railway in Boston, using locomotives drawing current from a centre conductor rail in the hope of saving the fortunes of a mule-operated line that had fallen into extreme disrepair. The railway was saved but the unsatisfactory performance of the central third rail caused future United States railways to use the side-laid conductor rail, conduits, or overhead conductor.

By the mid-1880s, in addition to important public demonstrations of electric traction held at great exhibitions (e.g. Chicago in 1883), there were several electric street railways in operation, usually with some version of the overhead trolley system as devised by van de Poele, Daft or Sprague. Conduits, conductor rail and battery traction were also used (Crosby and Bell, 1892; Whipple, 1889). Electrification extended rapidly in the 1880s, using many different systems, and the extent of American electrification contrasts with the state of affairs in Britain at this time, though many of the electric railways in the United States were short, with two to five miles being the common range of length. At this stage of electric traction development, the van de Poele overhead trolley arrangement was popular, though the Sprague system was to supplant it. One of the oldest traction companies to use electricity is described by Rohrbeck (1980). Typical was the electric railway opened in October 1886, at Port Huron, Michigan, two and three-quarter miles long, using the van de Poele system (Whipple, 1889).

Four motor cars are equipped with one motor for each car, one of the motors being of fifteen horsepower, and three of ten horse power each, the large motor being on a sixteen foot car, and the smaller ones upon fourteen foot cars. The total weight of the larger car is 9,500 pounds each, and of the smaller ones, 9,000 pounds each. A speed from twelve to fifteen miles per hour is obtained. The maximum gradient is three per cent, which is overcome at a rate of seven miles per hour. Fifteen tons have been carried up this gradient at that rate. The mileage of cars per day is from 90 to 100, the service per day being sixteen hours. A Westinghouse engine of 45 horse power drives a Van Depoele generator, which is of 30,000 watts capacity.

At this time, in both Europe and the United States, an average steam locomotive on main line railways could provide a steady output at the driving wheel rims of more than ten times the output of the generating station of the Port Huron line, so that main line electrification had to wait for electric traction technology capable of far greater power outputs, and operating and economic circumstances favouring the replacement of steam traction (Taylor, 1907). The post-1895 electrification of main lines in city limits and tunnels provided the first opportunities for developing main line electric traction (Bezilla, 1980; Condit, 1977, 1980; Middleton, 1977). Before then, technical and commercial considerations focused attention on street railways and rapid transit. There were isolated experiments with locomotives suitable for light railway duty or branch line service. In Germany, electric traction was introduced into non-gaseous mines by Siemens and Halske. The 'Ampere' was rebuilt and fitted with Detroit Electrical Company batteries, which powered Fisher-Rae motors. It is reported as achieving an average speed of 12 miles per hour, including stops, and a range of between 100 to 120 miles on one charge. This is outstanding performance if true, though battery traction was never a serious rival to the system that used trackside generators and distribution by trolley or rail. In 1889 Whipple wrote:

The opening of the year 1888 was the signal for the greatest impetus in electrical street railway construction it has yet experienced. The purchase by the Thomson-Houston Electric Co. of Boston, of all the Van Depoele patents, and the successful developments of some experiments that had been made by the Sprague Company, opened the eyes of the public especially the street railway world, to the fact that the electric propulsion of cars was no longer an experiment, and that the signs of the times were pointing toward electricity as the coming power.

The most important system opened that year was the Union Passenger Line in Richmond, Virginia, which used the Sprague overhead distribution system and the Sprague truck, which mounted the motors on the bogies in a manner that enabled electric cars to run over uneven track without excessive damage to rolling stock or street fabric. The axle-hung, nose-suspended motor became a world standard. Sprague combined components into a successful system which solved the problems facing electric street car lines. The Union Passenger Line was a relatively large one, with a total trackage of 12 miles, over which 40 cars were operated, to a scheduled running time of one hour for 7.5 miles, or 107 miles in a day of 18 hours. The generator station used three Edison dynamos of 40 kW capacity, wound for 500 volts DC. The railway was difficult to work, as it had steep gradients and sharp curves, with branch lines running over unpaved roads where draught horses got stuck in mud and snow. Electric traction made operations more successful. The Sprague system defined a reliable form for electric street railways. It used motorised cars riding on Sprague trucks, rather than locomotives, and it drew DC from an overhead distribution system. There were important variants, such as the centre-conduit system or distributions using both trolley and buried conduit. This combination system, employed in London, used a 'plough' running in the conduit, but the Sprague type demonstrated at Richmond became the archetype for street railways. This Sprague system was developed to serve heavy-duty rapid-transit lines, and for

Figure 3.1 Electric street car of Observatory Hill Passenger Railway, Allegheny City, Pennsylvania, USA, which began operations in January 1888 using the Bentley-Knight system. Supply was distributed through overhead conductors and sub-surface conduit. Each axle was driven by a 15 hp motor. Total weight was 5 tons.

Source: Out of copyright.

conductor-rail suburban and limited-length main line sections of electrified route. The alternatives to the Sprague exemplar remained limited to street tramways and did not lead to more advanced systems. The electric trolley system replaced horse-traction and steam-locomotion on light-duty passenger lines, though older traction modes survived on minor lines in country districts. The cable-hauled passenger railway was a more formidable rival, and it had unique advantages which prevented its complete disappearance, such as the ability to work very steep inclines. When local electricity supply was available, the steam winding engine was replaced with an electric motor, with the cable and rollers left in situ, as in San Francisco and on many mineral lines in hilly districts.

The demonstration of the Sprague system in Richmond in 1888 was followed by 25 years of intensive electric railway development in the USA, and by 1917 there were some 60,000 street cars or trams, running over 26,000 miles of street railway. At the same time, the heavier-duty 'interurban systems' ran some 10,000 cars on 18,000 miles of intercity electric railways (Middleton, 1961, 1967, 1974; Tassin, Nouvion and Woimant, 1980). There was an electric network about 44,000 miles long, worked by some 70,000 electric-powered cars. At the peak of operating efficiency between 1942 and 1944, the American main line steam railways of Class One category worked some 39,400 steam locomotives and 790 main line electric locomotives over a system with a route mileage of 250,000 (Stover, 1961). There were more electric motorised cars on the trolley systems in the United States than there were steam locomotives on Class One railways. If the electrified suburban networks of the steam railways are included, along with the elevated lines and the other rapid-transit routes, it is clear that railway electrification had taken place on a very large scale by 1914. Electrical engineering had built new railways, such as the interurbans, before main line electrifications got underway in Europe and the USA (Hilton and Due, 1960). The size of these pre-Great War electric railways is evidence of the important role traction played in electrification between 1880 and 1914. This preceded general electrification after the war, when the high-efficiency super-power generating station, and standardised grids, introduced the 'Electric Revolution' to wider industrial and domestic spheres.

The Sprague lines were the centre from which the modern electric railway grew. F. J. Sprague (1857–1934) worked closely with Edison, and others whose businesses were merged by Vaillard to form General Electric (USA). In 1882, Sprague visited the British Electrical Exhibition at Crystal Palace, London, where he was secretary to the jury testing dynamos and gas engines. During his stay, he experienced the discomforts of the steam-powered underground railway, and he considered applying electric traction to railways using overhead distribution and an under-running trolley. His first schemes for electrification concerned the railway that was to be the British proving ground for his ideas. Sprague resigned his commission in the United States Navy and served as an assistant in trials of electric traction which Edison began at Menlo Park in 1880, following his 1879 visit to Berlin where he saw the Siemens locomotive. In 1885, Sprague perfected the type of motor, mount, gearing and truck which is known by his name, and the Sprague truck, the Sprague trolley, and the Sprague system became the most important. The Sprague system was one of several, and Thomson-Houston, Brush, Bentley-Knight, Daft, and other companies marketed the complete electrical and mechanical apparatus of an electric street railway under the company name. By 1892, the number of companies was reduced by Vaillard's merger of Sprague and Edison to give Edison-General Electric, which merged with Thomson-Houston to give General Electric. The term 'system' was in common use and met with criticism:

> The word 'System' as applied to electrical street railway construction, is of little meaning. It is used in an abstract way, merely, in connection with a name to designate a particular manufacture of certain important appliances. Beyond this it has no value. Appliances of various makes work well together; methods of construction are similar or dissimilar, as the conditions necessitate. As commonly used, some 'systems' have merits not to be found in others, while others possess superior points lacking in some.

Whipple failed to see that engineers recognised in 'systems' a new way of analysing engineering ensembles, suggested by electrical engineering. The need for manufacturers to provide a complete electrical and mechanical ensemble signalled the advent of a new sort of technology: the integrated system. The Sprague system was recognised as a fine one, and became an important standard for heavy-duty rapid-transit railways. To this day, railway engineers refer to the motorised bogie, with DC axle-hung, nose-suspended motors, as a Sprague truck. In the 1880s, Sprague anticipated electrifying steam railways like the London Underground, the American elevated routes, and suitable main lines. The companies displayed little confidence in his proposals, probably because they feared that the new systems would be unreliable, and would fail under onerous operating conditions. This caution of the heavy-duty companies compelled Sprague to concentrate on street railways as the field of application, with cable-worked railways as candidates for replacement. Battery-electric traction was first used, but the Sprague overhead distribution, the under-running trolley and the Sprague truck were introduced, and by 1891 over half the two-hundred-odd electric systems in the United States were supplied by Sprague. In the 1880s, a series of important improvements were made and became standard features, such as the carbon motor brush introduced by van de Poele. Early motor brushes were brass, which wore badly and sometimes required two changes per trip. The carbon brush lasted much longer and it became universal.

3.2 The extension of the Sprague system

The street car railways spread in the 1880s and became a feature of American life. Their name came from the wheeled current collector which ran along the overhead distributor, and 'trolley' came to mean tramcar in common speech (Crosby and Bell, 1892). These railways, and the interurbans which developed from them, were opposed by the steam railways, and there were legal disputes

The electrification of street railways 29

Figure 3.2 The Sprague truck which in 1888 introduced the axle-hung, nose-suspended motor and drive which remained a standard form in electric and thermal-electric traction for almost a century.
Source: Out of copyright.

over rights of way, particularly where the trolley line wished to cross an existing steam railway. Nevertheless, by the early 1890s the electric street railway was well established, and the network grew for the next 25 years. A general history of these early systems is found in the following writings: CERA 91, 1980; 106, 1962; 107, 1963; 113, 1973; 115, 1976; Hilton and Due, 1960; Middleton, 1967; *Railway Gazette*, 1935; Schramm and Henning, 1978; Schramm, Henning and Dworman, 1980; Swett, 1975, 1979. From these first street lines general electric traction developed, not only on railways, but on public roads. The first trolley-buses were demonstrated in Germany in 1882, and the first battery buses were tried in the late 1880s in several countries, including Britain (CERA, 116, 1929; Keilty, 1979; Schultz, 1980).

The early electric street lines were the beginnings of the long-distance interurbans, and the electric rapid-transit railways (Autocar, 1905; CERA, 116, 1976; CERA, 118, 1979). The transfer of electric traction technology from the street railway to the heavy-duty rapid-transit system involved a great advance because rapid-transit duty was much more onerous than anything found on the trolley lines. In the United States, the heavy-duty rapid-transit system was typified by the steam-worked 'elevated' lines of Chicago and New York (CERA 113, 1973; 115, 1976), which had been constructed in the 1870s. These were operated by Forney 0-4-4 tank engines, built by Baldwins, which were practically standard for such routes. These locomotives proved as satisfactory as the British Class A 4-4-0 used in London, and a typical 28-ton compound-expansion model of 1893 could reach 25 to 35 mph with five cars after 350 to 400 feet. Twenty-five miles per hour in 350 feet means an acceleration of 1.9 ft/s/s, or 1.27 mph/s, which is very good. Calisch (1913) quotes typical electric train accelerations on suburban lines as 1.0 to 1.3 mph/s. The steam tank engine 'Decapod', built by the Great Eastern Railway in 1902, accelerated at 1.1 mph/s and orthodox steam traction provided 0.4 to 0.5 mph/s with trains heavier than the five coaches usual on the elevated lines (Ahrons, 1927). Steam traction on the elevated lines was unpopular, and smoke nuisance apart, there was a growing need to increase line capacity beyond what steam traction could provide. In 1895, this need compelled the Chicago Metropolitan West Side Elevated Railway to carry out the first full-scale electrification of a heavy-duty steam railway using the Sprague system (CERA, 113, 1973).

During 1893–94, Sprague had introduced multiple-unit control for elevators (lifts) in the New York Postal Telegraph Building, and the technique was used to control trains on electrically operated railways. Multiple-unit control became an essential feature of the electric railway, and it is impossible to overstate the importance of Sprague's contribution (Sprague, 1899, 1901). Equivalent systems were devised by Westinghouse, whose electro-pneumatic alternative to Sprague's electric system enjoyed great success (Agnew, 1937). By 1895, the two leading suppliers of electric traction systems were Edison-General Electric and Westinghouse. Each developed their own range of essential components, but agreement to use each other's patents was reached in the early 20th century. Down to 1910, electric railway traction in the United States was dominated by the Edison companies: thereafter the Westinghouse company gained equal status through its success in pioneering main line AC electric traction systems (CERA, 118, 1979).

The Sprague multiple-unit system was used in 1897 by the Chicago South Side Elevated Company, which was electrified using 120 cars supplied by Sprague under onerous legal conditions required to persuade the railway to try electric traction. The trials succeeded. Electric traction was established on this heavy-duty, short-range passenger railway, using the LVDC conductor rail system, a form closely linked to the Edison companies, and exported by them. In the period 1890–1910, the short-haul passenger lines were the most common form of electric railway and the Edison companies (General Electric) were dominant, but in 1907 Westinghouse installed a single-phase AC system on 33 miles of the New York, New Haven and Hartford main line. This introduced a system better suited to long-distance main line railways, and the Westinghouse company rapidly became as well established in the international railway traction business as the Edison combine (Westinghouse, 1910; Westinghouse Electric, 1929). The success of electric traction on the Chicago elevated lines was beyond dispute, and the Sprague system spread widely, with cars carried on Sprague trucks, controlled by multiple-unit apparatus. By 1903, all the New York City elevated lines were electrified, and the ability of electric traction to work all rapid-transit routes was proven (CERA 113, 1973; Condit, 1980; Cundahy, 1979; Middleton, 1974, 1977). The electric urban and suburban railway had found a reliable form worthy of investment.

These rapid-transit lines differed in important ways from the street railways. Much greater powers were required by the heavier rolling stock and higher accelerations. High voltages in trolley wires or conductor rails were not possible in the 1890s, and DC voltages were low, so that high powers meant currents in thousands of amps, which made third-rail distribution the best arrangement. Direct current systems, using DC power houses, enjoyed advantages in the pioneer period compared to AC systems. DC could be stored in accumulators to safeguard against generator station failure, to supply extra power at peak loads, and to improve load factor. The common form used final current distribution via lineside third rail rated at 600 to 1200 volts, with return through the running rails or an insulated fourth rail in special cases. This proved satisfactory for rapid-transit railways, underground systems, elevated lines, and some main lines once worked by steam. The LVDC system is still undergoing technical improvement and is widely used in the circumstances for which it was first developed (Moody, 1961). It reached its acme in the Southern Railway, and Southern Region of British Railways (Moody, 1963, 1979), and in the New York Central Railroad third-rail routes (Middleton, 1974, 1977), which developed electric locomotive haulage of heavy express trains and goods trains in addition to multiple-unit workings, though steam traction was retained for long-distance operations and many goods services. Without this LVDC railway, the underground and surface lines of London, and the elevated lines of Chicago and New York, would not have met the demand to increase carrying capacity. The ability of the LVDC railway to handle heavy loads, under adverse conditions, is described by Middleton (1974). He records that the Chicago electrified elevated system contained the world's busiest junction, at the Tower 'L' crossing, which handled 200 rush-hour trains totalling 900 to 1000 cars per hour. In March 1930, when snow hindered all other surface traffic, 21,270 elevated cars passed Tower 'L' in a single 24-hour period. Without electric signalling, control, track circuiting, and train position monitoring, this would have been impossible.

Electric power generation and electric traction came to London as consequences of American innovation and enterprise, and Britain's legitimate claim to important pioneering achievements (McGuigan, 1964; Pearson, 1970) does not affect the overall picture. London can claim the first electric underground railway in the world, the City & South London Railway, constructed in 1890, by British enterprise, and operated with locomotives drawing current at 450 volts from a third rail (Lascelles, 1955). This was a tube railway, passing in an iron cylinder under the River Thames, and, though an economic failure, it was a technical achievement of the first order. The next electrified tube railway was British in origin, though it used American rolling stock. This was the Waterloo & City line, built in 1898 with the backing of the London & South Western Railway. The engineers were Greathead, and the LSWR consultant, Galbraith. The loading gauge was that of the City & South London Railway, and the cars were imported from the United States. These British achievements were praiseworthy, but were not part of a plan by British industry to use electrical engineering to conquer markets at home and abroad. They were not consequences of a vision, or ambition supported by education, research and development. They were isolated achievements. A major event in British electric railway history was the construction of the Great Northern & City Railway, authorised in 1892 but not built until 1904. This line took main line rolling stock, and from the start many functions were electric, including signalling, safety devices and the American-type rolling stock. This 'American' line was a commercial failure, but was an important engineering development, being the first to use multiple-unit control. It is still in use, with dual-voltage AC/DC stock, taking 25 kV one-phase AC on the former Great Northern main line, and drawing on the 750 V DC third-rail supply in the tunnels.

The Liverpool Overhead Railway of 1893 predated the electrification of the steam-worked US elevated lines, and in the 1880s, there were several British electric streetcar lines. These were simple, light-duty railways, and the first heavy-duty electrification in Britain was on the 'cut and cover' underground lines of the Metropolitan, and District railways which operated the 'Circle' routes in Central London with Class A 4-4-0 steam tank engines. It was here that the American system triumphed in all its aspects.

American methods were displayed in power generation, distribution and traction, and in the financial, legal and managerial departments. The Metropolitan and the District railways electrification schemes were influenced by the control exercised over the District company by American financier Charles Tyson Yerkes, who rejected a proposal, put forward by Ganz of Budapest, to electrify the District portion of the 'Circle' lines using 3 kV, three-phase AC traction (Bruce, 1968, 1983). Yerkes favoured third-rail collection after the Sprague LVDC model, rather than the three-phase system advocated by Ganz in conjunction with Westinghouse. This found limited and lasting application in Italy, and limited but short-lived application in Central Europe and the USA (Tassin, Nouvion and Woimant, 1980).

In 1899, experiments were carried out on the Metropolitan Railway with multiple-unit stock. The four-rail system was laid down to prevent return currents interfering with and damaging buried telephone lines and gas or water mains. In 1905 and 1906, locomotive haulage with British Westinghouse equipment was tried (Bruce, 1983). Between 1902 and 1905, the Lots Road power station in Chelsea was built (Howson, 1986). This was a typical 'American Station', built on the American scale, and it set new standards of size, generating 11 kV AC transformed and converted to 550/600 V DC for third-rail distribution. By 1905, the heavy-duty electric railway was established in London to be a testing ground for new kinds of safety device and signalling equipment. Examples are the track circuits, using DC relays, designed by H. G. Brown of Westinghouse Brake & Signal Company for the Metropolitan & District Railway. Before 1910, a successful form of heavy-duty rapid-transit railway was established in Britain, which was steadily improved and extended to present times, when the Jubilee line represents the LVDC system in the electronics era. The evolution of the LVDC railway from origins can be charted in the writings of Agnew (1937); Benest (1963); Hardy (1986, 1987); Hardy and Connor (1982); Linecar (1947); Moody (1979); Prigmore (1960).

Chapter 4

Heilmann, Ward Leonard and the electric railway

4.1 Introduction

In the 1890s ways of electrifying main lines carrying general-purpose traffic were considered. No obvious best system was in evidence. The existing electric railways were LVDC lines in rapid-transit service, which were not general-purpose railways. There were doubts about the advantages of electrifying general-purpose railways. There were conflicting opinions about the design of important components, such as motors. There was no consensus about how best to mount a motor in a locomotive or car. There were disputes about supply, power station design, distribution networks, location of substations, and current collection methods. Successful practice favoured the LVDC form, but there were doubts concerning its suitability for general electrification. The range of opinions is evident in Austin (1915); Barbillon (1923); Barbillon and Griffith (1903); Dawson (1909, 1923); Manson (1925), and Tassin, Nouvion and Woimant (1980).

J. J. Heilmann systematically investigated various electrical systems and components with a view to general railway electrification. He worked at the same time as Sprague but he did not focus to the same extent on rapid-transit systems, though he considered them. He invented the thermal-electric locomotive, and several of those who worked with him were to contribute to electric traction and the use of the internal combustion engine on railways (Duffy, 1989).

4.2 J. J. Heilmann and electric railway traction

Between 1893 and 1897, J. J. Heilmann of Alsace investigated traction by a steam-power house mounted on wheels, and several forms of electric railway using lineside power stations feeding locomotives through conductor rails and overhead trolley wires (Du Riche-Preller, 1893; Tassin, Nouvion and Woimant, 1980). Within each system, Heilmann and his staff compared three-phase AC and low-voltage DC motors. They constructed the self-propelled coal-fuelled power house, the 'Fusée Electrique', in 1893, and two larger engines in 1897 (Elektrotechnische Zeitschrift, 1897; Engineer, 1897). Heilmann investigated both third-rail and overhead LVDC traction using a DC locomotive built out of the motor-bogies of the 1893 'Fusée Electrique', which was compared with the 1897 steam-electric locomotives (Leonard, 1892, 1894, 1895; Tassin, Nouvion and Woimant, 1980). He also investigated a control system, for locomotives, resembling that of Ward Leonard. These trials favoured LVDC traction using stationary electricity-generating stations and simple electric locomotives or cars supplied through conductor rails or trolley wires. At the time of these tests, little was settled about the technology of heavy-duty general electrification, and even the suitability of the DC traction motor was questioned by S. P. Thompson, though defended by W. von Siemens. Heilmann's work in the 1890s involved men who contributed to railway technology in the 20th century. There was Brown (of Brown-Boveri), and Durtnall, who was one of the staunchest advocates of the thermal-electric locomotive and of electric transmissions in transport. Durtnall pioneered petrol-electric and oil-electric locomotion in Britain between 1900 and 1930 (Durtnall, 1925, 1926), and designed electric transmissions for 'electric ships' built in the 1920s. Heilmann's experiments explored a variety of possible traction systems, including locomotives, and trains in which the vehicles were powered by a through-train busbar fed from the conductor rail or a power station on wheels. Heilmann began his researches before Sprague established the LVDC railway in heavy-duty rapid-transit service, and when the only electric railways were the light-duty street lines.

In the 1890s, Sprague himself questioned the value of electric traction for long-range duties. He regarded electric traction by locomotives as less promising than operations with multiple-unit trains (Sprague, 1901), and before 1900 proposed that rapid-transit services were the ones suited to electrification. Sprague may have been discouraged by his joint venture in 1893 into electric locomotive design with Dr Duncan and Dr Hutchinson of Johns Hopkins University. They designed a 60-ton, 1000 hp 0–8–0 locomotive for work on a Chicago goods transfer line which the Northern Pacific proposed to electrify. The engine was built by Baldwin, but bankruptcy of the Northern Pacific ended the electrification scheme (Middleton, 1974). Sprague (1901) argued that any line using electric traction needed to use multiple-unit working of passenger cars, which might be difficult on lines where goods traffic and long-distance sleeping car expresses were worked by steam locomotives at different speeds.

In the 1890s, electric traction was advocated for rapid-transit systems, lines across waterless deserts, and lightly used secondary routes like the interurbans or branch lines. Electric locomotion for main lines and heavy-duty traffic was in the development phase and, apart from the small electric tractors of Siemens, Edison, Daft and others of the late 1870s and 1880s, there were no models for electric locomotives capable of doing the work of contemporary steam engines. The 1893 Baldwin engine, designed by Sprague, Duncan and Hutchinson, was remarkable for its size, weight, power and good design considering the lack of precedent. In 1893 General Electric built its first electric locomotive capable of heavy railway work at the GE works in Lynn, Massachusetts. This was a four-wheeled 30-ton engine with two gearless motors supplied from a 500 V DC overhead trolley wire. A second engine of the BB type, weighing 35 tons and supplied from a 500 V DC trolley wire, appeared in 1894 and worked successfully for 70 years (CERA, 116, 1976). The Pennsylvania Railroad first installed electric traction in 1895, on a 7-mile branch line between Burlington and Mount Holly using an overhead trolley wire at 500 V DC to supply combination passenger-baggage cars, of which two had two 75 hp motors, and one had four 50 hp motors (Bezilla, 1980). The famous Baltimore & Ohio Number 1, made up of two semi-permanently coupled engines to form a BB unit, was built in 1895 for the 7.2-mile belt line through the Howard Street Tunnel, which used an overhead conductor, converted to conventional third rail in 1902 (Condit, 1977). In every sense of the word, Heilmann and his eminent staff were pioneers. Heilmann came from an Alsace family of manufacturers in a district noted for innovations in engineering methods. Perhaps the first quantified energy analysis of industrial steam engines was carried out in Alsace by Hirn (Cardwell, 1971). The Heilmann family were progressive, and well informed about the latest developments.

J. J. Heilmann was one of the first engineers, with connections in the steam-engine and electric power industry, to investigate electric railways in a systematic manner. He investigated the question in a general way, without limiting himself to the electrification of rapid-transit lines, or routes in tunnel. In the late 1880s and the early 1890s, few engineers believed that electric traction could do more than improve street tramways, or replace the steam engine on the underground or elevated lines. Some thought that electric trains would surpass steam traction in speed and put forward projects for new high-speed railways, though there was little evidence that electric traction could then do what its more enthusiastic supporters claimed. In the USA, Weems constructed a narrow gauge electric railway on which a small car reached 120 mph (192 km/h) in 1889 (Reed, 1978; Tassin, Nouvion and Woimant, 1980). The test vehicle had two axles, and picked up current from overhead wires. The rails were used as return, and the car ran at speeds up to 120 mph. This demonstration did not prove that full-sized passenger-carrying railways could be built to the same principle with economy, but it led to schemes for high-speed electric passenger railways, separate from existing steam railways, between the major cities in North America and Europe. Though premature, the 19th-century proposals stimulated discussion of high-speed electric railways at a time when none existed, and encouraged engineers to design electric traction systems very different from street tramways. Some proposals were probably fraudulent and raised money for schemes which made little engineering sense. Others were advocated seriously by visionary engineers. One 20th-century engineer, C. Kearney, made high-speed systems a central feature of a utopia, organised along engineering lines, which he described in a novel depicting a Britain transformed by enlightened technocracy (Kearney, 1943).

4.3 The 'Fusée Electrique' of 1893

In his review of the first Heilmann thermal-electric locomotive, Du Riche-Preller (1893) listed four categories of railway which used 'Electromotive Power' at high speeds. He classified the Heilmann machine as an electric locomotive which would enable railway electrification to be achieved cheaply without fixed works. None of the specific systems listed by Preller could outperform a good steam engine of 1890, let alone one designed according to the rational principles developed by Goss and Woodard between 1890 and 1910.

The first class of electric railways considered by Preller were high-speed suggestions rather than experimental models. They were all of impractical design and associated with dubious projects for linking, for example, St Louis with Chicago. They were inferior in concept to the high-speed electric motor cars constructed for the German Association for the Study of Electric Railways by Siemens and Halske, and AEG, and tested in 1901 on the Zossen-Marienfeld military railway (Lasche, 1902). These German trials heralded the arrival of high-speed electric locomotion in a way that the early model tests did not. The schemes reviewed by Preller were for high-speed railcars, feeding through overhead distribution networks from central stations via transformers which supplied low-tension current to overhead trolley wires. There was little discussion of the engineering problems of working at over 100 mph on these completely unproven systems.

In the second category were battery locomotives represented by the six-wheeled engine then under construction on the Nord railway in France, but heavy weight and limited range and power made these unsuitable for anything other than light duties or shunting.

The third category was the 'simple' electric locomotive, drawing power from a lineside supply. These were represented by the famous 'Ampere' of 1883, or the four-axled 87-ton, DC locomotives designed in 1893 for the Baltimore & Ohio Railroad Howard Street Tunnel line, or the Sprague-Baldwin 0–8–0 engine built for the Northern Pacific Railroad. In 1893, such locomotives could not replace the long-distance, high-speed steam locomotives of the 4–4–2 type, which were setting up new records for speed in Europe and the USA.

Of the three categories listed by Preller, class 1 contained high-speed units which were purely conjectural, and classes 2 and 3 contained engines which were unsuited to high-speed, long-range work or to the general-purpose duties required by a main line steam railway. Only category 4 made engineering sense at the time, and this comprised 'engines carrying their complete generating station on the frame, as is the case in the locomotive which bears the name of its principle promoter, Mr J. Heilmann of Paris'.

Heilmann's locomotive, the 'Electric Rocket', was the outcome of carefully co-ordinated research work conducted by eminent engineers with the facilities of the foremost engineering companies in Europe. Preller lists as Heilmann's 'coadjutors', Mr Drouin, Mr C. Brown 'formerly of the Winterthur Locomotive Works and now of Basle', and Mr C. E. L. Brown, 'late of the Oerlikon Works

(Zurich), and now of Brown-Boveri and Co., electrical engineers of Baden, near Zurich'. Preller states that:

the original intention was to equip a complete electrical train, in which each axle was to carry a motor actuated by energy derived from the generating plant of the locomotive; but that it was deemed advisable, first of all, to test the locomotive alone, with the object of utilising the existing rolling stock of railways, leaving the larger and more ambitious design to be carried out subsequently.

The 'Electric Rocket' was a contemporary Edison DC power house mounted on railway bogies driven by DC motors. It was an instrument for electrifying railways rather than for introducing a new form of steam locomotive. Between 1890 and 1920 there were advantages in putting the electric power house on a locomotive platform if long-range electrification was the goal. Before the mid-1920s, most coal-fired power stations were of a low thermal efficiency. The pioneer installations often lacked vacuum condensers and were most uneconomic (Parsons, 1939; Taylor, 1907). Losses in generators, motors and transmissions were high, so that thermal efficiency from coal feed to motor output was often less than that of a contemporary steam locomotive which could provide a thermal efficiency of the order 6 per cent from coal feed to train hook. There were no large-scale grid supplies, and the many local stations, feeding restricted areas, operated with different voltages and frequencies. A railway system, covering an area larger than districts supplied on a non-standardised local basis, would have to construct its own power station and network. This is what many of the electrified railway companies did, but it was expensive. A self-contained power house on wheels, able to take the place of the ordinary steam locomotive, would enable railways to be electrified cheaply without building their own power stations and supply networks. The overall thermal efficiency would be higher because losses in the transmission lines would be eliminated. Electric transmission would eliminate destructive loading of track and bridges due to unbalanced forces intrinsic to the reciprocating engine, and the use of DC traction motors would give the tractive effort versus speed characteristic proved desirable on the street tramways. The locomotive power house, or thermal-electric locomotive, offered the advantages of electrification without heavy investment in fixed equipment on any kind of line, main line, rapid-transit, suburban, passenger or goods. The problem of mixed working, with some lines electrified and others not, did not arise with this mode of traction. Tunnels could be worked with the fires banked and current picked up from a rail. The thermal-electric locomotive could burn coal and work on existing steam railways with minimum disruption, because the railways already relied on coal-fired traction. Existing coal and water facilities would be retained. No new passenger or goods rolling stock would be needed, and prototype units could be tested using the existing railways without building fixed works. The thermal-electric system could be progressively updated by retiring old units and building new ones with generators and motors of improved kinds, whereas the electric system using fixed works could not be updated as easily.

Figure 4.1 The 'Electric Rocket' of 1893, illustrated with bodywork removed to show Gramme dynamo, the 2-cylinder compound-expansion steam engine and the Lenz boiler with side water-tanks and coal bunkers. Each axle was motored.
Source: Out of copyright.

The Heilmann locomotive provided a flexible method for electrifying railways because it could work as a dual-mode tractor, able to pick up current from a lineside supply, and able to generate its own electric power on unelectrified lines. It could operate multiple-unit electric stock outside electrified areas, and work goods traffic, either as a tractor or as a bogie-mounted supply station feeding current to motors distributed through the trains. The 'thermal-electric' locomotives could pick up extra electric power from conductors laid down on steep grades to supplement the limited capacity of the engine's steam plant which sufficed for less arduous sections of the route. The universal, flexible tractor is a constant theme in traction policy, and Heilmann was one of the first to seek it.

In 1893, there was no commercial justification for building a network of new lines segregated from the existing railways for running trains at 160 mph, even if rolling stock capable of such a rate was built. It was wiser to increase the capacity of existing railways, without reconstructing them. The Heilmann locomotive was regarded (Du Riche-Preller, 1893):

... with a view to utilise existing railways as they are. M. Heilmann has hit on the via media of constructing an electrical locomotive which, carrying its own generating plant, requires neither overhead nor underground conductors, nor a contact rail, but can circulate freely on any part of any existing railway system, and drive express passenger trains at 60 to 70 (maximum 100) miles per hour, while at lower speeds it will be available as a most powerful goods engine.

Adverse critics said that losses would be greater in a Heilmann locomotive than a conventional engine. This was admitted, but supporters of thermal-electric traction claimed that the steam engine driving the generator would run at the rotational rate for optimum efficiency independently of the speed of the train. This was not possible with a conventional locomotive where the engine was coupled directly to the wheels. They claimed that electric transmission provided better tractive effort versus speed characteristics, and enabled every axle to be powered. The Heilmann locomotive ran in either direction without turning, though there was a preferred direction of chimney to the rear. The Heilmann locomotives suffered from limited capacity, because their boilers were small to make room for the electrical gear and auxiliaries, and not one of them outperformed a good ordinary locomotive in power and speed. During the early phase of his investigations, Heilmann directed comparisons between three-phase AC traction motors and the DC motor, in view of the success of the three-phase transmission system installed between Lauffen on the River Neckar and Frankfurt-on-Main. This three-phase system was designed by C. E. L. Brown during his time at Oerlikon Works, and between 1890 to 1910 Brown and others advocated three-phase motors in railway traction (Scott, 1899). Before selecting a motor type, Heilmann and his collaborators constructed a test bed, made up of the motor under trial and generators to supply either AC or DC. Motor powers were 60 hp normal, 100 hp max. A brake was made up of resistor banks and a DC dynamo driven from the shaft of the motor being tested. This electric dynamometer enabled the motor power, electromotive force and current to be measured at different speeds and related to motor losses and temperature rise. At the same time, Goss at Purdue University in the USA was subjecting the steam locomotive to the same thorough, systematic analysis.

C. E. L. Brown designed the three-phase induction motor tested by Heilmann. It had 54 bars of insulated copper forming an eight-pole drum winding connected in the star combination. The armature was mounted inside a steel drum, which was fixed to the rotating pulleys driving the brake dynamo. The tests revealed disadvantages which prevented three-phase motors from being used on Heilmann locomotives, because the 'dynamo' (alternator in modern terminology) and the motors had to be synchronised 'at least approximately'. The motors were directly coupled to the wheels, and the 'dynamo' was directly coupled to the driving steam engine. If train speed was varied, the driving engine would need to change speed in synchronism with the motors. This made the steam engine's rotational speed no more independent of the train speed than was an ordinary locomotive's engine, directly coupled to the wheels through a connecting rod, and removed one of the chief advantages of the proposed electric system. It gave the advantage to DC (Du Riche-Preller):

the continuous current motor proved to be preferable, for the minor disadvantage of commutator and brushes was far outweighed by its advantages – to wit: the ease with which starting can be effected ... together with the fact of the speed of motors, and hence of the train, being entirely independent of that of the dynamo, and hence of the steam engine. It was on these grounds, and fortified by the series of searching experiments ... that for this, the first locomotive of its kind, with travelling generating plant, and intended for the gradients and alignment of existing railways, the original intention of using the three-phase alternating current was abandoned in favour of the continuous current.

The Heilmann locomotive established the DC traction motor as the norm for all thermal-electric locomotives including diesels. It represents the first attempt to electrify a heavy-duty, general-purpose main line railway. It was a powerful electric locomotive as the output of its steam engine exceeded the power available to drive entire street tramway systems in the 1880s. The selection of DC motors in place of three-phase AC motors was crucial. The 1893 prototype engine was a contemporary electricity-generating station mounted on wheels. In his history of the early electric power industry, Parsons (1939) describes the pioneer stations of the 1880s. They were often Edison stations, of small capacity, with low-pressure shell boilers – sometimes of the 'locomotive' type – and with two-cylinder reciprocating engines of the Trevithick type. These could be of simple- or compound-expansion type, and they usually exhausted to atmosphere.

The Heilmann 'Fusée Electrique' consisted of a locomotive boiler with a corrugated firebox without a superheater; a horizontal compound-expansion steam engine directly coupled to a Brown-Gramme DC dynamo; an auxiliary steam engine and dynamo for excitation; water tanks and coal bunkers; and it exhausted to atmosphere through blast-pipe and chimney, as normal for railway engines. These major components were mounted on a single frame, which was carried on two four-axled bogies, with each axle driven by an axle-wound DC motor. The nominal rating of each motor was 45 kW, and the total electrical power was 360 kW nominal, with a maximum output of 440 kW, or about 590 hp at the wheel rims. The first Heilmann locomotive was 100 tons without fuel or water, and the weight in running order was 118 metric tons, equally carried on the eight axles, with a static rail loading of 14.7 tons per axle. This was the same as the axle loading of the largest contemporary European goods engines, such as the 12-wheeled 'Duplex' Mallet tank engine, used to work trains over the St Gotthard line, built by Maffei in 1890, and which weighed 85 tons. In 1893, the largest British express passenger locomotive weighed 50 tons without tender, or 90 tons with a four-axled tender in running order, and an American engine weighed about 10 to 15 tons more. These machines provided 1000 indicated horsepower maximum, with 500 hp at the wheel rims, so the power-to-weight advantage was with the orthodox steam locomotive. Without a vacuum condenser, the overall thermal efficiency of the Heilmann locomotive could not exceed that of the ordinary type, for any savings resulting from superior operation of the Brown compound expansion engine was offset by losses in the electric plant, and the boiler efficiency of the ordinary engine was probably higher than that of the Lenz boiler fitted to the 'Fusée Electrique'. This was before Goss and Woodard greatly improved the ordinary locomotive.

The Heilmann locomotive was introduced as an electric alternative to steam traction when the latter was deemed obsolete. The rational methods of Goss, Fry and Woodard rescued steam traction from obsolescence and introduced reproportioned machines, within the traditional form. These were full of development potential and could outperform the 1893 Heilmann locomotive and many early electric engines. The boiler and steam engines of the Fusée Electrique were constructed at the PLM Forges et Chantiers Works in Havre; the dynamo and motors at the Brown-Boveri and Co. electrical workshops in Baden near Zurich; and the bogies, frame and accessories at the works of the Compagnie de Materiel de Chemins de Fer, at Ivry in France. This subcontracting of the construction of components to several companies in France and Switzerland, at a time when most locomotives were 'built under one roof', resembles modern locomotive building practice, though it was normal throughout the electrical industry, and marine engineering, in the 1890s. The trials with the 'Fusée Electrique' were sufficiently encouraging for Heilmann and his collaborators to improve the design, and respond to adverse criticism of the concept. Engineers described the 'Electric Rocket' as underpowered and heavy, while the output of ordinary engines was being increased. They also said it was too expensive to convert to electric traction using Heilmann locomotives. Heilmann designed two larger and more powerful engines of the same general form as 'Fusée Electrique'. These appeared in 1897 as engines 8001 and 8002, which were compared with simple electric engines that picked up current from a third rail and an overhead conductor. The simple machines were built from the motors and bogies of the then-dismantled 1893 machine (Duffy, 1989; Elektrotechnische

Zeitschrift, 1897; Engineer, 1897; Stoffels, 1976). The Lenz boiler, with its corrugated, stayless furnace, gave way to a larger orthodox boiler, and the horizontal two-cylinder compound engine was replaced by a sextuple Willans engine, of the type which drove low- and medium-output electricity generating stations. Engine power was increased to 1000 kW. No superheater, and no vacuum condenser, were fitted to any of the three Heilmann engines, and they exhausted to atmosphere, as did many of the stationary power stations of the period. The electric transmission was DC, and there were eight axle-wound traction motors, each rated at 92 kW for a speed of 100 km/h. All the eight axles were driven, and the total weight of 124 tonnes in running order was available for adhesion. The eight powered axles were grouped into two bogies. The steam engines came from the British firm of Willans, and the electrical equipment was supplied by Brown-Boveri & Co. A maximum speed of 120 km/h was anticipated. The increased engine power overcame the objection that the original engine was underpowered, but there remained to be met the charge of excessive cost. Heilmann formed a company called the Société de Traction which funded construction of locomotives for letting out to railways, just as the Compagnie Internationale des Wagons-Lits hired out sleeping and dining cars to the Continental railway companies. The British journal *Engineer* (1897) remarked, 'The railway companies have therefore everything to gain and nothing to lose by the adoption of electric locomotives.'

This idea of Heilmann's anticipated the 'equipments obligations' system, used in the USA after the 1914–18 war to finance rolling stock re-equipment (Stover, 1961). Under its terms, equipment remained the property of the funding organisation. The operator hired apparatus, which was taken back on default of payment and hired out elsewhere. This demanded designs standardised for likely users. Goods wagons were standardised by the 20th century, and many were supplied under this scheme, but steam locomotives were not, apart from a few classes chosen for wartime building programmes. Generally, each builder, and many railway companies, had designs of their own. The steam locomotive makers never co-operated to produce a standard design for universal hire or sale within a nation despite the success of the wartime standard types. The hire-purchase of standard units was to be part of the General Motors strategy for eliminating steam traction from American railroads (Morgan, 1955). Heilmann, with his experience in progressive electrical industries, pioneered a similar outlook in the 1890s. The policy of his Société de Traction was to work with electrical industries to design, construct and supply the minimum number of standard types of thermal-electric locomotives for operating railways. Standardising the design would increase the numbers in production runs, lower the cost, and encourage operators to hire them. In the event, only two units, 8001 and 8002, were built and tested, despite talk of orders from European, Russian and American railway companies. Tests of the 1897 locomotives began on the Compagnie de l'Ouest between Paris and Mantes, a distance of 115 km, and both engines worked well, but though the railway seemed intent on using both machines on fast, heavy goods trains, they were withdrawn after little use and were not developed further. German railways, the Southern Railway of Russia, and the Ohio River, Madison and Central Railway in the USA showed interest in Heilmann locomotives, but no more were built. The ordinary steam engine was being improved by less radical innovations and was performing better. There was no reason for railways to buy or rent thermal-electric machines, which were heavy, complex and needed electricians to keep them in order. Contemporary researches into railway electrification, some carried out by Heilmann himself using parts of the 1893 locomotive rebuilt as an electric tractor, showed that the use of lineside power stations, fixed conductors and DC traction motors was a more promising way to electrify railways (Tassin, Nouvion and Woimant, 1980).

4.4 Heilmann and the Ward Leonard system

The 1897 experiments were carried out to find how to electrify lines; and how to control electric locomotives. They compared multiple-unit trains with locomotive-hauled trains and explored the role of electric traction on main lines. Heilmann produced two designs for simple electric locomotives, numbers 4001 and 4002, the latter being considerably larger than the former. Engine 4001 certainly was built, but there is doubt about how far construction of 4002 was taken (Tassin, Nouvion and Woimant, 1980).

(The) electric locomotives of the Western Railway of France (were tested) during 1897, on the route between Saint-Germain-Ouest and Saint-Germain-Grande-Ceinture. From a steam-powered stationary generating station, situated in the middle of the route, current was distributed towards Grande-Ceinture by a third rail (placed centrally between the running rails), and towards Saint-Germain-Ouest by an overhead line. The two Heilmann locomotives, (actually it is not clear that the second machine . . . was constructed), were carried on four-axled bogies, with four motors arranged in parallel: it is a question of an unexpected development from the first thermo-electric locomotive, 'la Fusee' (the Rocket) of 1893. The electric plant (conceived in the same design philosophy), had rotary converters, continuous-current to continuous-current, with pressure (voltage) varying through the range from 0 to 400V; the four-pole dynamos could supply up to 1,200 A at start: it was the Ward-Leonard principle with a direct current primary supply.

These engines were early examples of the motor-generator or motor-converter type, and they worked well, though the tests did not lead immediately to general electrification in France or elsewhere (Tassin, Nouvion and Woimant, 1980):

These machines operated electrically using an external supply, either constant 'pressure' or constant 'intensity' by means of rotary-converters, but the French engineers concluded that in spite of the interesting results obtained electric traction was only applicable in certain special cases, very limited, when steam traction couldn't furnish a satisfactory solution (to operating difficulties).

This remained the case until the 1950s. Between 1890 and 1920, electric traction engineers faced unresolved problems, such as how to achieve adequate speed control of railway traction motors. This caused Heilmann to abandon three-phase motors in 1893, and select DC, and his exploration of the forms railway electrification might take identified something very like the Sprague LVDC system using a conductor rail or trolley wire, which was the only practicable system in the 1890s for most kinds of railway service. The question of speed control of motors was pressing in the 1890s, and Ward Leonard's system was one solution (Yeomans, 1968).

For separately excited motors, speed control is obtained by varying the voltage applied to the armature. In the Ward Leonard system a constant-speed electric motor drives a d.c. generator whose field is separately excited from a d.c. exciter which is coupled to the same shaft. The output voltage of the main generator is controlled by a potentiometer in its field circuit, and is applied direct to the armature of the main motor. The field of the main motor is supplied from the constant voltage output of the exciter. The speed of the main motor may thus be controlled over a wide range by the potentiometer.

The system was invented by the American engineer H. Ward Leonard, who graduated from MIT in 1883, aged 22, and who worked with Edison as one of four engineers with special responsibility for introducing the Edison power station system. As related by Yeomans in his history of Ward Leonard drives:

The early experience of Ward Leonard gave him an approach to drive problems that differed from that of his critics. He was concerned much more with the economics of the power-station operation and distribution systems than were the electric-railway engineers, whose main concentration of effort had been in the design and operation of the street cars themselves.

Ward Leonard considered the electrification of railways much greater in extent than the short-run electric street lines and rapid-transit lines of the 1890s (Leonard, 1894). As Yeomans relates, the Sprague series motor had proved itself a sturdy and reliable machine on these pioneer lines.

... but the heavy currents drawn in starting or in crawling heavily laden up steep gradients caused severe fluctuations in the power-station loadings. In addition, power losses in the rheostats used for starting resulted in apparently un-avoidable inefficiencies. Many engineers had attempted to apply variable-ratio transmissions between motors and axles to achieve better efficiency at starting and at low speeds, but none had succeeded.

Ward Leonard first described his system in 1891 and 1892. He proposed a variable-ratio power converter, or motor generator, working on the principle that the motor would run at maximum efficiency when 'voltage varies as speed and torque as current'. The system would give high torque at starting, without creating a high demand for power from the supply station, and it promised regeneration of current during deceleration of trains or the descent of gradients (Leonard, 1892, 1896). It was admitted that the necessary gear would be heavier and more expensive than the controls on a simple Sprague locomotive, but Ward Leonard claimed that his system would be more efficient, and that because of the steadier load there would be savings in plant needed at the power house. He argued that, taking the system as a whole, it would be cheaper than Sprague's. Despite the demonstrated success of the Ward Leonard control in lifts and cranes, it was criticised. Sprague defended the switched-field series motor which he had developed initially for street tramways and he condemned the Ward Leonard system as impracticable and uneconomic. There were clashes in public meetings of the AIEE when Sprague quit the chair to take part in voicing his criticism (Yeomans, 1968). Ward Leonard was not deterred, and in 1894 read his major paper 'How shall we operate an electric railway extending 100 miles from the power station?'. During the meeting he demonstrated a model of his control system which impressed those present by the absence of sparking at the generator brushes, which Ward Leonard set at no-load neutral, rather than setting them to commute at full output with a highly saturated field. Ward Leonard visited France and Britain in 1895, and on his return to the USA commented on the lack of readiness in the United States to try out new engineering ideas (Leonard, 1895). He identified the interests of large companies as an obstacle to technological advance, perhaps feeling that the Edison-Sprague companies (General Electric) regarded his innovation as a threat within the traction field. Ward Leonard reported on the Heilmann locomotive of 1893, and Heilmann's use of the Ward Leonard system in his simple electric locomotive of 1897 was perhaps due to the American's visit to France in 1895. The Ward Leonard system never was a standard feature of electric locomotives, though it worked successfully for over 70 years in a wide range of industrial applications, including printing presses, electric lifts, electric cranes, rolling mills, electric excavators, and minehead winding gear (Burns, 1920). The variable-voltage Ward Leonard system normally had a direct-current motor driving the generator which supplied the traction motors. The DC motor could be replaced by various types of AC motor, or by steam, gas or oil engines. The Ward Leonard system was one point from which the motor generator locomotive was evolved, though these normally converted a single-phase supply in the trolley wire to a DC supply to the traction motors.

Ward Leonard did not restrict his system to a DC driving motor. His 1894 paper on long-range railway electrification contained a proposal for transmitting power at 20 kV to unattended transformer stations. These in turn fed an overhead trolley wire at 500 V AC, for supplying motor-generator locomotives in which the driving motor was of the single-phase AC type. The traction motors were DC. If the DC driving motor were replaced by a steam engine, one got the system used by Heilmann in his 1897 engines. In these, the Willans engines were governed to run at constant speed, and Heilmann may have considered driving the exciter off the main engine shaft, rather than by a separate steam engine, and so derived a steam-powered equivalent of the Ward Leonard system, with the constant-speed DC motor replaced by the constant-speed steam engine. Heilmann and Ward Leonard were often quoted between 1890 and 1910, but Sprague and S. P. Thompson criticised both. In 1902, Thompson rejected DC traction as obsolete (Swinburne and Cooper, 1902):

... the time for supposing that continuous currents may be seriously employed for heavy railway work has gone by ... all questions of 3-phase or 2-phase will eventually disappear ... No machine with a commutator on it has the slightest chance of surviving for the use of high railway speeds .... the future ... lies for single phase alternating work.

Thompson referred to the contemporary German high-speed electric cars on the Zossen-Marienfeld line, which used three-phase traction motors (Lasche, 1902), and after criticising the tendency to use the terms 'direct current' and 'continuous current' as if they were the same, he considered the Ward Leonard system. Thompson admitted it was an interesting and singular combination, but he argued it made no sense to carry around a whole substation of transforming gear on the locomotive 'in order to make the locomotive work with currents that are not the right ones to be supplied to it'. If this sort of complex solution was accepted, Thompson said, engineers might as well revert to Heilmann's scheme and mount an entire central station on board the locomotive (Thompson, 1902):

It was hardly more complicated than some of those round-about contrivances which Messrs. Swinburne and Cooper have suggested for doing that which ought to be done simply and directly. I do not think we shall arrive at that simple and direct system of driving until we have proceeded to that stage of further invention which I have suggested, namely, the perfection ... of the single-phase motor.

Professor S. P. Thompson's remarks did not go unchallenged, and Geipel (1902) defended the Ward Leonard system as possessing advantages over 'both the series system, and the ordinary parallel system' when it came to starting from rest, so that the motor part of the motor generator took much less power from the supply, and hence the power house, during the start up and acceleration of trains, than did the other systems. He argued that Ward Leonard locomotives could easily operate with single-phase AC supply in the trolley wire because (Geipel, 1902):

It is merely necessary with the Ward Leonard system to place on the locomotive a single-phase motor without regulating or controlling resistances. All that is necessary is a starting switch and a pair of high tension fuses. ... Objection has been taken to the weight of this motor generator, but, after all, it is but a small percentage of the total weight of the train.

Geipel argued that the Ward Leonard locomotive was more efficient than systems using fixed motor-generator substations placed along the line for feeding low-voltage current to the locomotives. In practice, economy was decided by many factors, including the length of the electrified route and the frequency of trains. Motor-generator locomotives were used on electrified sections of steam railways where traffic was infrequent, and 'putting the substation on the locomotive' was an economic thing to do. The Great Northern Railway in the USA provides an example. Geipel argued that there was no cause to go to the extent Heilmann had done and put the entire generating station on the locomotive, as this led to excessive weight, '100 tons is one thing and 20 tons is another', and he favoured locomotives similar to the two simple electric tractors which Heilmann tested against the 1897 thermal-electric units.

Heilmann's work showed up the deficiencies in the coal-fuelled thermal-electric locomotive, though two similar engines were built in Britain in the 20th century (Duffy, 1989; Stoffels, 1976). The Heilmann locomotives made sense at a time when it was unclear which traction motor and power supply system best suited railway work. Newly installed equipment could become obsolete after a short time. Heilmann investigated railway electrification principles when nothing was settled, and he pioneered motor-generator locomotives, as did Ward Leonard. He built the first thermal-electric locomotives and used them to experiment with long-range heavy-duty main line electric traction between 1893 and 1897. Heilmann deserves recognition for his sound method and wide scope, as a pioneer of electric traction.

Chapter 5

Electrification of British rapid-transit lines

5.1 Electric railways and the power supply

The electrification of American horse-operated street railways and steam tramways in the 1880s relied on LVDC. The power often came from batteries or cells, or was supplied from a dynamo driven by a low-pressure reciprocating steam engine or a water wheel. The Gramme dynamo found an early use in this role. Alternating current machinery was not sufficiently developed for common use in the 1880s, and direct current had many advantages. It could be stored in batteries, which improved the load factor of power houses, and which provided a supply if the generators failed. DC traction batteries could be carried on self-propelled vehicles. The DC motor was reliable, and had good characteristics for traction, such as providing a high torque on starting. LVDC current could be distributed from the power house to the vehicle via a lineside rail, or an overhead conductor, without difficult insulation problems. During this first period of electrification, when there was a lack of reliable AC transformers, generators, motors and distribution, the advantages were with the system in which all the components were LVDC (Mordey and Jenkin, 1902). This 'simple' DC electric traction system was used on many late 19th-century and early 20th-century street railways (Hopkinson, 1903), and on the first short sections of heavy-duty electric railway (Mordey and Jenkin, 1902; Calisch, 1913). In the early years there were limitations set by the ability to generate DC in large amounts. Heavy-duty rapid-transit services demanded high powers for accelerating trains from frequent stops. Low-voltage supply meant that currents were high on starting, which caused marked voltage drop and high losses in the conductor rail. This was not too serious on a short line of a mile or two. Electric working could still prove economic even when overall thermal efficiency was poor, which it would be when the driving engine was a low-pressure reciprocating non-condensing type. Even with condensing steam engines, compound expansion, and watertube boilers, the systems efficiency was low before 1920 (Parshall and Hobart, 1907; Duffy, 1997). The early British electric lines used this simple DC system

Figure 5.1 Coal-fired power house built in 1890 for City & South London Railway, with four 450 hp non-condensing compound-expansion Fowler reciprocating steam engines which drove through belts the Edison-Hopkinson dynamos which supplied 450 A at 500 V DC.
Source: London Transport Museum.

Electrification of British rapid-transit lines 53

(Vickers, 1986), including the City & South London Railway (Lascelles, 1955); the Liverpool Overhead Railway (Woodward, 1993; Jarvis, 1997); and both the Waterloo & City Railway and the Great Northern & City Railway (Linecar, 1949).

Later, these lines were supplied from AC stations when the electric railway network grew sufficiently large to justify large, centralised power houses and when the advantages of AC generation and transmission were recognised. Many electric street tramways continued to be supplied from DC stations until they were replaced by motorbus services, or were linked to the national grid after the 1939–45 war.

In Britain, the first DC electric lines ran in tunnel under great cities and there was a fear that return currents passing through the ground might damage buried cables and pipes, and an insulated fourth rail was often provided to carry it. It helped to meet the Board of Trade requirements that voltage drop be kept within limits, and was also used to leave the running rails free to carry track circuit currents for working automatic signals. This four-rail system was used on the Mersey Railway and on the London underground lines, though at first the Central London Railway did not use it. It became standard in London Transport days, though Calisch remarked in 1912 that no American railway used it, and he believed it to be quite unnecessary.

5.2 Alternating current power stations and transmission

Extended electrification would have been hampered if the only source of power had been through the simple DC system. In the 1890s, the AC system developed, which increased overall systems efficiency and enabled routes 10, 20 or more miles in length to be electrified with economy (Snell, 1905). The most important component was the large power station, usually generating three-phase AC between 6000 V to 12,000 V for transmission over longer distances, with less loss than was possible with the simple DC system (Bowers, 1982; Parsons, 1939; Snell, 1905; Trotter, 1907). Large-capacity stations became possible when AC generators run in synchronism could provide a regular, sinusoidal output (Parsons, 1939). Each of these new 'high-pressure' or 'super' stations supplied an area which would require several separate stations under the simple DC system. This 'centralising' of the stations was a major factor in increasing systems efficiency. At the same time, efficient transformers were introduced for increasing transmission line voltage far above that of any DC line, and for stepping it down at points of delivery to customers. This reduced transmission losses. The invention of the motor-generator enabled the desirable features of the DC railway to be retained, when supply came from a centralised AC station. The motor-generator, or motor-converter, was a combined AC synchronous motor and DC generator. It took AC supply via a transformer from the high-tension transmission line, linked to the power station, and converted it into LVDC which was fed into the railway's conductor rail distribution (Parshall and Hobart, 1907). These

54 *Electric railways 1880–1990*

Figure 5.2 City & South London Railway electric locomotive of a type introduced in December 1890 to work the first underground passenger railway. A conductor rail between the running rails carried the supply at 500 V DC.
Source: London Transport Museum.

units were reliable and long-lasting. Though gradually displaced by mercury-arc rectifiers after 1930, some sets lasted into the 1970s. Several of the units installed by General Electric under New York Central Station in 1912 survived until 1989, when they were replaced by solid-state devices. The system of centralised AC station, transformers, three-phase AC transmission line, motor-generator converters, and conductor rail distribution (which enabled large currents to be passed) enabled the LVDC railway to be extended far beyond the limited geographical range of the first street railways, tunnel lines and short-run elevated railways. The new system could be used with storage batteries, fed with DC through motor-converters, so that DC was available for feeding the conductor rail when generator supply failed, and for improving load factor and voltage regulation. The system needed manned substations every 12 miles or so of route, and sometimes electro-mechanical boosters were required to maintain voltage in long lengths of conductor rail between substations (Duffy, 1997). It proved a very reliable system, and could be integrated with sections of railway supplied from an older 'simple' DC station. It worked well, and supplied the networks which grew in and around New York, Chicago, London, Liverpool, Paris, Berlin, Vienna and other great cities before 1914 (Linecar, 1949; Parsons, 1908; Tassin, Nouvion and Woimant, 1980). It supplied London Transport railways, and the extensive network of the London & South Western Railway which became part of the huge system associated with the Southern Railway, and the Southern Region of British Railways. It supplied the Lancashire & Yorkshire system round Liverpool, and between Manchester and Bury, where 1200 V DC was fed to the conductor rail. In the USA it was used to supply the elevated railways and subways of New York and Chicago (Parsons, 1908) Its role in world railway electrification has been charted by Tassin, Nouvion and Woimant.

Electrified rapid-transit networks were extended between 1900 and the 1922 amalgamation of Britain's railways into four large groups, and this required AC power stations of a size in excess of any 19th-century station. Large centralised stations were opened at Lots Road and Neasden (London Transport); Durnsford Road (London & South Western Railway); and Stonebridge Park (London & North Western Railway). During this period, the high-voltage DC railways system emerged in the USA, with overhead wires carrying 1500 V DC, or 3000 V DC (Tassin, Nouvion and Woimant, 1980). This HVDC system was suitable for general-purpose railway electrification, and it became a world standard rivalled only by the single-phase AC system developed in Germany and Switzerland in the first 20 years of the 20th century (Jenkin, 1906). On rapid-transit railways, the standard form remained the LVDC system, supplied from an AC station via converters.

5.3 The electrification of the Mersey Railway

The Mersey Railway was a short passenger line, 4.75 route miles, linking Liverpool to Birkenhead. It opened on 1 February 1886, and was extended to link

Birkenhead Park, Rock Ferry and Liverpool Central between 1888 and 1892. It operated a shuttle service in tunnels with gradients of 1 in 27, using large, heavy tank engines of the 0–6–4 and 2–6–2 type. Constant mechanical power was required by drainage pumps, ventilation fans, and the hydraulic system which worked lifts and hoists. This came from individual steam engines of low thermal efficiency, and operating costs were very high. The company directors considered electrification as a means to reduce operating expenses. The example of the City & South London Railway (1890) and the local Liverpool Overhead Railway (1893) showed the engineering and economic advantages of the LVDC system, but the Mersey Railway's declining fortunes ruled out electrification funded from its own resources. The company was losing considerable funds each year, was facing bankruptcy, and no British investor was willing to meet electrification costs. Electrification was undertaken by the American G. W. Westinghouse, to demonstrate the benefits of electrification to British railways, and to win orders for his own companies. Throughout the 1890s, Westinghouse established his companies in the USA and Europe as reputable manufacturers of electrical engineering systems, including power plant and railways. Originally founded to make the Westinghouse air brake, the company diversified into railway signalling, and electric traction. Westinghouse pioneered AC power and traction, and in the 1890s his railway engineering work was overshadowed by the achievements of Edison and the General Electric group of companies. By the early years of the 20th century, the Edison and Westinghouse groups of companies in North America or Europe manufactured all major DC and AC traction systems in their own name or through subsidiaries. In 1899, Westinghouse founded the British Westinghouse Electric & Manufacturing Company, at Trafford Park, Manchester, to make ready for an anticipated British railway electrification boom. Sprague had vindicated his system by electrifying a steam-worked elevated line at his own expense: Westinghouse chose the Mersey Railway for the same purpose. The Mersey Railway was a much heavier-duty line than were the City & South London Railway in London or the Liverpool Overhead Railway. Its locomotives (Ahrons, 1927) were as big and heavy as a main line goods engine, and they worked trains of standard gauge 'steam railway' size and weight. Electrifying the Mersey Railway would show that electric traction had the potential to work many of the services provided by the ordinary steam railways. Westinghouse further intended to show that this could be done more economically with electric traction. Westinghouse's proposal that the Mersey Railway be electrified at his expense was accepted by the company, which had little choice, as financial decline was acute and seemingly unstoppable. It was the first conversion, in its entirety, of a British steam railway to electric traction.

Westinghouse chose the simple LVDC system, following the example of the first electric street railways, the City & South London Railway (Lascelles, 1987), and the Liverpool Overhead Railway (Jarvis, 1997). This simple system was well suited to a route length of only 4.75 miles. In 1902, there was no local electricity supply, and a railway power station was built. The current was generated at 650 V DC for distribution via a live outer rail, with return through an insulated centre rail. The simple LVDC system was used with storage batteries which improved load factor and provided emergency supply for extracting trains from the tunnels in case of power failure. It avoided the expense of transformers and rotary converters. The conversion to electric traction was carried out without interrupting steam services. Cost is variously quoted. Linecar (1949) gives £386,000 for 4.75 route miles. Calisch (1913) quotes £414, 368. Gahan (1983) gives £300,000. The variations may be due to inclusion or exclusion of costs of refurbishing stations, which was carried out as part of the electrification programme.

The Mersey Railway power station, which followed established Westinghouse practice, was erected next to the pumping station in Shore Road, Birkenhead, near Hamilton Square. It had a total capacity of 7735 kW (Shaw, 1910). There were nine Stirling water-tube boilers with a pressure of 170 lbs per sq. in. which supplied three vertical cross-compound engines, each of which drove a 1200 kW DC generator, rotating at 90 rpm and providing traction current at 650 V DC. In addition, there were two 200 kW sets for power-station lighting, and for powering the motors which replaced the steam engines that drove the hydraulic lift pumps. All circuits were controlled from a 19-panel switchboard. The boiler house reflected contemporary best practice. The Stirling boilers had mechanical stokers and ash removers that loaded ashes into wagons positioned on a siding which ran into the power station off the dock railway on Shore Road. After electrification the ventilation blowers and steam extraction fans were not needed. The blowers were removed, but the disused steam extraction fan at Liverpool Central station was left in situ until the 1970s. The steam pumping engines, used to prevent the tunnels from flooding, were retained unchanged during the first electrification programme, but in the 1920s the great beam engines were replaced by electric pumps. One set was scrapped, but two were maintained on a stand-by basis until 1959. One of the original pumping stations has been preserved. Steam traction ended at midnight on Saturday, 2 May 1903. Electric traction began on Sunday, 3 May 1903 at 1 p.m. The Mersey Railway advertised the new system as offering the travelling public seven advantages:

Elimination of the smoke and steam nuisance; greatly improved ventilation using electric fans; increased number of trains running on such short intervals (three minutes headway) as to render timetables unnecessary; greater acceleration of trains and higher running speeds with shorter journey times; improvement of the lift services due to electric pumping of the hydraulic system; clean rolling stock of the latest American pattern; greatly improved illumination of stations and carriages by electricity.

The rolling stock consisted of 24 motor coaches and 33 trailers, fitted with Westinghouse electro-pneumatic multiple-unit control. The 1903 motor cars had four 100 hp DC motors mounted two per bogie, and they followed the pattern set by the American elevated railways and interurban lines in the 1890s. The bodies were wooden, with clerestory roofs and rectangular windows above 'matchboard' lower bodywork, where the joints of the vertical 'matchings' provided good drainage for rainwater. Many of the first generation of British electric railway vehicles were to be of this type. The cars were 60 ft long. The bodies were carried on American underframes, and on diamond frame bogies. Each motor car had a cab at one end only. The end platforms were open, except for the driving end of the motor cars, which was totally enclosed. Open ends, and the use of gatemen to work the protective lattice gates, were features of several early electric railways. When labour costs began to rise, this uneconomic practice was eliminated by enclosing the ends and forming vestibules with entry doors. This was done between 1904 and 1912. The doors were not driver-controlled but were opened inwards by the passengers. Buhoup combined centre buffer and coupling was fitted. Air sanding gear was used, though it was later removed, and Gahan (1983) reports that motormen were then expected to sand the rail by hand from a supply-bag on the rare occasions when it was needed. The bogies were of the Baldwin-Westinghouse type, from the Baldwin Locomotive Works in Philadelphia. Gahan writes that the car bodies were built by Milnes, of Hadley, Shropshire, and that four additional trailer vehicles were built in 1908 by Milnes Voss & Co., Birkenhead, with the trucks constructed by Mountain & Gibson of Burnley, East Lancashire. All electrical equipment and control gear was manufactured by Westinghouse, and all fitting out and completion was done at the Trafford Park works

The electrified Mersey Railway used multiple-unit control from the start. Multiple-unit control of trains was demonstrated by Sprague in the 1890s, and Westinghouse introduced his own system. This was fitted to Mersey Railway stock so that a train with a motor car at either end could be driven from one cab. Controls were fitted to selected trailer cars from 1904 onwards to enable trains to be divided in slack periods and run as short trains driven from either end. The drivers controlled the trains using a small, vertical handle, which regulated the main drum-type controllers in the compartment behind the cab. These main controllers regulated power supply to the motors using the Westinghouse electro-pneumatic apparatus. Gahan (1983, p. 45) relates that the electro-pneumatic control apparatus caused a 'miniature electric storm' in the front and rear control compartments, so that: 'These compartments were lined with asbestos slate and so were parts of the underframes in the vicinity of equipment likely to heat or flash.'

A remarkable feature was the lack of train-mounted compressors for the Westinghouse air-brake and pneumatic control gear. The air reservoirs on the vehicles were charged between trips by the guard and the motorman using a flexible air-hose, which they linked to an air-main supplied from an underground compressor. This system was not replaced until 1938.

5.4 Economic benefits of electrifying the Mersey Railway

The Mersey Railway electrification attracted a great deal of attention, and its economic fortunes after 1903 are summarised in a detailed review by Shaw (1910). There was no doubt that electrification of traction and auxiliary services had saved the company, but there was dispute concerning methods for accounting the savings arising from electrifying a line. These disputes concerned all electric railways, not just the Mersey company, and are found in papers published between 1900 and 1910. Electrification costs depended on traffic density, route length, electrified track mileage, working schedules, train weights, gradients and distances between stops. These varied from one line to another. By 1903, there were many different kinds of electric railway at work or projected, and the lessons of one line, such as the Manhattan division of the Interborough Rapid Transit Company, New York City, did not necessarily apply to a simple, short route like the Mersey Railway, the Liverpool Overhead Railway, or the City & Great Northern Railway in London. Electric traction was reliable, and it increased the number of passengers carried per mile, but there was dispute over economic analysis and comparisons with steam railways. Advocates of electric traction argued that costs of new rolling stock and station improvements should not be included as electrification costs because they would be required on any line being upgraded, regardless of the traction mode. Calisch (1913) listed the principal items making up total capital cost as the power house; the transmission system; and the electric traction equipment in the rolling stock. He warned that (pp. 85–86):

Figures relating to the cost of these items for another railway are no guide unless one is fully conversant with the local and exact operating conditions of the line to which they relate.

Calisch (1913) quotes typical capital costs as power house: 20–40 per cent of total cost, with an average of 30 per cent; transmission lines, substations and track equipment: 35–55 per cent with an average of 45 per cent; and electrical equipment on rolling stock: 20–30 per cent with an average of 25 per cent. Estimates of capital cost of electrification per train mile depended on the interest charged on capital. Taking 4 per cent as a typical value for the period 1900–10, Calisch quotes the capital charge per train mile for five British electric railways as follows: District Railway (London), 4.5d not including costs for power station; Lancashire & Yorkshire Railway, 2d excluding rolling stock costs, 2.5d including rolling stock costs; London Brighton & South Coast Railway (London Bridge to Victoria section) 5d not including costs for a power station; Mersey Railway, 4.3d for the complete electrification scheme; North Eastern Railway, 1.93d not including costs for a power station.

However much argument there was about accounting methods and how conclusions might be drawn concerning general railway electrification, it could not be disputed that the Mersey Railway electrification was a success. In a paper to the Institution of Civil Engineers (1909), Mr Joshua Shaw stated that with reference to the Mersey Railway:

... with electric traction 1 lb of coal costing 8s 9d per ton will move one ton of load 2.29 miles at an average speed of 22.25 miles per hour, whereas with steam traction the same weight of coal costing 16s per ton would move the same load 2.21 miles at an average speed of 17.75 miles per hour.

The power stations of electric railways could burn much cheaper coal than could steam locomotives. Before electrification, the yearly coal bill for the Mersey Railway was £7100; after it was £5246. Electric traction worked faster trains, and fuel costs per train mile were reduced from 5.49d to 1.67d. Electric rolling stock worked higher mileages per day than did steam locomotives and steam-hauled stock. J. Aspinall, the Chief Mechanical Engineer of the Lancashire & Yorkshire Railway, and later its General Manager, stated that the average mileage worked per year by a company steam locomotive was 20,000, whereas an electric motor car, or an electric locomotive would work about 50,000 or 60,000 miles. Annual mileages for Mersey Railway stock, before and after electrification, were quoted as follows: steam locomotives: 17,274; steam-hauled coaches: 22,672; electric motor cars: 48,064; electric trailers: 36,453.

Calisch quotes maintenance costs in pence per train-mile for the Mersey Railway, before and after electrification, as 3.47d for steam stock and 1.67d for electric stock. He remarks (p. 94):

The annual cost of maintaining a motor car is, however, somewhat more than the maintenance of a steam locomotive, but owing to the greater number of miles run per annum by the former the cost per mile run is considerably less.

The reduction in traffic expenses was from 14.50d per train mile with steam traction to 6.35d with electric traction. Electric traction greatly increased train mileage with less rolling stock, which was more economically operated. The following comparison between steam working on the Mersey Railway in 1902 with electric working in 1912 shows the improvement. The figures are for the half-year ending 1902; and the half-year ending 1912. Gross earnings increased by 102 per cent from £29,470 to £59,651; working expenses increased by 11.3 per cent from £28,081 to £31,245, but net revenue increased by 1950 per cent from £1388 to £28,406. Total paid-up capital increased by 13.9 per cent from £3167,563 to £3609,108. Operational costs per train mile showed considerable reductions and are quoted in the writings of Shaw and Calisch. The case of the Mersey Railway attracted much attention, and perhaps it encouraged too sanguine expectations about the benefits of electrification which were not realised. The extraordinary increase in net revenue was due to the abnormal conditions found on the railway immediately prior to electrification. Net revenue had fallen from £6074 to £1388 in the last complete half-year of steam traction, and it fell still more to £216 in the following half-year when four months were worked by steam and two by electric traction. Electric traction raised the net revenue from £216 to £28,406 for the half-year ending 1912, and after paying interest on the additional capital, the net half-yearly revenue was increased by £19,902.

5.5 The prospects for electric traction in Britain after the electrification of the Mersey Railway

The Mersey Railway demonstrated the technical and economic advantages of electric traction in heavy-duty service. It encouraged the replacement of steam traction on systems where steam engines could not meet increased demands for higher operating speeds; more frequent services; decreased costs per train mile, and elimination of smoke nuisance. The simple LVDC system established electric traction as superior to steam traction in rapid-transit operations on systems like the Liverpool Overhead Railway, the Waterloo & City Railway, and the Mersey Railway. It suggested that LVDC would meet the needs of other rapid-transit railways considering electrification, such as the Metropolitan and District lines in London, and elevated and underground systems in the USA and Europe (Calisch, 1913; Parshall and Hobart, 1907; Shaw, 1910). However, it did not establish electric traction as superior to steam for all urban and suburban rapid-transit operations, and it definitely did not indicate any general superiority of electric over steam traction at that time. In the most intensively worked rapid-transit services, in tunnels or on viaducts, electric traction was technically superior. Steam-worked rapid transit within cities involved frequent stops, with small locomotives accelerating and decelerating relatively heavy trains over routes with sharp curves, steep gradients, and wet tunnels, and here electric traction effected great economies. The innovations that improved the main line steam locomotive between 1910 and 1930 could not be incorporated easily into the small machines used in rapid-transit work. On these lines steam traction could not meet higher performance levels which could only come via electrification. On main lines, the improvements to the steam traction system sufficed, and electrification was limited to special circumstances before 1930.

In the 19th century, encouraged by the successful advent of electric traction, a general displacement of steam traction was anticipated by engineers like Westinghouse, Edison, and Webb – the Chief Mechanical Engineer of the London & North Western Railway. Speaking at the Jubilee of the Crewe Mechanics' Institute in January 1896, F W Webb said:

In 10 to 15 years from now, trains moved by electricity will run from all the large centres of the country at a rate of speed which can hardly be realised, probably 100 mph.

This did not happen, and those who anticipated extensive electrification of British railways in the early 20th century were disappointed (Mordey and Jenkin, 1902). The plans made by Westinghouse for his Trafford Park factory received a major setback, and the site passed under British control. Only slowly did it become apparent that one could not argue the case for general railway electrification from a few successful examples drawn from rapid transit. Electrification was expensive and was undertaken if there were convincing technical, economic or other reasons. In some cases, such as New York, the law abolished steam traction from within city limits and the railways had no choice but to electrify. The threat of competition from tramways and motor buses drove

Figure 5.3 Lots Road power station built in 1905 beside the Thames at Chelsea, London, to supply the District Railway. It was then the largest power station in Europe and the biggest traction power station in the world.
Source: London Transport Museum.

companies that could afford electrification to implement it. Many companies could not afford electrification from their own resources, and lacking state aid carried on with steam traction as best they could. The lack of a standard electricity supply grid covering a wide area discouraged railway electrification. Without a public supply, railways had to build their own stations, which added greatly to cost of fixed works. Overall thermal efficiency of stations was low before 1910, after which date the gradual introduction of water-tube boilers, high-steam pressures, large turbo-alternators and vacuum condensing improved fuel economy. Outside the range of hydroelectric stations, coal-fuelled electricity supply was relatively inefficient and modern steam locomotives were often more economic, and if they could cope with traffic, they were used. Low power station thermal efficiency was compatible with technical and economic success on the Mersey Railway, or the District line in London, but not on a long main line of 200 miles. Between 1910 and 1930, the steam traction system was considerably improved. Superheating, wide fire grates, piston valves and rational proportions transformed the locomotive following the work of Goss, Lomonossoff and Chapelon (Ahrons, 1927). These did not require great investments, and no retraining of the workforce was needed to exploit them. Applying American rational management to the steam traction system improved utilisation of equipment so that between 1910 and 1930 a well-managed steam railway could outperform an electric railway in most instances. On many urban and suburban services out of the great conurbations, on semi-fast medium-distance trains, steam traction was good enough. On secondary lines, cross-country services, goods trains, for shunting, and for stopping trains there was no alternative to the steam engine before the diesel established itself circa 1940 in the USA.

Extensive electrification of railways needed a standardised supply grid and state involvement to compel groups of private railway companies to agree on a common system. During the years of economic crisis between the wars electrification without some state help was not likely. Before the amalgamation of 1922, there was little chance of a national electric railway network being developed as there were too many independent companies, and no national supply grid. Extended electrification came with nationalisation (1948) and may have ended with privatisation. The first phase of railway electrification was largely confined to extending the LVDC railways associated with the underground, elevated, or inner-city networks of the great cities. In Britain much was done before the amalgamation of 1922, when several companies installed limited networks of LVDC out of their major termini. The growth of the Lancashire & Yorkshire Railway electric lines out of Manchester Victoria and Liverpool Exchange is an example. The London & South Western Railway system out of London Bridge and Waterloo became the core of a network which was further developed by the Southern Railway, and the Southern Region of British Railways. This 660/750 V DC conductor rail system, together with the similar systems found in the London area, working on the four-rail principle out of Euston (LNWR, LMS), or operated by London Transport, became one of the biggest electric railway networks in the world as it underwent steady expansion after 1920. It became

64   *Electric railways 1880–1990*

the largest LVDC system anywhere, and its initial success was owed to the general use of centralised AC stations, AC distribution and motor-generator substations feeding the conductor rails.

Westinghouse demonstrated on the Mersey Railway what electric traction could do, but there were engineering and economic reasons why electrification could not be extended, at that time, beyond those rapid-transit routes where steam traction could not deal with a rising demand which the railways determined to meet. The most important of these in the history of British electrification were lines making up sections of the London Underground.

5.6 Electrification of the London Underground

The electrification of the London Underground was a major event in the history of electric railway traction. In Britain it was as significant as Sprague's first use of electric traction on the Chicago elevated lines. The transforming effect of electric working on the London 'cut and cover' lines from 1900 onwards is charted by Howson (1986). These were not small-scale railways worked by light equipment inferior in quality and performance compared to the main line steam railway; the London Underground lines were intensively worked railways moving loads which required rapid acceleration and deceleration along tightly curving routes with frequent stops. Electric traction had definitely established itself as a reliable alternative to steam traction once it displaced steam locomotives from the busiest sections of the London Underground. The first sections selected for electrification were those forming the Inner Circle Line, which linked main line termini north of the Thames with the City and Westminster. Most services were operated by the Class A engine, which remained the standard from its introduction in 1864 to its disappearance after electrification of the lines that used it. This 4-4-0 tank engine remained at work in London Transport days, on non-electrified lines, and as a civil engineer's locomotive. On the Metropolitan lines, many survived until 1919, the date of their general withdrawal, but the last one worked as a branch-line goods engine until 1948. These classic engines were built by Beyer-Peacock and were practically perfect in every way so that they required few modifications. They determined the general level of performance and efficiency of the steam-worked underground from 1864 until electrification. They were designed by Sir John Fowler and W. Adams, and over 120 were supplied between 1864 and 1886 (Ahrons, 1927). They represented best practice for the 1860s and worked exacting duties demanding high-power outputs. Howson (1986) provides a brief account of what steam operation on the Inner Circle Line was like in 1893 when the 13 miles of the circuit were completed in 70 minutes, inclusive of 27 stops and one engine change – a praiseworthy performance compared to the 50 minutes required with electric traction in 1986. Boiler pressure was 9.4 bar, and weight in working order was 42 tonnes, of which 31 tonnes were on the driving wheels. They were as big as a main line 4-4-0 express engine (without tender) built for work between 1860 and 1900. As late as 1900 these 4-4-0 tank engines were large, powerful units by any standards, in a completely different power, weight and speed category from the electric street car or the light trains on the Liverpool Overhead Railway or the City & South London Railway. Any electrical system taking their place would be an advance beyond the engineering equipment of a street trolley line, however well-proven the latter. Sprague had faced such a challenge when electrifying the Chicago Metropolitan West Side Lines (1895) and the South Side Elevated Lines (1897). The necessary engineering systems marked such an advance that the directors of the latter only allowed Sprague to proceed after he agreed to make good the consequences of failure. If electric traction could displace the Class A tank engine, it could compete with steam traction on the short haul, heavy-duty, densely used railways working in the great conurbations.

Electrification of the underground and elevated lines did not simply replace the steam engine by an electric unit, and leave all else the same. The electrification of rapid transit resulted in the close integration of components, and the optimisation of systems. It was quickly recognised that an electric railway system could be closely monitored; and that there could be automatic, semiautomatic and centralised control protected by safety devices to an extent not possible with steam traction. Techniques which had failed to realise their potential on steam-worked lines became vital components on electric railways. Colour-light signalling, electric control, powering of points and track circuiting were used on steam railways, but there was never that continuous integration of all subsystems and components into a single whole that electrification promised from the start.

5.7 The influence of American electrical engineering

George Stephenson believed that (Johnson and Long, 1981, p. 201) 'the day will come when electricity will be the great motive power of the world', but when the new motive power did come to British railways it was largely the consequence of American enterprise in an age of industrial endeavour dominated by American practice. The City & South London Railway (Lascelles, 1955) was the first electric underground line in the world, running with locomotive haulage and 450 volts third-rail distribution. This was a noteworthy British achievement, and though a commercial failure, its technical success encouraged further electrification schemes in the 1890s, including those conducted by the Metropolitan Railway in 1898. The 1898 Metropolitan tests were exploratory in nature, to define the elements of the future system which would eliminate smoke and increase capacity (Bruce, 1983).

The Metropolitan Railway carried out its own electrification experiments at Wembley Park, but in 1898 joined the District in financing a joint experiment on the latter company's line from High Street, Kensington to Earl's Court. Power was provided at 600 volts d.c. and supplied to two conductor rails arranged one on either side of the running rails .... The provision of two conductor rails was probably arranged for two reasons.

First it was to avoid the need to improve the conductivity of the running rails through the fish plates and crossing connections and secondly, to avoid interference with block signalling apparatus by stray earth currents if the conductivity of the running rails was bad.

The four-rail variant was deemed necessary in metal tunnels and in other rights of way which were in close proximity to cables and telephone lines, gas and water mains, and where ground currents had to be kept low to reduce interference and corrosion. In the early Metropolitan Railway experiments, converted steam passenger stock was used, and rebuilt 'steam stock' was widely employed on a regular basis on the underground and surface electric railways running into London. Converted steam stock was used well after World War Two, and its use did lead to a form of multi-unit train distinctly different in structure from the standard American rapid-transit form. The set of cars used by the Metropolitan and the District Railways for the 1898 trials did not follow any American example, and was newly built by Brown, Marshalls & Company, with Siemens electrical equipment. This train was not completed until 1899, when it took part in the series of trials started in 1898. Its basic form was of steam-stock passenger coaches with electrical gear installed in the luggage compartment, rather than the American type motor car which became more usual shortly afterwards. The experimental train had two motor coaches and four trailers, with each motor coach having four gearless motors, one per axle (Bruce, 1983):

Although a bus line was provided through the train connecting all the collector shoes together on both motor coaches, no through control lines were provided, to that only the motor coach at the leading end was powered when the train was in motion. It is understood that one of the original conceptions of the proposed electrification scheme for the Inner Circle line was the provision of only one motor coach per train, fitted with a driving cab at both ends. This vehicle would then have been required to run round the train like a locomotive to reverse.

This latter feature marked a radical departure from the Sprague multiple-unit norm, and was not established. The experiment was sufficiently successful to encourage the Metropolitan Railway (Baker, 1951), and the District Railway (Lee, 1956), to go ahead with electrification. Nine tenders were offered to the company, which indicates the diversity of design philosophies, though the Sprague form met the challenge of the alternatives, some of which were ill suited to rapid-transit work. An offer was received from Ganz, for a 3 kV three-phase system, using two overhead trolley wires per track with the running rails taking the third phase. Ganz quoted a low price. They reasoned that motor cars could haul existing steam stock – whereas multiple-unit trains would need new carriages – and claimed that manned substations would be unnecessary with the triphase system: hence the low estimate of costs. The Ganz tender was accepted despite the novelty of its engineering equipment, but the American financier and rapid-transit 'tycoon' Charles Tyson Yerkes acquired control of the District Railway, and rejected the Ganz offer after pressure from his American advisers. This led to dispute, arbitration and a decision in favour of the low-voltage, direct-current, third-rail system (Barker and Robbins, 1963, 1974). In the event, the four-rail system was used to meet the legal requirement laid down for street tramways that the voltage drop in the running rails, when used to take return current, should be limited to seven volts to prevent leakage currents of a strength sufficient to damage buried mains and telegraph cables. The four-rail system was successful, and the last steam train ran round the Inner Circle on 22 September 1905. The two independent companies, the Metropolitan and the District, which operated their own trains round the Circle agreed to extend the four-rail LVDC system elsewhere.

Yerkes' control of the District Railway, and his rejection of the Ganz scheme in favour of LVDC systems after the 'Sprague' model, established American engineering as usual before the completion of electrification of the Circle Line. American practice guided later District Railway schemes for electric traction. The former Ealing & South Harrow Railway was incorporated into the District company in 1900, and by June 1903 was electrified after the American model using Class A stock (Bruce, 1983):

The A class comprised two seven-car trains built by the Brush Electrical Engineering Co. of Loughborough. The cars were 50.25 feet long and were basically of American design, bearing a striking resemblance to cars constructed about the same time for the Interborough Rapid Transit Company of New York.... One train was equipped with BTH equipment and GE66 traction motors similar to the multiple unit equipments which had already begun to operate on the Central London Railway. This train, however, incorporated a 'dead man's handle' type master controller, following American General Electric practice already established on a number of railways operating in Chicago and New York.... The BTH equipment consisted of individual electromagnetic contactors controlled directly from the 600-volt traction supply without an accelerating relay so that hand notching was under the direct control of the driver. This system was known at the time as the Sprague-Thomson-Houston System.

Another train was fitted with Westinghouse equipment for comparison, and thus the A stock incorporated work by nearly all the great pioneers of electric traction in the United States: Brush, Sprague, Thomson, Houston, Westinghouse and Edison. The British Thomson-Houston (Sprague) equipment proved the more reliable and was standardised by the District Railway, though the Westinghouse brake became the general standard, rather than the Christensen brake fitted to the British Thomson-Houston train. Unfortunately, the useful sanding apparatus fitted to improve wheel adhesion on rail had to be discontinued as the sand interfered with the track circuits of the newly installed automatic signalling system. Automatic signalling was successfully demonstrated in Britain on the District Railway by engineers working for the 'traction financier' Tyson Yerkes, who were electrifying by American methods.

In 1906, the electrified District Railway installed new equipment that represented an advance at components and systems level. This was the Train Describer and Recorder-Receiver which helped to control and monitor the movement of trains. Signalmen were aided by track circuits and the illuminated track diagram which showed train position on a schematic representation of the route. This was first used in 1905. Signalmen used the Train Describer and

Receiver to send train destination, coded as a particular pattern of movable metal studs arranged round a drum. This electromechanical memory stored information in a four-digit binary code and was decoded by a Combinator fixed above the Receiver. Between 1890 and 1910 new devices such as this were used to integrate complex systems, with self-regulation and automatic action to a degree not possible on the steam railway. The Strowger switch used in automatic telephone exchanges is another example of an innovation dating from this period which heralded a new sort of engineering (Pierce, 1981). The Strowger company produced an early form of cab-mounted warning signal known as the Strowger-Hudd system.

The electrification of the District Railway was an exercise in penetration of a foreign market by systematic organisation of financial and engineering strategies, using new systems that displaced the native equipment which was outclassed from the technical and economic standpoints. The 'New American Methods' in technology and management were too potent to ignore. Automatic signalling was a case in point (Bruce, 1983):

This was a development of that first introduced in 1901 on the Boston Elevated Railway. Track circuit signalling was materially assisted by the use of the fourth rail traction system, which left the running rails clear of traction current so that they could be utilized exclusively for signalling requirements. The system required the running rails to be divided into insulated sections by means of block joints. A train or just a pair of wheels on any section caused the two running rails to be shorted together detecting the presence of a train. Incorporated with the track circuit was the train stop and trip cock. When a signal was at danger the train stop was raised and should the approaching train overrun, the trip cock would come into contact with the train stop, the brakes then being automatically applied. This system, apart from the problems caused by sanding, was very successful and was adopted throughout the District, over the congested parts of the Metropolitan, and also on the Tube lines.

These cars were fitted with the American 'Ward' mechanical coupler, which became the standard coupling for the Underground group of companies until replaced by an automatic coupler, which not only linked the cars – as did the Ward coupler – but made the electrical and pneumatic connections as well.

The American influence set the pattern for the Underground group of companies. The vehicles were known as cars instead of carriages and for a time the bogies were known as trucks. The American term 'motorman' was adopted for the driver. In addition the practice generally adopted in New York of having only one motor truck per motor car became standard on the District for a long time.

The B classification stock of the District Railway resulted from the 1903 decision to electrify the entire system, and to eliminate steam traction completely, following the success of the Ealing & South Harrow electrification. The order was for 420 new cars, to provide 60 seven-car trains, 12 made up with four motor cars, and 48 with three motor cars. Bruce reports that they were of the basic American design with BTH traction equipment and GE69-type traction motors.

Much to the chagrin of the British car-building industry the order for 280 of these cars was placed with a French syndicate . . . The 140 cars built in England were constructed by Brush of Loughborough and the Metropolitan Amalgamated at its Ashbury and Lancaster Works.

The cars, completed in 1905, carried Sprague-Thomson-Houston control gear. One set on each motor car controlled two GE69 traction motors placed together in one powered bogie. The other truck was unmotored, and carried the collector shoes. This stock was successful, though the original American cast steel motor bogies suffered repeated fractures and were replaced by plate framed riveted construction which became standard. The basic components of the new electric railway were first integrated on the main line scale of loading gauge on the Great Northern & City Railway. This company was formed in 1892 to link the Great Northern Railway at Finsbury Park with the City of London at Moorgate by a railway laid in a tube capable of taking main line rolling stock through tunnels of 16 ft diameter instead of the usual 11 ft 8.25 in diameter. It was only 3.5 miles long, but it was an important line because it unified several key components of the traction system into a main line-sized prototype which worked well. Severe financial problems delayed the start of construction until 1901, when the British Thomson-Houston Company won the contract for all electrical work. This included 11 seven-car trains and three separate motor cars held as spare (Bruce, 1983):

The Sprague-Thomson-Houston 'train control' system was specified. This was a pioneer decision as far as England was concerned, since this type of equipment had only just passed beyond the experimental stage in America. Although the Central London Railway had the honour to have been the first to operate multiple-unit trains in Europe, the Great Northern & City made the decision to use such equipment some time earlier. But since passenger service on the line did not begin until 14th February 1904, the railway could not claim that it had the pioneer installation of this equipment.

The Great Northern & City Tube Railway was amongst the first in Britain to use track circuit automatic signalling. A four-rail system was installed with both the positive and the negative current rails outside the running rails, which avoided interference between traction current and the track circuit. Before 1906, the Sprague-Thomson-Houston controllers were not provided with a dead man's handle, so that assistant motormen were carried on full-length trains, though two-car units were fitted with it and could be driven by one man. Before train doors were closed by the guard or the driver, underground trains on certain lines carried six or seven men simply to close the end-platform gates of the cars and to check that the train could be started without injuring anyone. This could be tolerated when labour was cheap, but the rise of labour costs after the Great War hastened the introduction of labour-saving devices wherever possible. Apparatus for operating doors from one point by lever or push button is an example. The bogie design of the Great Northern & City Tube Railway followed the pattern set by the McGuire Manufacturing Co. of Chicago, which used one traction motor per bogie, and two bogies per car. This did not become standard

70 Electric railways 1880–1990

Figure 5.4 Steel bodied rolling stock built in 1904 by Brush for Great Northern & City Railway, London.
Source: London Transport Museum.

practice on the London Underground railways and the single-motor bogie was not used again until 1935, when it was fitted to an experimental streamlined tube train of that year. Transverse seating arrangements within the GN & CR cars were 'in the American fashion'.

The Great Northern & City line was not a financial success, because of the high outlay required by the large loading gauge and the expensive tunnel work, and it was bought by the Metropolitan Railway in June 1913. In this year, the Metropolitan Railway placed orders for 23 motor cars and 20 trailer cars which departed from the basically American type of electric rolling stock that then dominated much of the Metropolitan and District systems, though District Railway rolling stock constructed between 1910 and 1913 showed American influence. The origins of British electric railways in the rapid-transit systems, elevated railways, and interurban networks of the United States were always evident, which is natural granted the dominance of the British market by American companies, or their British subsidiaries. British rolling stock development is described by Benest (1963), Bruce (1968), Hardy (1986, 1987) and Prigmore (1960). Westinghouse influenced Britain through the Metropolitan Carriage and Waggon Company, which later became part of Metropolitan Vickers (Leupp, 1919). There were also the Edison and General Electric groups which became Edison-General Electric, and later General Electric (Reich, 1985). Sprague, who had been assistant to both Farmer and Edison, formed a company which merged with the Edison group. Elihu Thomson and E. J. Houston founded Thomson-Houston, which took over the van de Poele enterprise in 1889, and then merged into General Electric. Brush founded a company which likewise went into General Electric. Such eminent electrical engineers as Steinmetz and Batchelder worked for General Electric at this time, and the considerable number of influential and powerful men who had worked for Edison at some time included Insull, the traction magnate who controlled much of the interurban network: he had been Edison's secretary. Ward Leonard had also worked for Edison. Henry Ford worked for Edison, and was encouraged in his own researches by him (Ford and Crowther, 1922, 1926). Ford acknowledged that he took Edison as an example, and he remained a close friend until Edison died. Tesla also worked for Edison, but became adversely critical and went to work for Westinghouse, with whom he developed the fundamental technology of AC systems. Bernard Shaw worked for the London offices of the Edison company, and he used the experience in his satirical attacks on the 'new man' being created by 19th-century culture.

There was considerable German expertise in electric traction in the 1880s and 1890s, and a demonstration of Siemens' pioneer electric locomotive in Berlin in 1879 had roused Edison's interest in this subject (Tassin, Nouvion and Woimant, 1980). The lack of a sustained German effort to penetrate the British market is striking. By the end of the 19th century, American practice was unrivalled, and any native British design tradition emerged slowly and gradually on the Metropolitan Railway surface lines, and on the suburban routes of the London & South Western; the London, Brighton & South Coast; and the

London & North Western railway companies. Even here, American influence was strong in the early years. Werner von Siemens' 1878 demonstration of a small electric locomotive suitable for mine railways was a notable achievement, and the public trials of this system in Berlin in 1879 were a great contribution to electric traction by Germany, but it was the energy and effectiveness of the American 'Edison Faction' after 1880, in engineering, finance, management and publicity, which was remarkable (Dickson and Dickson, 1894; Hughes, 1976; Josephson, 1959; Reich, 1985). Between 1900 and 1910, disciples of American methods established the electric railway in the British capital and penetrated the growing electrical industry and the financial world. British managers of genius modelled themselves on the American pattern. Albert Knattries was born in Derby in 1874, and emigrated to the USA where his name was changed to Albert Stanley (Ellis, 1959). He made a career in public transport, particularly with the Detroit electric street cars (Schramm, Henning and Dworman, 1978, 1980), and became General Manager of the New Jersey Street Railways in 1907. He returned to Britain and was appointed General Manager of the District Railway by Chairman George Gibb (Lee, 1956), who was responsible for organising the once-independent railways into the Underground Electric system and the London Electric Railway. Stanley followed Gibb, and became Managing Director of the Underground (Barker and Robbins, 1963, 1974; Day, 1974; Howson, 1986). He was knighted in 1914, and created Baron Ashfield of Southwell in 1920. He was instrumental in creating the London Transport of the inter-war years, and he did much to modernise equipment, including the reconstruction of Lots Road power station between 1932 and 1934 (Howson, 1986). Like Henry Ford, he got an engineering education in the Detroit electrical industry.

Chapter 6

Electrification 1900–20

6.1 The Great Eastern Railway and the Liverpool Street Station experiment

The success of the first electric railways in London suggested to the general-purpose railway companies that electric traction might suit the extensive and heavily used urban and suburban networks radiating from the larger London termini, particularly Waterloo, London Bridge and Liverpool Street. In 1900, few British railway companies were willing to undertake expensive works unless they were certain that the money would be recovered. The underground lines were practically compelled to electrify but the general-purpose companies were not dependent on rapid-transit services for their entire income as were the Metropolitan and District railways, and similar systems. Only a limited number chose to electrify parts of their system, where steam traction worked near its limit. The London & South Western Railway, the South Eastern Railway, the London, Brighton & South Coast Railway, and the London & North Western Railway chose electrification, but the Great Eastern Railway elected to retain steam traction after improving its performance by using methods learned in the United States. This was done by reorganising the use of existing equipment, and the senior management of the company paid much attention to contemporary 'American methods'. The GER manager, Sir Henry Thornton, earned a reputation for applying such methods, and he organised the Canadian National Railways in the 1920s, and became an early advocate of oil-electric traction. In 1913, there were several electrified heavy-duty suburban lines in operation or under construction (Hardy, Frew and Willson, 1981; Frew, 1983; Beecroft, Frew, Holmewood, Rayner and Stevenson, 1983). There was considerable variation in voltage and distribution systems. The Mersey Railway from Liverpool Central station to Birkenhead Park started electrical operations in 1903 with 650V DC third and fourth rail distribution. The Lancashire & Yorkshire Railway 650 V DC third-rail North Liverpool network grew between 1904 and 1914. In 1913, at Holcombe Brook, near Bury, the L &Y tested a 3.5 kV DC overhead distribution which was converted in 1918 to the 1.2 kV DC third-rail system. In 1908, the Midland Railway opened its electrified line from Lancaster to Heysham using 6.6 kV, 25 Hz AC with overhead distribution, the first AC railway in Britain. The North Eastern Railway electrified system on Tyneside grew between 1904 and 1909 using 600 V DC third rail (Hennessey, 1970; Hoole, 1987, 1988). The Great Central Railway's Immingham Tramway, a British interurban, opened between 1912 and 1915 using 500 V DC overhead distribution. The London, Brighton & South Coast Railway, with Dawson as consultant, used 6.7 kV, 25 Hz AC, with overhead distribution. This was installed and extended between 1909 and 1912, with further additions down to 1925. Amalgamation into the Southern Railway took place in 1922, and by 1929 the LB&SC AC system was converted to LVDC third rail in the interests of Southern Railway standardisation (Bonavia, 1987). The South Eastern & Chatham Railway projected an extensive network of 1.5 kV DC third-rail lines using multiple-unit sets and locomotive working of long-distance passenger and goods trains. This was never implemented, and the former SECR lines were eventually electrified in Southern Railway days using the 660/750 V DC 'London Standard' (Bonavia, 1987; Moody, 1979). The London & South Western Railway 750/850 V DC third-rail network was developed after 1905, and it gave rise to the 750 V DC standard which formed part of the 660/750 V DC system operated by the Southern Region of British Railways, which is the biggest third-rail system on earth (Moody, 1961, 1963, 1979). The candidates for electrification circa 1913 were those suburban and interurban lines which took heavy traffic at high speed with frequent stops, and those urban, elevated or underground systems taking heavy traffic at moderate speeds with frequent stops.

In the 1920s British engineers like Dawson (1909, 1923), Raven (1921–22) and O'Brien (1920) proposed extending electrification, and they outlined the first schemes for main line electrification in Britain, supplied from coal-fuelled stations. The schemes, whether for rapid-transit or main line services, met with opposition from the senior managers of the home railway companies. The economy and fuel efficiency of electric railways were still contentious issues at this time. Calisch reported in 1908 that the overall thermal efficiency of the electric rapid-transit system, from power station coal supply to motor car driving wheels, exceeded that of rapid-transit steam traction, and he recorded that an electric train needed 49 lbs of coal per train mile, and a steam locomotive on similar duty needed between 80 and 100 lbs per train mile. This was in rapid-transit work, where steam traction was inefficient due to short runs (no time to build up superheat – hence superheaters were seldom fitted) and when relatively long periods were spent halted in stations, not only burning coal, but increasing condensation losses as cylinders cooled. The need to steam hard during acceleration caused firethrowing and the loss of imperfectly consumed coal. Electric traction on the underground increased speeds by 30 to 40 per cent and effected considerable savings in power station coal, though the figures quoted by Calisch did not apply to comparisons with main line steam traction of modern design.

After the 1914–18 war the Great Eastern Railway needed to increase capacity on the steam-worked suburban lines out of Liverpool Street station in London. This system was a candidate for electrification, but the company could not afford it. There were two alternatives to electrification, assuming that things could not be left as they were. New steam locomotives could be introduced for working the suburban trains, with acceleration and tractive effort equal to those of electric units. The mechanical engineer Holden explored this possibility in 1902 and built a three-cylinder simple-expansion 0–10–0 tank engine, the famous 'Decapod', which could accelerate loads as quickly as contemporary electric traction if not as economically. This solution was rejected (Ahrons, 1927). The civil engineer's department objected to the heavy loads imposed on the track and bridges by 'Decapod' and replacing existing locomotives by large numbers of the new type would have been expensive. The other solution was to increase capacity on the lines out of Liverpool Street by making better use of existing engineering equipment with the help of automatic signalling and some reorganising of facilities. If this failed, then electrification would have to be faced, or loss of traffic to tramways and motor buses accepted. There was one factor which limited any proposed solution: the Great Eastern Railway was not prepared to rebuild the Liverpool Street terminus. This would involve an expense which the company could not meet. The second solution was the one which was attempted and it proved successful. It attracted much attention, not least from managers anxious to increase performance without electrifying lines. Considerable expense was involved though much less that electrification would have demanded. New layouts were installed at Liverpool Street, and separate engine docks were built for each platform road, located so that locomotives could use them without interfering with train movements. Engines could be watered, and coaled at the terminus, and other services – such as oiling or fire-cleaning – were done at suburban stations. Every effort was made to service engines during stops when trains were loading or unloading. Other improvements were effected by relocating signals and changing track layout.

The Liverpool Street reorganisation was sometimes interpreted as proving that steam traction could rival electric traction in rapid-transit duty. It is important not to misinterpret this example of an older engineering system staving off replacement by a new one which had established itself elsewhere in London. The GER retained steam traction as the short-term solution to operating problems because it was too poor to afford electrification. In consequence, it carried on with a system which became inadequate over much of East London following the increased use of the motor vehicle, the electric tramcar, and the trolleybus in the 1920s. When this happened, the steam operated lines were closed, the services were withdrawn, or electrification was carried through as part of post-nationalisation state-funded policy. The Great Eastern Railway Company enjoyed excellent management and used American methods. It was these methods rather than American electrification which raised line capacity by 50 to 75 per cent in 1920. The full story is told in the paper 'The Last Word in Steam Operated Suburban Train Services' published in the *Railway Gazette* of 1 October 1920. The title is accurate: it was the last word, and nothing more could be done to improve the steam traction system in rapid transit. The most relevant extracts are as follows (*Railway Gazette*, 1920):

... a 50% increase in train frequency has been attained through the cumulative effect of a number of small operating and mechanical improvements brought about by the close co-operation of the Traffic and Engineering Depts., resulting in a two-minute service of Chingford, Enfield and Palace Gates trains over two lines between Liverpool Street and Bethnal Green. ... and into four platform roads at Liverpool Street terminus.

The anonymous author of the *Railway Gazette* paper stressed that it had been almost axiomatic to accept that the GER had done everything possible in its handling of its suburban traffic in and out of Liverpool Street. There were implications for electrification policy (*RG*, 1920):

The scheme was not conceived in any spirit of antagonism to electric traction, but rather as a makeshift or stopgap to surmount an immediate difficulty. On underground lines electric traction is obviously essential. ... but for surface lines it would seem that the success of the Great Eastern Railway experiment will have the effect. ... of making it a much simpler matter for a railway authority to decide when to electrify and when not to electrify. It narrows the issue very considerably. It has practically reduced the deciding factor to the elementary one of £.s.d.

The article argued that it was an error to suppose that increased frequency of services could only be got through electrification, and claimed that the GER experiment showed that the limiting factor was not service frequency, nor mode of traction, but the speed with which trains and platforms could be emptied. The reformed steam operations using existing rolling stock could provide the frequency required, but adding electric traction to these reformed operations would yield no benefits because the station entrances, doors, passages, stairs and walkways couldn't cope with the crowds anticipated if electric operations were introduced. The experiment was valuable because it showed which improvements were due to reformed operations, independent of traction mode, and indicated the likely consequences of electric traction. Certain benefits thought peculiar to electric traction could be achieved through operational reform with steam traction. The exercise arose from a study to find out what would be the benefits of electrification plus automatic signalling. Even with automatic signalling, steam traction couldn't handle the traffic possible on an electric railway with automatic signalling, but as mentioned above, the consequences of electric traction and automatic signalling would be to choke the station with passengers and cause chaos. This study was done at a time when the pressure on suburban services was increasing. Some years later, with motor buses, and tramcars threatening railway traffic, electrification and automatic signalling were regarded as techniques for stopping passengers leaving the railway for other modes of transport (Gordon, 1919). There can be no doubt that the revised steam operations were a 'phenomenal success' and they coped with increased traffic through the 1920s. They created an example of the steam hauled rapid-transit system brought to practical perfection, but the very success of the 'Liverpool Street Experiment' signalled finality, because there was little that could be done to improve its performance, whereas the contemporary electric systems were in the early stages of evolution, and capable of considerable development. The GER used well-organised steam traction to solve a pressing problem cheaply, but electrification out of Liverpool Street station, plus improvement to station passageways, would have been a better long-term investment. The steam services out of Liverpool Street were adversely affected by trams, trolleybuses, and motorbuses brought by new roads like the Silvertown Way into areas where the GER enjoyed a monopoly until the mid-1920s. One can but speculate on what would have happened to transport, and civic development in North East and East London, if electrification had taken place in that part of the city, on the scale it did on the Southern lines immediately following the 1914–18 war.

6.2 Electric rapid-transit railways and general railway electrification

The success of electric railways like the District Railway, and the Great Northern & City Railway, suggested that electric traction might be used beyond the limits of rapid-transit networks in the vicinity of large cities. Between 1900 and 1920, the common type of electric railway was the low-voltage, direct-current, three- or four-rail system developed for heavy-duty rapid-transit service. This was found in an advanced form in the London Underground lines, and in the electrified surface lines of the Metropolitan Railway, and those companies which became the Southern Railway in 1922. These electrified lines formed the 'London Standard' network of third-rail LVDC, with supply voltage between 660 and 750 V DC (Agnew, 1937; Hardy, 1986, 1987; Linecar, 1947; Moody, 1979; Prigmore, 1960). The entire network never was converted throughout to the higher standard voltage, and present-day rolling stock is designed to work between 660 and 750 V (Frew, 1983).

A brief summary of British railway progress in electric traction, in this transition period, is apposite. In Britain, railway electrification did not evolve out of an electric street tramway industry because there was none, and British railway electric practice was imported from the United States in parallel with British electric tramway practice (Jackson-Stevens, 1971). The British Isles were, however, the scene of notable pioneer ventures, (Hardy, Frew and Willson, 1981). The City & South London Railway, opened in 1890, was the first electric underground railway. The Liverpool Overhead Railway, opened in 1893, was the first overhead electric railway in the world, and the second urban electric railway in Britain. It was the first to use electric automatic signalling (with semaphores), and in 1921 it saw the first use in Britain of daylight colour-light signals. The line was extended in 1894 and 1896, and was operated with equipment built by Brown and Marshall (later Metro-Cammell), and Dick, Kerr & Co. of Preston. Stock could not be operated in multiple which limited trains to three-car formations. This important venture did not herald a major British initiative in electric traction, and the system was closed in 1956 and dismantled when the viaduct required extensive renewal.

The Waterloo & City Railway, opened in 1898, marked the beginning of the London & South Western Railway development of electric traction, leading to a great network of 750–850 V DC conductor rail lines. As noted in the previous chapter, the Mersey Railway, opened with steam traction in 1886, was electrified by Westinghouse in 1903, using four-rail distribution (converted to three-rail in 1955), with cars erected at the Trafford Park plant, from which the Westinghouse Company hoped to mount its programme for electrifying British railways and industry. The larger British companies were not slow to consider extending electric traction beyond the range of the short-run 'shuttle' service pioneer lines. The London & South Western Railway was a technically progressive company which commenced electrification in 1898 using imported American equipment on the underground line under the Thames between Waterloo and the City of London. An underground LSWR power house at Waterloo supplied LVDC until 1915 when a new surface station was built near the main terminus. In 1899 five double-cab single cars were supplied by Dick, Kerr of Preston to supplement the original American four-car sets, and in 1922 the LSWR built trailers for the line at its Eastleigh works. The success of this pioneer line influenced the LSWR to choose LVDC, with third rail, following its 1913 plan to electrify suburban services supplied from the railways own power station built in 1915 at Durnsford Road, Wimbledon. For this network, the side-contact conductor rail used in the Waterloo & City Railway gave way to the top-contact arrangement. One of the world's largest electric railway networks was brought about by the 1922 amalgamation which took the LSWR, LB&SC, and the SE&CR into the newly formed Southern Railway, which implemented a policy of standardising on LVDC because it was the most widespread system. The London & North Western Railway decided to electrify suburban services out of Euston Station, London, in 1911 using 630 V DC, with three- and four-rail distribution as necessary. Additional track was laid between Watford Junction and Camden Town, and the LNWR power station was built at Stonebridge Park near Willesden. Electric services started in November 1914, using four three-car sets supplied by Metropolitan Carriage & Waggon Co. The LNWR section of the North London Railway from Broad Street Station to Richmond and Kew Bridge was electrified in October 1916, by which time the LSWR section of this joint railway was already electrified for District line services. This operation of several company's trains over one joint railway compelled standardisation of electric systems from the start, and is one reason why the LVDC conductor rail system became the 'London Standard', The South Eastern & Chatham Railway planned to electrify a considerable part of its system, with general electric working of passenger and goods services, using 1500 V DC in a protected third rail, but this plan foundered when permission was refused to build a company power station at Angerstein Wharf on the Thames.

An important exception to the LVDC schemes was the London, Brighton & South Coast Railway plan to use overhead distribution with a single-phase AC supply, 6700 V, 25 Hz. In 1903, authority was given to electrify from Victoria to London Bridge, and public service started in December 1909. In 1925, the lines as far as Coulsdon and Sutton were electrified, with trains operated by 21 motor vans built by Metropolitan Carriage, Waggon and Finance Company. After absorption into the Southern Railway in 1922, the decision was taken in August 1926 to convert this AC network to third rail. Conversion started in June 1928, and was completed by September 1929 (Bonavia, 1987). Much AC rolling stock was converted for DC use in May 1929, and some sets worked until the mid-1950s. This line provided Britain with its only home-based AC experience on the large scale until the 1950s when British Railways accepted the 25 kV, 50 Hz AC system in place of the HVDC standard. The Midland Railway, in many ways a technically conservative company, opened a short branch line from Morecambe to its harbour at Heysham, working a shuttle service with a steam motor. The railway had built a power station to supply the electrical equipment of its harbour, and this was used to feed an AC traction system installed on the branch. This was the first use of AC traction on British railways. The equipment was similar to that used on the LBSC, the supply being single phase, 25 Hz, 6600 V. Electrification was authorised in 1906, for 9.5 miles from Lancaster, to Heysham via Morecambe. Electric working of local services began in 1908. Goods trains and boat trains were steam worked. Siemens and Westinghouse equipment was used in three motor coaches. In 1951, the frequency of the line was altered to 50 Hz, and the system used to investigate single-phase electrification using industrial frequency. The original sets, disused since 1940, were converted at English Electric (1953) and Metro-Vick (1957), using rectifiers and 750 V DC traction motors. One set ran in 1955 with a germanium rectifier (Johnson and Long, 1981; Webb and Duncan, 1979). This line closed in 1966, and though the British Railways experiment of the 1950s was a great success, the Midland Railway venture did not result in further electrification of the company's lines. In 1904 the Lancashire & Yorkshire Railway electrified the Liverpool Exchange to Southport commuter line using 650 V DC with three- and four-rail distribution, and electric systems were developed in the Liverpool and Manchester areas. Locomotive traction was tried in 1912 using a 2–4–2 steam engine frame, fitted at Horwich Works with four 150 hp motors, collector shoes, and a bow collector for use with overhead distribution in sidings (Mason, 1974). It saw service working coal traffic in the Aintree area, but was not duplicated though converting steam locomotives to electric drive using similar methods was occasionally proposed in this period. In 1913, the L&Y installed 3500 V DC, using overhead distribution, on the Holcombe Brook branch near Bury. This test was carried out in co-operation with Dick Kerr who anticipated foreign orders for HVDC (Vickers, 1986). In 1916, the decision was taken to convert this branch to 1200 V DC third rail following the start of passenger services on the nearby Manchester to Bury line which used 1200 V DC in a third rail. In 1917, before conversion, the 3500 V DC supply failed, so 1200 V DC was fed to the overhead wire and to the motors of a 1200 V DC set through pantographs and a cable, using the Bury line as a power source. Conversion to 1200 V DC was completed by March 1918, but electric traction was abandoned in March 1951, and the Holcombe Brook branch line closed in May 1952 after a brief spell of steam working. The amalgamation of the LSWR, LBSCR and SECR into the Southern Railway led to standardisation of systems in the London area using LVDC. Likewise, the merging of several undertakings to form the London Regional Transport system (Barker and Robbins, 1963, 1974) preserved the dominance of the same LVDC system throughout London Transport railways. The growth of a single company in London through successive stages should be noted: Underground Electric Railways (1902); London Electric Railways (1910); London Passenger Transport Board (1933); London Transport Executive (1948); London Transport Board (1963) and London Regional Transport Board (1984) (Day, 1974; Howson, 1986), and, at the time of writing, London Underground Limited. Any privatising of the system following the example of the Conservative Government's policy for the former British Rail will of necessity leave the LVDC standard in place.

A 'systems technology' like electric traction benefits from the grouping of smaller enterprises into a large, centrally controlled unity, and engineers must regret privatisation within the British context. For many engineers and managers between 1900 and 1920, 'electric railway' meant a low-voltage direct-current system, with conductor rail distribution, and sometimes fourth-rail return. This was developed by the Metropolitan Railway (Baker, 1951), the London & South Western Railway, the New York Central Railroad, and the Pennsylvania Railroad, to serve their London or New York termini on a scale which involved main line working of heavy passenger, and sometimes goods, trains over routes of the order 10 to 20 miles long. The LVDC railway which Sprague did so much to establish was seen as a possible model for general electrification, but it never became a norm for main lines in the 1920s, due to the low overall thermal efficiency of the system. Nevertheless, the success and reliability of this LVDC system influenced railway managers, directors, shareholders, and engineers in the matter of main line and general railway electrification. This was the case in Britain where there were no electrified networks to provide experience other than the LVDC systems, which were far more extensive than the AC examples of the Heysham branch in 1908 and the Victoria to London Bridge line in 1909 (Agnew, 1937; Moody, 1961, 1963, 1979; Vickers, 1986). Between 1900 and 1930, a considerable number of managers, directors and engineers failed to distinguish between electrification based on extension of the 'London LVDC Standard' and electrification using either high-voltage DC, or some AC system. Granted the widespread establishment of the London Standard, it was questioned whether or not there was a role for any new electrical system on British railways. Warnings were sounded about establishing several, non-standard systems of electrification, which would create problems equivalent to those which resulted from building railways to different gauges in the early 19th century.

It was suggested that LVDC networks should be developed round the larger British cities. Those serving Liverpool, Manchester and London should be joined by networks round Birmingham, Newcastle and elsewhere. The obvious way of creating a British electric railway network would be to link up these existing and projected systems by relatively short conductor rail links. There were practical objections to this argument. The LVDC third-rail railway was a passenger carrying system, usually worked by multiple-unit trains rather than by locomotives. Several authorities argued that electrification was only suitable for this sort of work, because the working of non-passenger trains, such as loose-coupled goods trains, had to be locomotive hauled (Sprague, 1901). This would render general electrification uneconomic, by lowering operating speeds, or by necessitating mixed working with steam engines used on goods trains, for shunting, and for trains working through electrified sections between stations beyond its boundaries. It is striking how many projects put forward before 1956 accepted the need for mixed-mode traction which always diminished the efficiency and economy of electric traction.

6.3 Locomotive working

Locomotives were demonstrated early in electric railway history, examples being the units of Daft, Edison and Sprague, but locomotives were only needed when main line electrification emerged after 1910 and required the through working of trains made up of steam stock under circumstances where steam locomotives could not be used. Large, powerful locomotives had been built earlier, and more were to be built between 1910 and 1920, but these worked relatively short electrified sections in the mountains, or in tunnel (Dawson, 1909, 1923; Haut, 1969; Tassin, Nouvion and Woimant, 1980). Before 1910, locomotives gave way to multiple-unit sets, or served in an auxiliary role. The City & South London Railway opened in 1890 with four-wheeled electric traction locomotives, and by 1901, 52 were at work supplied by Siemens, Mather & Platt, Beyer-Peacock, Crompton-Parkinson, and other makers. This did not become general practice, and multiple-unit working followed the reconstruction of the original line to become part of what is now the Northern Line. The Waterloo & City Railway (LSWR), which opened in 1898, used a Siemens shunting engine (preserved in York museum), but the passenger stock was not locomotive-hauled. In the same year, General Electric (USA) had a contract for 30 locomotives to be supplied to the Central London Railway based on the 1895 design of engine used in the Baltimore Tunnel electrification. The CLR locomotives were 44 tons Bo-Bo machines of 349 kW (468 hp), compared with the 1080 hp of the Baltimore & Ohio Railroad locomotives. Their performance was inferior to a Class A tank engine at its best so the Central London Railway developed multiple-unit trains for later work. The District Railway used Bo-Bo locomotives, and a particularly interesting series of electric locomotives was introduced to the Metropolitan Railway with the British Westinghouse class of 1904. Their use on surface sections, in express service, in goods working and on maintenance trains did not prevent multiple-unit sets from becoming the norm for most services. A clear

82 *Electric railways 1880–1990*

Figure 6.1 Motor car for Central London Railway built in 1903 to replace locomotives with which operations began in June 1900. In the background is Wood Lane power house which fed substations with a 25 Hz AC supply. Rotary converters supplied conductor rails with 500 V DC. Wood Lane Depot traction supply was at first via overhead distribution, later changed to ground-level conductor rails. Note wooden cooling tower on right.
Source: London Transport Museum.

line of development can be traced from the Metropolitan locomotives, through the work of John S. Raworth, of Raworth's Traction Patents to the British-Thomson-Houston locomotives of 1906, and the Metropolitan Vickers locomotives of 1922, which have been described by Benest (1963). These latter engines were amongst the first British main line electric locomotives, and worked on Metropolitan Railway express trains between Baker Street and Aylesbury. They were rated at 1200 hp for one hour, and at 1780 hp at 26 mph, and were fitted with four MV339 motors, the same type which were installed in the Southern Railway 4-SUB multiple-unit sets. It is probable that the design influenced A. W. Raworth, Electrical Engineer of the Southern Railway after 1925, who extended the Southern Railway third-rail network and designed the very successful Co-Co DC third-rail electric locomotive, no. CCl (Haresnape, 1983; Tayler, 1995). This was a 1097 kW or 1470 hp type introduced in 1941 for mixed traffic working, which originated in a Maunsell-Raworth 1938 project for a four-axled 1500 hp engine. It showed the form which express passenger, goods and mixed traffic locomotives would have taken had low-voltage third-rail DC electrification been extended and become the norm, and had British railways retained a great volume of goods working over electrified sections. The evolution of this locomotive from earlier types, all within the 'London Standard' system, and from an American source (the General Electric Baltimore & Ohio Railroad design of 1898) is highly significant. Eventually three were built, with bogies designed by Bulleid, and they saw mixed-traffic working on the Southern Railway and the Southern Region of British Railways, including express service on Newhaven boat trains. They were succeeded by the Bo-Bo Class 71 of 1959, and present types for working Channel Tunnel goods trains over the LVDC network in Southern England.

In the period 1900–20, there was British experience with overhead distribution DC locomotives. These formed an important group, which owed much to American practice, and the advent of the HVDC system described in a later chapter. The best-known examples were built for the 1.5 kV DC North Eastern Railway mineral lines between Shildon and Newport, opened in 1914 but closed in 1935 (Hoole, 1987, 1988; *RCTS*, 10b, 1990); and those built for the much-delayed 1.5 kV DC ex-LNER Manchester-Sheffield project, which dated from 1935, when the Shildon lines were de-electrified, and for which civil engineering work started in 1940 (Haresnape, 1983). Completion was delayed until 1954 by the war, by economic difficulties in the late 1940s and early 1950s, and by collapse of the tunnel at Woodhead. Many questions arise concerning the slow growth of British main line electrification after 1920, and the extent to which a distinctly British school of electric locomotive design developed out of American or Continental origins. How many sources of origin has the modern AC main line system, and the high-voltage main line DC system? Did American expertise overshadow British electrical engineering expertise to such an extent that Britain suffered a permanent loss of place as far as the new form of traction was concerned? Did a distinctly British tradition emerge in the 1920s, with the Metropolitan Vickers DC locomotives, and the Maunsell-Raworth project.

84   *Electric railways 1880–1990*

Figure 6.2   LVDC locomotive of a class built by Metropolitan-Vickers in 1921–23 for the main line surface routes of the Metropolitan Railway (later London Transport). Four 300 hp traction motors were fitted.
Source: London Transport Museum.

However these questions are answered, they concern the development and extension phase of electrification which followed the 1895–1905 decade when the electric railway established itself following American practice, (Agnew, 1937). With respect to the Waterloo & City electrification of 1898, Vickers remarks (Vickers, 1986):

The electric lines built in the next few years were to be British applications of American technology, presaged by the Waterloo & City Railway's purchase of American car bodies ... Because the American market had become overcrowded, profits and prices were falling, so an aggressive search was on for new fields to conquer. As the world's richest city London was the prize, and was to receive a flood of American expertise, capital, and businessmen to revolutionise its transport.

Why British engineering expertise, management skills, capital and business enterprise did not revolutionise British transport and launch the 'Electric Revolution' on the basis of British initiative is a question which awaits the analysis which it deserves, but part of the answer may lie in Britain's failure to move away from a paternalist model of society to a more democratic one. The British 'take-off' of the industrial economy, and the date by which it matured – 1870 according to Rostow (1977) – were separated from the onset of the mass-production, mass-consumption society in Britain by many decades. There was a long period of stagnation between the maturing of the steam-power-based industrial economy, with which Britain dominated the first two phases of industry-based global development, and the importation of the mass-production society from the United States. The United States maturation of its industrial economy coincided with the 'Electric Revolution' and was followed by the immediate introduction of mass consumption and a phase of accelerated growth. The United States transformed its methods of marketing, advertising, planning and financing industries, which expanded as the nation underwent a joint ideological, industrial and social transformation which was welcomed and identified with national progress (Odum and Odum, 1976; Rostow, 1977). There was no such commitment in Britain, but rather a devotion to conservatism. A fear of social change was displayed in the late 1890s and the early decades of the 20th century. This was coupled with a loss of technological initiative and vision, and a failure to comprehend that the changing nature of engineering required extensive changes in planning, organising, managing, funding, researching and developing all branches of engineering, not least education (Barnett, 1986; Urwick and Brech, 1951, 1953). As a result, Britain could do little to prevent American expertise dominating new technologies such as electric traction.

6.4 LVDC lines and main line traffic

There were circumstances where LVDC systems, developed for rapid-transit operations, handled main line traffic for short distances. The electrification of the tunnel and terminus sections of the Pennsylvania and the New York Central railways following the rebuilding of their New York stations and the abolition of steam traction from the city by law resulted in LVDC traction for heavy passenger and goods trains. Locomotives of great power and size were needed, as engineers developed the LVDC system to meet main line conditions. Some engineers argued that no case for general electrification could be based on experience with terminus and tunnel lines which had been created by legal fiat rather than engineering demand. No less an authority than Sprague declared (Sprague, 1901):

> I think it safe to say that the electric locomotive will not generally take the place of the steam locomotive. If a steam-railway adopts electric traction, it must radically change its service, it must adopt smaller train-units, complete within themselves, operated independently, or in combinations making up longer trains then one must ... use a multiplicity of motors ... distributed on the different cars or motor-car units.

At this time the electric railway was usually a low-voltage DC system with conductor-rail distribution and manned substations. It might be an AC railway, with single-phase commutator motors, or a three-phase system with two wires per track and traction motors running at a limited number of set speeds. Where supply came from a coal-fired power house, the overall thermal efficiency was low. Between 1900 and 1930, steam locomotive design was greatly improved by Goss, Woodard and Fry, and fuel economy and power output were increased. New types of engine were introduced, like the Mallet compound-expansion articulated locomotive which was used in heavy goods and pusher service, and it could not be assumed that electric traction would be superior to steam locomotion in every field. In 1906, J. Muhlfeld, the Motive Power Superintendent of the Baltimore & Ohio Railroad, compared steam and electric traction in goods service, and Condit (1977) remarks:

> As a consequence, Muhlfeld was far from convinced that electric traction was appropriate to the movement of freight, and when the B&O placed the first Mallet locomotive in American service on 6 January 1905 he was sure it was not.

Muhlfeld reported the supremacy of the new Mallet engine:

> The results ... obtained from this ... locomotive ... cannot be duplicated by other single units of steam, electric or internal combustion locomotives now in use on American railroads.

There was good reason for this view: the steam locomotive was equal in power to the electric, and its operating and maintenance cost was $0.245 per mile, or little more than two-thirds that of the earlier (electric) machine tested by the Baltimore & Ohio. This view changed between 1900 and 1920. As Muhlfeld was comparing traction modes, the first trials took place between electric and steam traction on the extended terminus lines of the New York Central Railroad, and the Pennsylvania Railroad. These showed conclusively that low-voltage DC conductor-rail locomotives, working the heaviest passenger trains in adverse weather conditions, could outperform the best 4-6-2 type express steam locomotives, though with a lower overall systems thermal efficiency (Bezilla, 1980; Gibbs, 1910; Middleton, 1974; Westing, 1978). Muhlfeld believed that the steam locomotive could be developed to provide greater powers with improved thermal efficiency. He experimented down to 1933 with high-pressure boilers, water tube fireboxes and double- and triple-expansion compounding, but failed to establish a new form of steam engine to compete with oil-engined and electric locomotives (Stoffels, 1976; *Locomotive, Carriage & Waggon Review*, 1933).

The first electric lines carrying main line traffic were electrified sections in tunnels, or railways within city boundaries from which steam traction had been abolished by law. Within conurbations, electrification had generally used the LVDC system for rapid-transit services, and in New York this conductor-rail system was extended to an engine-changing station outside the prescribed zone, on both the Pennsylvania and the New York Central railways. Electric traction had been used previously, for example in Baltimore, where electric locomotives hauled trains through the under-city tunnel with the steam locomotive still attached. The New York systems were on a larger scale, and resulted from the law of 1903 which banned all steam railway traction from New York, south of the Harlem River, after 1 July 1908. This compelled the New York Central Railroad, and the New Haven Railroad to introduce new systems (CERA, 116, 1976; CERA, 118, 1979; Gibbs, 1910; Middleton, 1974; Sloan, 1957). Likewise, the New York City Tunnel Extensions of the Pennsylvania Railroad, completed in 1910, could be nothing else but electrified, and this railway – which was to call itself 'The World's Great Standard Railway' – conducted a series of comparative trials to identify which was the best electric system for the railway being constructed to serve the new Pennsylvania terminal station. This Tunnel Extension Railway linked Pennsylvania Station, and several yards in Manhattan and Long Island, with Manhattan Transfer station, where steam traction took over from electric working on the main line. Trials were carried out with experimental DC and AC locomotives (Bezilla, 1980). Alternating current traction was recognised as potentially superior, but LVDC was installed, using lineside conductor rail for most of the system, and overhead conductor rail, supported from massive gantries, at complex junctions and in the terminal yards (CERA, 116, 1976). Alternating current equipment was undergoing rapid development by Westinghouse in the USA, and by Oerlikon in Switzerland, but there was insufficient time for the Pennsylvania Railroad to test it before the legally binding last date for steam working. The Pennsylvania Railroad chose LVDC because it was a reliable system, proven on tunnel railways. Furthermore, direct current could be stored in batteries. This improved the load factor of the generating stations and enabled power to be reserved in large numbers of cells which could keep the railway running for two hours in the advent of a total power failure, and would prevent trains being immobilised in the river tunnels.

Third-rail DC was laid down between the new Pennsylvania Station, and Manhattan Transfer, where steam traction took over. The distance between Pennsylvania Station and Manhattan Transfer was 8.78 miles, and the total length of the Tunnel Extension Railway, between Laurel Hill Avenue (Long

88 *Electric railways 1880–1990*

Figure 6.3 Baltimore & Ohio Railroad No. 1 built by General Electric (USA) in 1895 to work the electrified Howard Street tunnel section in Baltimore. Supply was via an overhead system built from conductor-bars, later changed to the conventional third-rail arrangement.
Source: General Electric (USA).

Island) and Manhattan Transfer was 13.41 miles. However, the heaviest trains needed to be worked and very powerful locomotives were designed. The DD1 class was the most outstanding of the early types. They were two unit machines, each unit having a four-wheeled unpowered bogie and four large-diameter driving wheels, like a steam locomotive. Tests suggested that the asymmetric wheel arrangement, such as the 4-4-0 layout common for passenger steam engines, worked the rail less severely than did symmetric arrangements, particularly in the lateral direction (Westing, 1978). Large DC motors were placed in a box-cab body, with drive through rods and jack-shaft. The general design reflected the theory that a high centre of gravity shifted load onto the outer rail on curves, and pressed it against the tie plate. This would reduce lateral strain below the excessive levels experienced under many early electric locomotives. The Pennsylvania Railroad Type DD1 was an outstanding success. It showed that it was possible to work heavy passenger trains using electricity at express speed (Carter, 1922; Dawson, 1923). Economy and reliability were better than with steam traction, though the limitations of the LVDC third-rail system prevented long-distance application at a time when manned converter substations were usual when supply came from an AC station, and where overall thermal efficiency was low when supply came from a low-voltage DC station. Long-distance electrification of steam railways awaited high-voltage DC systems, or AC systems. The DD1 type was built between 1909 and 1911 and was a powerful machine. The first two units, numbers 3998 and 3999, were each fitted with two Westinghouse 315A motors rated at 2000 hp, or 4000 hp in total, providing 3160 hp or 2357 kW continuously, which outperformed any express steam locomotive. The type was so reliable and cheap to run that units were still at work in the early 1960s. They exerted a tractive effort of 79,200 lbs force and could haul loads exceeding those pulled by the largest Mallets at the time of their construction, though Bezilla (1980) pointed out that they:

... did not represent much of a technological innovation. The locomotive not only embodied conservative mechanical design; the whole low-voltage d.c. system was soon rendered obsolete by advances in a.c. and high-voltage d.c. systems. Nevertheless, the DD1 did provide the low-cost, dependable rail transportation that the Pennsylvania had been seeking.

The DD1 drew current at 600 V from the lineside conductor rail via a collector shoe, and used a small overhead pantograph under the overhead conductor rail gantries. The first two units were more powerful than the rest of the class, and there is some controversy concerning the rating of the later machines. Carter (1922) describes the 1910 engines as having two 1,250 hp motors, but other writers have quoted each motor as providing 1065 hp continuously. At the time of their construction, the other major DC locomotives in use on LVDC systems were the New York Central & Hudson River Railroad terminal section locomotives built in 1906. These had 1D1 wheel arrangement, and four gearless motors of 550 hp drawing current at 625 V DC from a conductor rail beside the track, with overhead sections in sidings. These locomotives were the first in the world

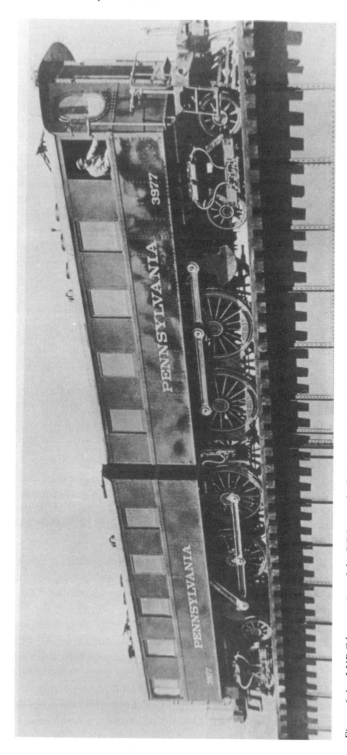

Figure 6.4 LVDC locomotive of the DD1 type, built for Pennsylvania electric services between the New York City terminus and Manhattan Transfer. They were built at the Juniata shops of the Pennsylvania Railroad between 1909 and 1911, with Westinghouse electrical equipment.
Source: Westinghouse (USA).

to be fitted with multiple-unit control, having been constructed with Sprague apparatus. They were rebuilt with outer bogies as 2D2 types, and were joined in 1914 by BBBB machines fitted with eight individual, gearless motors of 325 hp. All used the 625 V DC conductor rail supply. The New York, New Haven & Hartford terminal section locomotives, of the 1BB1 type, used four gearless quill drive motors of 350 hp, and were fitted with power equipment to work from a single phase source, at 11,000 V AC, 25 Hz, or from a DC source at 625 V. The other major American types then working included four three-phase BB locomotives for the original Great Northern Cascade Tunnel scheme, which used a 6000/6600 V, 25 Hz supply to four induction motors rated at 500 hp each. There were the six BB units built in 1910 for the Michigan Central Detroit River Tunnel, which drew 600 V DC from a third rail, lineside and overhead, and which were powered by four motors of 275 hp each. There were the two B&O locomotives of type DD, built in 1903 for the Baltimore Belt-line tunnel, which drew DC at 625 V from an overhead conductor rail to supply eight individual geared motors of 200 hp each. The Grand Trunk St Clair River Tunnel employed six locomotives built in 1908, of the C wheel arrangement, powered by three individual Westinghouse geared motors of 240 hp each, supplied from a single-phase overhead wire at 6600 V, 25 Hz. Descriptions of these engines are found in Carter (1922), Dawson (1909, 1923), General Electric (USA) (1929), Middleton (1974), Westing (1978). The North Eastern Railway, in Britain, was then using two BB shunting engines on the Newcastle on Tyne Quayside branch, which drew DC at 550/600 V from an overhead trolley wire, to drive two individual geared motors of 90 hp each (Hoole, 1987, 1988; RCTS, 10b, 1990), and the ten locomotives built in 1907 for passenger work on the Metropolitan Railway, London, which were supplied from a lineside conductor rail at 550 V DC, to drive four motors of 200 hp each (Benest, 1963). The original Simplon Tunnel electrification was a three-phase system, and the two locomotives of 1906 drew three-phase supply at 3000/3300 V, 15 Hz, to power two induction motors, each of which gave 400 hp when working with 12 poles and 550 hp working with six. Contemporary practice is described by Austin (1915); Carter (1922); Dawson (1923); Haut (1969); Talbot (c. 1920); Tassin, Nouvion and Woimant (1980). This is by no means an exhaustive list, as only the outstanding designs are mentioned, but the 33 DD1 engines of 1910 represented advanced practice and were the most powerful locomotives in the world.

The LVDC conductor rail system ceased to be a model for main line electrification circa 1910, but the LVDC electric street tramway, which Sprague did so much to develop, was the starting point from which the overhead trolley wire DC railway evolved, and contributed to evolution of the high-voltage DC railway as well as the heavy-duty electric rapid-transit systems using third-rail distribution.

6.5 The electric railway and internal-combustion engined traction

The LVDC electric railway, in the form of street lines, heavy-duty rapid-transit lines, and 'interurbans' played an important if minor part in the history of internal-combustion traction. Lightly loaded rapid-transit routes and branch lines needed light passenger and goods vehicles, capable of being self-propelled or run in multiple. The steam-motor, formed by uniting a small steam locomotive with a carriage body met with limited success and was generally replaced by light tank engines working in the 'push-pull' mode. Goods trains were usually worked by old engines unsuited for work elsewhere. However, the success of electric vehicles on street railways and rapid-transit lines suggested that better modes of traction could be developed for branch lines and minor railways, including electrical haulage. On lines where traffic did not justify electrification, some other form of locomotive traction was needed (Keilty, 1979). The LVDC railways and heavy-duty rapid-transit lines used bogie-mounted vehicles, with electric drive, often capable of multiple unit operation. Branch lines could be worked by such vehicles, using petrol-electric or oil-electric drive, battery power, or mechanical drive from an internal combustion engine. This general form of light railway traction developed from several sources, including powered road vehicles. The need to convert the steam street tramway to cleaner modes of traction involved many experiments, such as the unsuccessful trial in 1896 of the Connolly oil engine in a small locomotive on a London tramway. Petrol-electric omnibuses were tried on city streets after 1900 and the combination was also tried on railways. Steam railcars with mechanical transmission were used in service with some success, but those with electric transmission were a failure, and these coal-fuelled units proved no better than the small conventional locomotives which were used (Abbott, 1989; *RCTS* 10b, 1990). The heavy-oil engine was too heavy for general railcar use before 1930, though a 56 kW (75 hp) motor was fitted on the Mellersta and Sodermandlands Railway in Sweden in 1913. The usual internal-combustion motor in service on trolley, interurban and branch lines, was the petrol engine which kept weight low. The petrol-engined and oil-engined railcar was developed by railroad operating and engineering departments, with automobile engineers contributing to a lesser degree. The diesel locomotive proper, capable of supplanting any steam locomotive on any duty was to be in the early 1930s through the co-ordinating efforts of H L Hamilton (later of Electro-Motive Division of General Motors), who raised funds for his pioneer experiments by supplying branch lines and light rail systems with self-propelled gasoline-electric railcars. The internal combustion engined railcar appeared on branch lines and interurbans, and some were fitted with electric transmission by suppliers of trolley and interurban equipment, who drew on established practice of car construction and electrical gear. A survey is provided by the review articles which appeared in Locomotive Cyclopedia (USA), edited versions of which have been reprinted (1974). The Union Pacific Railroad tried a 100 hp branch line motor car with mechanical transmission in 1905, General Electric built an oil-electric railcar for the Delaware & Hudson

Railroad between 1904 and 1906, and between 1904 and 1914, 155 of the famous McKeen petrol-electric cars were constructed. General Electric produced some 90 petrol-electric cars between 1910 and 1917, after which date they concentrated on more urgent wartime contracts, and neither McKeen nor General Electric resumed their production of internal-combustion railcars when the 1914–18 war was over, though General Electric was to become, and to remain, a major builder of diesel-electric locomotives, despite losing initiative to General Motors in the 1930s. General Electric's co-operation with the Wason Coach Building Company produced a petrol-electric vehicle suited to interurban, street railway, or branchline use. It was constructed by established carbuilders; it ran in either direction with controls at each end; and it often formed a 'Combination Car' with passengers, luggage, and mail carried on the motor car which could haul trailers (Lind, 1978; Locomotive Cyclopedia USA, 1974; Stephenson Carbuilders Co., 1972). It seems that failure to develop a unified method for controlling the engine and the generator led to breakdowns which discouraged further General Electric endeavour in this field at a time when the company was engaged in all departments of electrification in the USA. Electro-Motive was originally an independent company before it was taken over by General Motors. It was founded by Hamilton, an ex-Southern Pacific employee, and it began as a maker of gasoline-electric railcars. Hamilton had worked as the Western Sales Manager for an automobile sales company. He noticed the failure of attempts by bus- and lorry-makers to put their technology on rails. Initiative to enter the railway market was lacking and attempts failed through poor design, so Hamilton recruited a small design team to produce a commercially successful gasoline-electric railcar. From the start he identified the financial, managerial, and technical problems of the design project, and solved them so that his product succeeded where others failed. In 1922, he raised money to build a single railcar, completed in 1923, which was then sold and the profits used to begin construction of the next. Between 1923 and 1930, some 500 were built, a typical model being the M300 of 1924, which weighed 35 tons, was powered by 175 hp engine, and could carry 44 passengers. Petrol-electric cars developed in the late 1920s and early 1930s were powered by 550 hp engines and hauled up to five trailer cars to provide the same sort of service as the electric interurban express trains. These trains represented one of several lines of development which resulted in the internal-combustion engined 'streamliner' trains of the 1930s, and they discouraged extended electrification of light railways during the 1920s and 1930s by promising 'cheap electrification' with petrol-electric or oil-electric traction (Locomotive Cyclopedia, 1974; Morgan, 1955; Reed, 1978; Stover, 1961). The technological significance of the street tramways, the interurbans and branch lines diminished when traffic declined in the face of motor transport until the German-inspired revival of light rail transport began in the 1970s. The heavy-duty rapid-transit networks remained important and did not decline despite the growth of road transport, and their role in growing conurbations demanded continuous engineering improvement. From their inception they used heavy rolling stock, and worked under onerous conditions, so that engineers took them as models when main line electrification was considered after 1890. As late as the 1920s, some planners regarded the LVDC system as applicable to extended main line electrification. The dismissal by the London Midland & Scottish of the eminent engineer O'Brien is attributed by some writers to his advocacy in the 1920s of conductor rail distribution for electrification of the main lines out of Euston, rather than the overhead HVDC system.

Chapter 7
Track circuits, describers and electrical signalling 1890–1920

7.1 Introduction

The first electric rapid-transit railways, whether elevated on viaducts over streets as in Chicago or New York, or running in tunnels as in London, faced peculiar problems in the matter of signalling, position monitoring and control. The problems were acute with the tunnel lines, but they could be severe on the twisting elevated lines which curved sharply between the high walls of tenements and warehouses. There was no long-distance sighting ahead of trains, as there was on main lines, and braking distance often exceeded sighting distance. The need to work intensive services between stations located close together, with rapid acceleration and deceleration, made operations exacting, and on any extensive urban network there were junctions and crossings, some of them in tunnels or on viaducts. Lineside signalling was essential to maximise efficiency and capacity, and it is remarkable what was achieved with a manually operated system using the electric telegraph, interlocking, the block system and semaphore signals reduced in size to fit in tunnels. Signalmen worked in tiny, cramped cabins in smokefilled tunnels; bracketed from the walls of deep cuttings; placed on gantries over the track; perched atop the tall trestles of the elevated lines and showed what the manual-mechanical system could do on the most heavily worked steam railways in the world. Steam traction and manual-mechanical signalling reached their most advanced forms on the Inner Circle lines in London, and on the elevated railways in New York and Chicago, shortly before electrification. These systems were always under pressure to increase capacity as cities grew, and as higher standards of living made the population mobile. Competition from street tramways and omnibus services, and from the suburban services of main line railways threatened the rapid-transit companies which were largely dependent on passenger traffic, and which could not offset losses by profits made on general-purpose operations.

In the early 1890s, the steam-worked rapid-transit railway had reached a

96 *Electric railways 1880–1990*

practically unsurpassable limit, and electrification was the only means of upgrading performance. This demanded an improved signalling and control system. Manual-mechanical signalling systems were improved and used on electrified rapid-transit railways, but the trend was to integrate signalling, communications and control more closely with the traction system and power supply. Automatic operations were introduced, and the survival of manual working became dependent of electro-mechanical aids and the powered working of distant mechanisms. This was made easier by ready access to electricity supply which was available in the large cities where rapid-transit railways were located. Electrification of all services was not universal on early electric lines, and oil or gas lighting of carriages, stations and signal lamps long survived. Electric supply for lighting and other services did not always come from a company supply, but might be drawn from a public main. Signalling had long relied on cells for supply in routine duties, and in emergencies. In the 1890s, the electrification of all services: traction, signalling, control and communications was a practical possibility on rapid-transit lines, offering many advantages. Expertise in low-current, low-powered electrical engineering was available from work with telegraphs and communications on the steam railways. In some cases the engineers and 'electricians' who pioneered the railway telegraphs in the 1840s and 1850s played major roles in the general electrification of rapid-transit lines, and concerned themselves with traction in addition to signalling. An outstanding case is Spagnoletti (1832–1915) who acted as telegraph superintendent of both the Great Western Railway, and the Metropolitan Railway, and who installed electric lighting in several London termini. He designed what was the largest electrical device in Britain – the three tons weight show board for Ascot Racecourse. It was on his advice that the City of South London Railway, for which he acted as consulting electrical engineer from 1889, chose electric rather than cable traction. He served as consulting electrical engineer to the Central London Railway, the Metropolitan Railway, the District Railway and the London Electrical Omnibus Company. Sir William Preece was another eminent pioneer of telegraphy and telephony who was interested in all aspects of electric traction. There was a body of men well versed in electrical telegraphy, signalling and communications, and who understood the requirements of railway operations, who gave the new electric railways the signalling and control networks which they needed. Indeed, the development of advanced signalling, communications and control techniques within the new electric railways was relatively trouble-free.

7.2 Pioneer systems

The City & South London Railway which opened in 1890 was the world's first electric underground railway, and the pioneer deep-level tube railway in London (Lascelles, 1987). It opened with mechanical signalling by Dutton & Co., of Worcester, using small semaphores to fit the restricted clearances. As related by Lascelles, the line used no distant signals, and each station was protected by a home signal, with the blocks extending from starting signal to starting signal, or terminus stop. The Spagnoletti 'lock and block' apparatus was used, working over two line wires, one wire ringing the signalling bell, the other (of opposite polarity) working the block release. There were two single-needle 'speaking' telegraph circuits for transmitting messages, linked to the railway's own power station, and telephones were installed to communicate between signal boxes soon after opening. Power for the signalling circuits came from primary batteries. Almost from the opening of the line, it was realised that block sections would require shortening to increase line capacity, and additional home signals, and electrically interlocked treadles, were fitted – for example at the Oval and Borough in 1892. When a train activated an 'electric treadle' in advance of an outer home signal, a signalman could set that signal to danger, work the block instrument, and release to clear the starting signal in the rear. An electric locking instrument controlled the section of line from the outer home to the starting signal. Lascelles describes the 'last vehicle' treadle installed at Stockwell in 1893 at the down outer home signal. A conducting brush on the last bogie of a passing train made contact, completing the block release circuit through the rails. Lascelles claims that this was the first use of 'positive last vehicle action' on an underground railway, though it was first used earlier that year (March 1893) on the Liverpool Overhead Railway, where automatic control of signals was achieved using the Timmis system. At first, the signal lamps on the City & South London Railway were oil, and then gas, but both types were susceptible to being blown out by the draught produced by passing trains, and in 1895 Mott and MacMahon installed electric signal lighting. A safety circuit was designed for repeater signals which illuminated the lamp only if the aspect of the semaphore agreed with that indicated by the signal-box lever. Lascelles reports that a red and green repeater was fitted to indicate the signal aspect to the signalman, and to serve as a pilot light showing that all lamps were on. This became a standard arrangement, used on the Central London line to which Mott went as engineer. Further block sections were created, and the signalling could handle 28 trains per hour until it was replaced by a modern automatic signalling system shortly after the 1914–18 war.

The pioneer electric elevated railway was the Liverpool Overhead Railway. Construction began in 1889, and the line opened in 1893. The significance of this railway in terms of engineering innovation has been described by Woodward (1993) who draws attention to the important advances in signalling arising from the high estimated costs of conventional signalling. This led to the use of automatic electrical signalling on the Timmis system, proposed in 1885, which worked signals and points using 'long-pull' electromagnets designed by Currie to operate lower-quadrant semaphores (Woodward, 1993, p. 70). The 1885 method was for working semaphores which would fail safe if the supply were cut off, with a counterweight returning the arm to danger. This was developed further on the Liverpool Overhead Railway, where the trains set the signals via a striker carried on the rear bogie which opened and closed contacts alongside the track. The Timmis system worked well, and is reported by Woodward to be the first of its type to meet the Board of Trade's strict requirements. Without Timmis signals, 13 signal boxes would have been needed to work the line, with at least two men in each box. Much later, in July 1921, the railway became the first in Britain to use daylight two-aspect colour-light signals, installed by the Westinghouse Brake & Signal Co. The City & South London Railway and the Liverpool Overhead Railway were simple systems, working shuttle-services over moderate distances. They showed that electrical automatic signalling and interlocking (albeit of a primitive form) were possible, but extending these techniques to main line general-purpose railways was a more formidable undertaking.

7.3 Automatic signalling and related innovations

Automatic block signalling increases line capacity. It can be used with steam, electric or diesel traction, and with mechanical or colour-light signals. It was suggested in the 1870s, by several innovators including Dr W. Robinson who proposed using a track circuit to denote the presence or absence of a train. His system, suggested in the USA in 1872, was designed to be fail safe and any failure of battery, circuit or rail set the signals to danger. The track circuit was a 'vital' circuit, which meant that it could give an erroneous reading through malfunctions such as a relay sticking in the closed position. Hence components were and are designed to prevent this happening and to be lightning-proof (Armstrong, 1978). In Britain, Preece and Spagnoletti produced systems which were tried on both steam and electric railways. Preece designed a system of electric block-signal instruments to be used with electrically controlled semaphores, available in three-wire and single-wire versions. Matters of detail were subject to endless experiment. The actual working of the signal, within an electrical system of control and interlocking was by rods, wires and weights; by compressed air; by hydraulic power; by electricity; or by a combination of these. The signal might be a large or miniature semaphore, or a spectacle frame of coloured glass moved before a lamp. The lamp might be oil, gas, electric-arc (rare), or electric incandescent light. The role of electricity within a system varied from exercising control to both controlling and physically moving everything. For physically shifting signals, or points, electricity was normally used in conjunction with pneumatic or hydraulic motors, though all-electric systems were tried from the start (H. R. Wilson, 1900, 1908).

Design of signalling bells attracted much attention, and a stream of patents attested to the need to devise bells, buzzers and klaxons of different sounds and power to enable separate sound signals to be distinguished from each other in noisy signal cabins or control rooms. In signal boxes several sets of apparatus might be sounding at once, with locomotives shunting wagons close by, or electric trains passing the box which might be inside a metal-lined tunnel. For this reason, most apparatus had a visual signal associated with the bell. All manner of batteries, bell strikers, bell-pushes and selectors were on the market in the 1890s and after. More importantly, the need to provide visual indication in signal cabins and control centres resulted in electric display boards to show train position, and to indicate signal aspect.

Signals were modified to suit underground use. Miniature semaphores (mechanical or electro-pneumatic), colour-light signals and signal repeaters were introduced. Oil and gas signal lamps were found on electric railways, but once reliable electric lights were available, able to withstand the shock of passing trains, electric signal lighting was adopted, usually in the colour-light form. For locomotive headlights, electricity began to supplant oil on a large scale in the USA circa 1915. Until then electric headlamps, supplied from a small impulse turbo-generator on the locomotive, were too unreliable due to filament breakage and oil was employed, though some railways used arc-headlamps. In many countries, the oil-fuelled portable lamp remained in use until the 1980s. Reliability was the key issue. Much railway equipment, like signal lamps, had to be left to work well without attention for days in all weathers. One failure could result in great financial loss even if the equipment failed safe. Simple mechanical devices, like the oil lamp were reliable and well understood. Electrical systems were accepted only when proved to be equally reliable. Electric traction established itself as reliable before electric signalling did. After reliability, economy was the deciding factor.

Automatic working of lineside signals and signals on trains through contacts completed by the train was tried on steam railways in the 1870s, using both electrical and mechanical devices. In the 1890s, the first successful prototypes of systems destined for wide use were installed on electric rapid-transit railways. In the first decade of the 20th century, during electrification of London's railways, Spagnoletti installed automatic signalling on the Central London Railway and the Great Northern & City Railway. With truly automatic signalling, the trains completed circuits which controlled the signal setting. Various methods were introduced for converting the control signals into mechanical actions including electro-motors, electro-pneumatic engines, and electro-hydraulic systems. These systems were later used on the general purpose railways, with all modes of traction. Generally, the electrification of a main line was accompanied by modernisation of the signalling system. The Pennsylvania Railroad electrification of the main line between Philadelphia and Paoli, authorised in 1913, is an example of this. There was considerable exchange of ideas between Britain and North America, and British railways like the District Railway in London set the standard for advanced signalling and control practice, which was copied in the USA. Electric traction engineering was dominated by big international combines and credit for improvements must go to several countries. British and American engineers co-operated closely in electrical engineering projects after 1880, including the Niagara Falls scheme, and the electrification of New York Grand Central terminal. Automatic signalling of the American Westinghouse pattern was installed on the Baker Street & Waterloo ('Bakerloo') line and similar equipment was used on the Metropolitan & District Railway in 1905 (Horne, 1990). These employed energised track circuits and pneumatic power to move the spectacle plates before the signal lamps. These moving-spectacle signals were eventually replaced by two-aspect colour-light signals in the 1920s. Trainstops were employed – which are still used – and semi-automatic signals were installed at cross-overs, where intervention by the signalman was part of the programme. Miniature lever frames for controlling the pneumatic signalling, and illuminated track diagrams showing train position on a layout diagram, established new standards in signalling which became widespread. However, automatic signalling and electro-mechanical control were not universal even on lines newly constructed in the early 20th century. The 'Central Line' in London (Horne, 1987) used mechanical signalling with Sykes lock and block equipment. There were 16 signal cabins, controlling semaphore and sliding spectacle signals. At first, the proposal was to work trains with a locomotive at each end, linked by a power line, but this was condemned by the Board of Trade as a potential source of fire, and the line was run with conventional electric locomotives, and multiple unit stock. In the early system, a bogie mounted brush released the locking for a section. Improvements were carried out in 1909 when the automatic train stop, and the 'dead man's handle' were installed. The decision to install automatic signalling was taken in 1912, and between 1912 and 1914 this was set in place by McKenzie, Holland and Westinghouse. In 1913, a power frame was erected at Shepherds Bush. The system included track circuits, with impedance bonds and the (largely) automatic operation of signals (and repeaters). Two aspect signals were used: colour-light in tunnels, and electro-pneumatic semaphores in the open. Compressed air was used to work train stops and the points at the Central Line stations at Liverpool Street and Wood Lane. Most remarkable of all was the electro-pneumatic interlocking with signals of a movable platform extension, which fouled the track in certain positions.

Electro-pneumatic equipment was applied to the rolling stock, for working multiple-unit control gear, and for operating doors in 'air-door stock', built between 1919 and 1922. As related by Bruce (1968), an order was placed with Cammell Laird & Co. in 1919 for 20 trailer and 20 control cars, the first tube stock to be designed with air operated doors. These were supplemented by 20 French-built cars of 1906 which were reconstructed as air-door stock. This stock reduced train crew to driver and two guards, and abolished the conductor who worked the gates on each car by placing one man in charge of door control. This reduced station stop time and cut labour costs when wages were rising. Each door leaf was driven by an air engine, supplied with compressed air at 30 lbs/sq in, controlled by an electrically operated valve (Bruce, 1968, pp. 37–38) which exhausted or filled a large diameter cylinder to allow the door to open, or to close against the action of a small diameter cylinder. Door control positions were provided at the non-driving ends of motor cars and control trailers, and circuits were supplied from the 600 V auxiliary line from the motor car. The guard controlling the doors could signal the driver to start using a push button and cab light. Bruce reports that each door leaf was provided with an interlock which made contact only when the door was correctly closed. These interlocks were in the starting light circuit, so that the cab lamp could not signal 'start' if any door was improperly closed. Miniature exterior semaphores, of bright

Figure 7.1 Interior of Signal Cabin E, Camden Town, Northern Line in June 1907, showing Westinghouse electro-pneumatic lever frame.
Source: London Transport Museum.

yellow colour, signalled which doors were open, and the front guard – who gave the starting signal – could not do so until these dropped into the 'doors closed' position. Eventually, pilot lights replaced the semaphores. If a train halted short in a station, isolating switches prevented the doors opening of the cars still in tunnel. This is just one example of how the new signalling and control techniques, dating from the period 1890 to 1910, were transforming every component in the traction system. Many other examples could be found. One has only to consider the routine, and relative lack of safeguards, for starting up a steam-hauled suburban train made up of 'slam-door' stock, by flag and whistle to appreciate the advantages offered by interlocked electrical systems controlling powered mechanisms.

By 1910, the London electric railways used signalling, control and communications systems, which were becoming more closely integrated, though there was much manual intervention and human oversight. Electro-pneumatic and other systems were used in automatic train control; automatic train stops and automatic signalling (Talbot, c. 1910). The automatic signalling was fail safe, with electro-pneumatic restoration of signals by the train, and four colour signalling was in use: red, green, orange, plus purple for use in brake tests. Electric fog signals were available for placing detonators on the rail in accordance with signal position, or – less common – for sounding cab signal bells or air-whistles. Manual cabins with small levers actuating electro-pneumatic systems for working signals, points, and train stops had proved their reliability and operational superiority to the traditional mechanical system of large levers working through wires, rods, cranks and weights. Economic considerations made it impossible to replace the latter by the former even where it was technically desirable to do so. There were too many mechanical boxes in place, some new with expensive equipment, and provided skilled labour was cheap these worked well. Mechanical boxes, of the orthodox pattern, were installed down to the 1980s (Signalling Study Group, 1986). Other modern aids to railway signalling and control which became normal equipment in electric railways in the first decade of the 20th century were the telephone in traffic control; the electric indicator board on stations, the train indicator in signal cabins; and the signalman-controlled indicator board. There were devices such as the combinator, which formed part of the receiving apparatus, and which stored information about train destinations and order of running. This used electro-mechanical means – such as stud position in a rotating drum – to control indicator board displays.

The reconstruction of the Grand Central terminus in New York, and the electrification of the former steam railways within the city, was the result of a law abolishing steam traction after 1908, following an accident when smoke obscured signals in tunnels on lines serving the station. The New York Central and allied railways had no choice but to electrify the terminal sections for several miles, and the decision was taken to rebuild the old station, and erect buildings for sale or rental over the former engine and carriage lines which in steam traction days had to be left open. The Pennsylvania Railroad followed much the same policy. The new Grand Central station, New York, was a remarkable

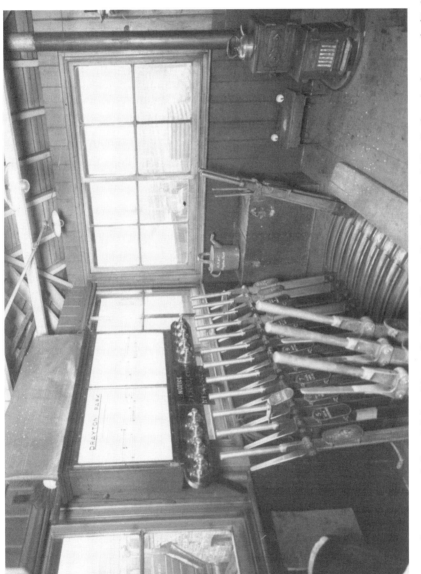

Figure 7.2 Interior of Drayton Park Signal Cabin, Metropolitan Line (former Great Northern & City Railway), photographed in November 1933. The old GN & C section was worked by the Metropolitan Railway 1913–39, and afterwards by the Northern Line.
Source: London Transport Museum.

mechanical, electrical and civil engineering achievement and a fitting witness to the importance of railways at that time in New York (Condit, 1980) and in the American economy (Stover, 1961). There were two levels of track, with main line and suburban operations separated, and a lower level for storing rolling stock. All trains were worked by LVDC using multiple unit trains and locomotives drawing current from a lineside conductor rail, and an overhead conductor rail at complex junctions. There were 113 tracks controlled from 5 interlocking stations or signal boxes, with a main 400 lever station as the centre. An all-electric system was used, with semaphore signals worked by electricity. Contemporary descriptions (Talbot, c. 1912) claim that the system was modelled after the automatic signalling system used on the District Railway in London, but it was not as advanced as the latter. The American system required the constant intervention of a human train director because there was no time for inter-box telegraphy, and an error on his part could result in collisions. Automatic indication of trains to the public was provided. The system used on the Metropolitan and District lines in London had to deal with trains running at intervals of 75 and 90 seconds, and some tracks carried 50 trains in one hour. The policy was to eliminate regular steam traction entirely on the underground sections, and to 'rehabilitate' the District line through electrification and re-equipment. Automatic signalling was installed to minimise the human element using electro-pneumatic equipment from Mackenzie, Holland and Westinghouse. Points and signals were worked by compressed air, with electromagnetic valves controlling the air flow. As usual in automatic signalling, the trains controlled the signals, of which there were several types, with different arms. The track circuits worked the illuminated track diagrams displayed in the signal cabins, the first use in the world of such diagrams in connection with signalling. All signals were controlled by track circuits. Automatic train stops were installed, and automatic fogging signals which sounded a warning by whistle. Lock and block using treadles and electrical locking was fitted. The system worked very well, and in less advanced form was used in Grand Central Station, New York.

In Britain, the standards setting innovations in signalling and control were not limited to London. Electric railways in Liverpool was the scene of both improvement and innovation. Both the Mersey Railway and the Liverpool Overhead Railway witnessed noteworthy developments.

The Mersey Railway began operations as a steam railway using semaphore signals installed by the Railway Signal Company, Fazakerley, Liverpool. The system followed typical late 19th-century block system practice, with manned signal boxes at each station and a cabin in the middle of the river tunnel which was used during the morning and evening peak periods to shorten the tunnel-section block and thereby increase line capacity. The Manchester firm of John Lavender supplied the telegraph system for interlinking the signal boxes. Many pioneer electric railways opened with similar signalling systems and the installation of electric signals, worked by track circuits, came several years later, often in the form of electric semaphores rather than electric light signals. The semaphore signals on the Mersey Railway survived until 1921. They were then replaced by Westinghouse two-aspect colour-light signals, automatic and semi-automatic, operated in most locations by track circuits with train stops installed to brake any train which overran a signal set at danger. At the larger stations, where the trackwork was relatively complex, the manned signal boxes were retained. These were found at Liverpool Central, Hamilton Square, Rock Ferry and Birkenhead Park. This Westinghouse system employed both automatic and semi-automatic working of signals, with electro-mechanical interlocking, and it followed the practice developed for semaphore signals on the pioneer lines, and later developed on electric railways like the Central London Railway, the Great Northern & City Railway, and the Metropolitan and District railways. In 1923, points-setting operations at Liverpool Central station were made automatic using timing relays to control trains arriving in a known sequence from Hamilton Square. The timing relays, in conjunction with track circuits, set signals and worked points so that trains could shunt and reverse to change platforms at the terminus. The train itself set signals to red behind it, which only cleared automatically when the train passed into another section and set signals further down the line to danger. The train controlled the signals through the track circuits with additional protection from train stops. Such systems became common throughout the London Transport system. The Mersey Railway signalling and control system worked well, and was not replaced until 1977. If the automatic system at Liverpool Central failed, manned operations could be conducted from two emergency signal boxes. James Street also had an emergency box in case the automatic installation at that station failed. These emergency boxes were seldom used as the automatic system proved reliable.

The Liverpool Overhead Railway used automatic semaphore signals from the start of operations in 1893. These were of the Timmis design and were controlled by a make-or-break contact worked by a striker on the last coach of the train. Linecar (p. 111) relates:

Each station was signalled by a home and a starter, and the latter also acted as the distant signal for the next station along. Working was by absolute block, thus: On leaving the station the train put the starter signal at danger by actuating a breaker. Fifty yards beyond, the train passed a maker, which completed track circuit clearing the section in the rear as far as the starter, which still remained 'on', protecting the train. On passing within the next home signal, the train completed the track circuit clearing the preceding starter. The current for these signals was supplied by the battery used to light the stations.

The first patents for automatic signals working according to these principles were taken out in the 1860s, and were intended for use in steam railways, but they did not become a success until electric traction was well established.

7.4 Main line signalling developments

Powered signalling was needed for any full and effective automatic system, and pneumatic, hydraulic and electrical systems were devised. Powered signalling was not always automatic, and manually controlled powered systems were employed to operate signals, switches and other devices situated beyond the range of mechanical wire and rod linkages. By 1910, pneumatic, hydraulic and electrical systems were available for working signals and points with both push button or lever operation under all railway operating conditions (Wilson, 1900, 1908; Dutton, 1928; Kichenside and Williams, 1967). The pneumatic signalling installed by the Great Central Railway at London Road, Manchester, is an example dating from this period. The demand for power-signalling increased as labour costs rose in the 20th century, and E C Irving of the British Pneumatic Railway Signal Co. claimed that one signalman with power-signalling did the work of three with a purely manual system.

There were proposals to eliminate the manual control of shunting yards using levers, rods and wires and install powered operations using a pneumatic supply at 15 lbs. per sq. in. gauge, though the shuntsmen in many yards relied on non-centralised hand operated lineside levers, and handsignals, into the 1960s. After 1900, the sorting of wagons by gravity using a 'hump' become common in Britain. The new marshalling and sorting yards had centralised signalling and switching and were suited to powered operations. Good examples were at Wath on the Great Central Railway, and the later yard at Feltham on the Southern Railway (Lamb, 1941, p. 100):

> The points leading into the sidings at both these yards are power operated from a signal-box, those at Wath on the electro-pneumatic, and at Feltham on the all-electric principle. The signal-man is thus relieved of a good deal of manual labour, while the points can be actuated more speedily than would be possible with an ordinary mechanical frame.

Miniaturised, mechanical interlocking of operating switches by a two-part movement of the signalman's lever became standard. In large shunting and marshalling yards and at complex junctions near busy main line termini, power signalling reduced staff to one quarter of the number needed for a manual system.

One of the first electro-pneumatic signalling systems was installed in 1913 at Bolton, on the Lancashire & Yorkshire Railway. In the early years of the 20th century, several main line railways installed test sections of automatic signalling on their routes, to signal steam-hauled traffic. The London & South Western Railway, a technically progressive company, and the North Eastern Railway installed automatic signals on important and heavily used sections of main line. In 1902, the LSWR installed them, using equipment which survived in use, on site, until 1966. The NER system used 40 automatic signals on a section between York and Darlington. The Great Central Railway, another progressive company, installed automatic block signalling, with electrical signals, over two block sections near Woodhead and conducted trials with white and red electrical signal

Track circuits, describers and electrical signalling 1890–1920   107

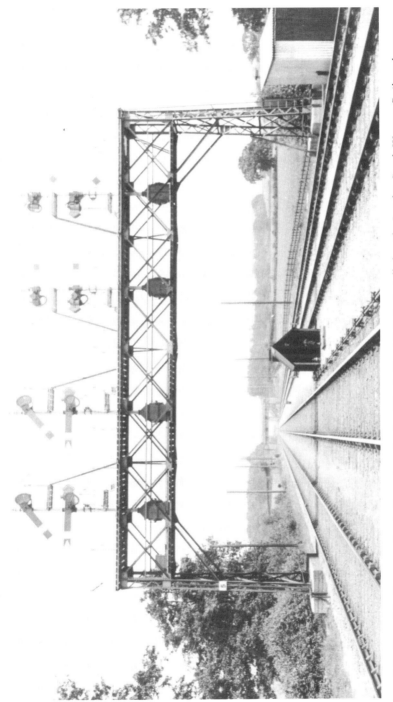

Figure 7.3   One of the first uses of main line automatic signalling in Great Britain, installed on the London South Western Railway between Basingstoke and Woking in 1904 using the electro-pneumatic system. The system remained in use until 1966.

Source: British Railways/National Railway Museum (York).

lamps mounted in the locomotive cab. Similar trials were carried out in the USA and Continental Europe.

In Britain, the extent to which automatic signalling spread was decided by installation cost at a time when the railways faced financial difficulties and were cautious about schemes demanding large amounts of capital. The manual-mechanical system operated via levers, wires and rods from signal boxes was already in place and largely paid for by 1900. It was reliable and proven. The automatic signalling system was expensive, particularly where it included automatic, powered points. If points were excluded, signalmen were needed to work them, and in most situations they could work the signals also. Most companies found little difficulty in recruiting diligent, self-disciplined, responsible crew to staff the signal boxes. In the USA, good-quality labour was relatively more expensive. The sorting of wagons much bigger and heavier than were found in Britain was pioneered in the 1890s using the 'hump' or artificial mound up which wagons were pushed to run down the other side towards the sorting sidings. In the USA automatic signalling, centralised control and powered operations were more economic, and were shown to reduce working expenses in marshalling yards. Automatic signalling assisted operations in fog, because cab signal-whistles, or track detonators ('torpedoes') could be incorporated into the system. Techniques which were proved in 19th-century manual-mechanical signalling were incorporated into automatic, powered networks. This main line use of automatic signalling with steam traction proved perfectly satisfactory.

In 1912 the Pennsylvania Railroad decided to electrify the heavily used main line from Philadelphia to Paoli using single-phase AC, at 25 Hz, 11,000 V instead of low-voltage DC or high-voltage DC. This programme was announced as heralding something new, even revolutionary, for it involved the joint electrification of traction and signalling over a vital section of a railway which represented the best engineering practice in the USA. The project was on the large scale. It planned for construction of the railway's own power stations, distribution systems, substations, rolling stock and a new signalling system. It was innovative, bold and successful, and marked the end of the low-voltage DC system as a candidate for main line working. The methods introduced for building the fixed works were also innovative, and the Pennsylvania Railroad made first use of construction trains which enabled the overhead system to be erected without interrupting the steam-hauled traffic on the section. This became standard practice. The signalling was of revolutionary design, and one technical author (Talbot, c. 1912) described it under the significant heading 'Adapting the electric sign to railway signalling'. A new, automatic block system was chosen, using position light signals which employed techniques found in the large illuminated signs which the Americans pioneered. In these signs, arrays of electric lamps displayed letters and patterns created by switching individual bulbs on or off. Words and sentences giving the latest news could be created, and given the appearance of being displayed on a moving band. These were a familiar sight in large cities in Britain into the 1960s. The same techniques were used in the Philadelphia to Paoli electrification, though on a smaller and simplified scale. Arrays or 'boards' of seven or ten electric lamps indicated signal setting with lights representing semaphore arms. This was made possible by the availability of electric lamps powerful enough to be seen in broad daylight. The bulbs used in the Paoli scheme enabled a light signal to be seen at 4000 feet in bright sunlight. The lamp power was increased during the day to a 'daylight voltage' of four times the night voltage. The trains set the signals automatically over blocks of 3500 feet. From this time on, the oil or gas lit semaphore was steadily replaced by the electric light signal, generally of the colour-light form. The semaphore signal survived, in part by being fitted for powered working and electric lighting of the moving spectacle plate, and examples of each type (powered or unpowered) survive to this day in North America or in Europe. Skegness station, Lincolnshire, is still signalled by semaphores controlled from a box at the end of the platform.

After the Paoli electrification proved that the electric light could signal clear indications in brightest daylight it began to spread, with some rivalry from the powered semaphore signal. By the time war broke out in 1914, automatic electric block signals were working on the Union Pacific Railroad, and automatic 'flashlight' (colour-light) signalling was installed on the famous Langen suspension railway between Barmen and Elberfeld, in the Wupper valley, in Germany.

7.5 Wireless telegraphy in train operations

Wireless telegraphy in train operations is used for general-purpose communications between administrative centres and local offices, and for the exchange of messages between control centres and moving trains. The spread of wireless telegraphy and wireless telephony between 1900 and 1920 supplemented the existing, comprehensive railway communications network based on the telegraph and telephone. The use of wireless to communicate with moving trains did not become standard until after World War Two, when portable radio transmitter-receivers became available, suitable for use on trains. During World War One wireless telegraphy and radio-telephony were much improved, and were employed in general railway communications to a limited extent in that era. There were attempts before 1914 to communicate by wireless telegraph between a land station and a moving train, and the first trials began in 1898. The signals sent were in Morse and similar codes. Wireless telephony in which speech was transmitted was tried before the Great War, though its rapid development and general use occurred afterwards. Communicating with trains by wireless telegraph proved unexpectedly difficult, and it took 15 years of experimenting to achieve success. Many projects were abandoned before the first train order was given by wireless telegraph on 18 November 1913 by a despatcher on the Lackawanna & Western Railroad. The United States led the way in this form of communication for railways because the need for it was pressing in that country. In Europe, where research into wireless telegraphy was active, trains were under constant observation from signal boxes, from stations, and from the many towns and villages through which they passed. Their passage was constantly monitored by land-line telegraphs and telephones. Vast open spaces, crossed by single tracks lacking signals and stations were not as common in Europe as in the USA and Canada. Wireless communication with moving vehicles was studied with military application in mind, but research was not carried out by railways in Europe as it was in North America. In the United States before 1910, there were times when bad weather brought down the telegraph and telephone wires, disrupting communications over a wide area. Trains could be lost in snow drifts for many hours. Bad weather, involving heavy snow or floods, sometimes immobilised all modes of transport which might be used to find a lost train, such as motor cars, the first aeroplanes, horse drawn vehicles, horses or other trains. Telephones to local farms and villages were scarce – even if they survived storms – and roads were poor in country districts. If a long length of railway was blocked or washed out, so that a search by rail could not be mounted, a train might be lost for a day or more. Even if a farmer found it, a message could take hours to get through to railway headquarters if landlines were down, and the countryman had to walk through drifts to the nearest working telephone or telegraph. An aeroplane might find a lost train, but bad weather could ground all flights, and organised searches from the air are more a matter of the post-Great War period. Wireless telegraphy solved the problem that could not be ignored, because it enabled a transmitter-receiver to be put in the train. The potential of wireless communication in railways went far beyond finding trains trapped by snow or floods. It revolutionised 'T&TO' by establishing a direct link between the central train control office and train crew in place of messages sent by land-line telegraph to be handed to the crew in written form. The latter practice has survived, though supported by radio communications which now link crew to control centre, and interlink members of a crew distributed along a train. In Europe, where lineside signalling from boxes linked by land-lines is less liable to disruption by weather, the advantages of wireless telegraphy were fewer.

After several trials were made in the period 1898 to 1909, sustained experiments began in the latter year, with the Lackawanna & Western Railroad leading the programme. In 1909, equipment was judged too primitive for good results, but encouraging tests were made in 1912, and success claimed in 1913 when train operation by wireless became practical. The Lackawanna & Western Railroad trials used wireless telegraph stations at Hoboken, New Jersey, and Scranton, Pennsylvania, which were 146 miles apart by rail. Trials over this section showed that it was possible to communicate with a moving train from the control centre, without resort to landline telephone or landline telegraph. Wireless communication was especially useful if there was signalling failure, or if an engine broke down in a remote place, far from a signal box or telephone. It had advantages in despatching train orders, where passing written orders by hand to crew was the usual practice, because wireless enabled orders to be changed and revised information to be given to trainmen on the move. The wireless pioneers faced restrictions imposed on aerial size and form by the railway loading gauge, which determined the cross-section of rolling stock. Tunnels and bridges set severe limits and the aerials were wires run along the train, close by the roof, one on each side, supported in porcelain insulators in the carriage flag brackets, with a third wire down the centre line. The wireless system on board the train was in the charge of a skilled operative who was given a special compartment in one of the passenger cars. The source of power was the normal axle-driven dynamos which generated low-voltage DC, mainly for lighting but which was powerful enough to work a motor-generator if necessary. Earth connection was made through the rails, though this caused anxiety about upsetting the automatic signalling circuits then being installed on the main lines of the first class US railways. However, the Lackawanna tests showed that the lighting dynamos were sufficiently powerful for wireless telegraphy and telephony, and that track signalling circuits were not upset by using the rails to provide connection to earth. Wireless telephony was used in the Lackawanna & Western trials. Four fixed stations were set up, and wavelengths of 3000, 1800 and 1600 m used over different sections. The power capacities of the transmitters were 5 kW and 2 kW. The prestigious express passenger trains, known as 'Limited' because of limited accommodation and necessity for advanced booking, carried transmitters of 1 kW, working on a wavelength of 600 m. Communication was possible with a moving train from a single fixed station over a range of about 130 miles. With several stations, communication could be maintained to and from a train throughout its 400-mile run along the Lackawanna & Western main line. The trials provided valuable information about the effect of motion past lineside structures on reception and transmission. On 21 November 1913 the first regular train to be equipped for transmission and reception began its service – the Lackawanna Limited. Apart from the wireless telegraph this was a typical steam-hauled express train of its day. The wireless was used to exchange service messages concerning number of passengers, and to arrange for the train to be met by ambulances, police or relief crew in emergencies. It proved its worth when severe storms wrecked telegraph and telephone land lines over hundreds of square miles in 1914. The damage was so severe that there was chaos for ten days, and all railways were disorganised apart from the Lackawanna & Western. Other railways had to revert to using train orders despatched by hand, but without the support of telephone and telegraph, so that operations were very slow. The ability of the Lackawanna Railroad to maintain traffic by wireless was noted by other companies in North America and elsewhere. The 'Great Storm' which struck Britain during the Great War, and which wrecked land lines across the country indicated the advantages of wireless telegraphy to the railways, the military and the government. On the Lackawanna Railroad the wireless room was next to the traffic manager's office in a land station so that the wireless telegraph system could be switched into traffic service if the telegraph and telephone land lines were brought down.

In the early years of railway wireless telegraphy and telephony, it was regarded more as a standby to be used when the existing systems failed, but gradually it became a major component of the system itself. In the 1920s, good-quality radio telephony was introduced into regular use. In Britain, tests with wireless communications to trains followed closely the US experience and there were successful demonstrations before 1914. The Great Northern Railway and the North Eastern Railway encouraged its use, and it became established in the 1920s after the amalgamation of most of Britain's railways into four big companies. Its use was in general communications, and not with moving trains. Throughout the age of the steam railway, and into the period of dieselisation, train crew could only communicate with control by stopping the train and telephoning from a signal post instrument, or by throwing a written message – usually wrapped round a lump of coal in an engine cloth – to a signalman alerted by engine whistle. On routes where signal boxes and stations were frequent, this was adequate. After 1960, with operating speeds rising, and with signal boxes disappearing as extensive centralising took place, and with stations being closed, radio communication with moving trains became essential.

7.6 Telephony and railway communications

From the start in the 1880s, railway telephony faced problems peculiar to itself. Telephones needed a connection by wire from the exchange to the point of use, which might be in an isolated signal box or remote station. In towns or in industrial districts it was expensive to set up the wires because independent lines were required to interlink signal boxes, stations, warehouses, engine sheds, substations, switch points, shuntsmen's huts, and the many administrative offices found in the vicinity of a big station like Manchester Victoria or London Bridge. Even with a free right of way alongside company tracks, it was costly to set up the land lines and the cheapest system was generally chosen. This was the 'omnibus' system which had disadvantages but reduced the number of independent lines required to cover an area. One early example was installed in 1880 on the South Eastern Railway between Charing Cross Station and Cannon Street Station, London. The range of the first systems proved to be limited by the power of the bell-ringer. Audible signals were used which had the striker driven by a clockwork motor, or a motor powered from a local battery supply. These supplied adequate local power to audible signals at a time when there was no mains supply. They enabled telephone bells to be rung at a distance but added to complexity and expense. The electrical engineering texts of the day reveal much interest in the design of bells and sounders. The availability of mains electricity largely overcame this problem, though all railway telegraphs and telephones required standby-power in case there was a failure of mains. Little could go wrong with cells placed in the signal box, which could be changed as needed by train or cart, and this form of power supply to the traditional signalling and communication systems was reliable and safe.

From these first applications there grew extensive networks, grouped into 'community of interest' circuits at the disposal of particular departments. In the simplest systems, all telephones would be set ringing at once by an incoming call, and the loud noise caused annoyance, misinterpretation of messages, and discomfort, and was a reason for introducing the selective traffic control (STC) telephone. After 1920, manual switchboards became usual until they were replaced by automatic telephone networks in the 1960s and 1970s.

Chapter 8
Evolution of the electric railway 1920–40

8.1 Introduction

Between 1890 and 1920, several examples of electric railway were constructed to serve different needs. There were rapid-transit railways, terminus lines, mountain railways and electrified main lines. There was no standard system of electrification and power supply, and no way of telling which system was likely to meet future circumstances better than the others. The relative advantages and disadvantages of the several kinds of motor, and of systems for converting power supply to traction current were matters of debate. There was considerable development of systems for converting AC into DC, and for controlling and mounting traction motors and transmitting their power (Dickinson, 1927; Harding and Ewing, 1911, 1916, 1926). The advantages of using high-voltage AC for the power supply from generating stations to the railway feeder points was recognised in the 1890s. The good traction characteristics of the low-voltage DC motor were equally evident. The attempts to combine the most efficient power supply with the traction motor with the best characteristics for a particular service resulted in several systems for electrifying railways. Before the mid-1920s the mercury rectifier was not available for large powers, so that conversion and frequency changing was by rotary converters mounted in lineside substations or on the locomotive. These methods defined the form and performance of the systems dependent on them. The main systems will now be reviewed in turn.

8.2 The three-phase railway electrification system

The polyphase electrical distribution system became established in the 1880s to transmit power over long distances, with relatively low losses. It played a crucial part in the growth of electrical engineering after 1890, following the success of the two-phase transmission used in the Niagara Falls hydroelectric scheme (Adams, 1927; Scott, 1899; Stillwell, 1901). In the 1880s, Tesla, Westinghouse and Steinmetz invented the three-phase transformer or 'secondary generator', and the first reliable three-phase motors and generators, capable of industrial service. Three-phase systems were excellent for distributing power over long distances, but three-phase induction motors were of limited value when there was no control mechanism for changing speed over a wide range. At this time there was no reliable single-phase motor suitable for use at industrial frequencies. The three-phase motors built before the 1970s ran at speeds set by the supply frequency and control was effected by pole-changing and concatenation, which gave four or five speeds. This discouraged the use of three-phase motors in general-purpose railway traffic unless constant speed working of heavy trains over mountain ranges was sought. Here the fixed speed characteristic controlled trains on downhill gradients and helped to maintain schedule uphill.

The three-phase railway used three-phase supply to the locomotive via two trolley wires, and three-phase traction motors. It is credited to C. E. L. Brown, who installed the three-phase transmission between Lauffen on the Neckar, and Frankfurt on Main, when he worked on the staff of Oerlikon. Later he worked with Heilmann on the 'Fusée Electrique' project. The development trials of the 'Fusée Electrique' showed that the three-phase motor was light for its power, and that elimination of the commutator reduced maintenance. It enabled regenerative braking to be used, and formed a component in a power system with an overall efficiency greater than that of the low-voltage DC traction systems of the 1890s. Brown advocated the three-phase railway, despite its need for two trolley wires and it found use on main line railways through difficult terrain in the mountains of Italy, Switzerland and the USA.

The first railway to use three-phase supply and three-phase induction motors was the Lugarno tramway, on which Brown installed a 40 Hz system in 1896. In 1897, the narrow gauge railway up the Gornergrat mountain was electrified, and in 1898 the electric Jungfrau railway opened. Both railways used three-phase supply and motors. The first standard gauge railway to use the three-phase scheme was the Burgdorf-Thun line, electrified in 1899 (Haut, 1969; Tassin, Nouvion and Woimant, 1980). This was an ideal railway for the three-phase system. It was 45 km in length with gradients of 1 in 40 up which trains of 100 tons were to worked at 32 km/h. The electrification was carried out by Brown-Boveri & Co., of Baden, Zurich, using 750 V, 40 Hz. The original B type locomotives had two motors of 150 hp each, driving the four wheels through gears, jackshafts and side-rods, and they weighed 30 tonnes. They were joined in 1910 and 1918 by more powerful units of the B-B type, each fitted with two 260 hp induction motors, capable of hauling 220 tonnes goods trains at 14–42 km/h, and 120 tonnes passenger trains. Pioneers of three-phase traction systems include Kando of Ganz, and Dolivo-Dobrowski, who was born in St Petersburg and who died in Heidelberg after developing three-phase systems at AEG. Ganz of Budapest formed links with Westinghouse of the USA and manufactured several railway electrification systems including three-phase schemes. Its eminent engineer, Kalman Kando, developed DC locomotives and several kinds of AC system, but his name is linked with a design for taking single-phase, high-voltage AC current from the catenary at industrial frequency and using electromechanical converters on the locomotive to give three-phase supply to the traction induction motors. Speed control was effected by pole-changing and concatenation. Concatenation or 'connection in chain' became the technical name for coupling the rotor of one asynchronous motor to the stator of the next. It was sometimes called 'cascade connection'. Ganz also developed the pure three-phase system, similar to that of Brown, with two trolley wires, three-phase supply and three-phase motors.

In 1896, Ganz built a Kando three-phase electric locomotive, and in 1899 the company constructed a three-phase test line in Budapest which was 1.5 km long, and electrified at 3000 V. In the late 19th century, Ganz tendered to electrify part of the 'Circle' lines of the London Underground using a 3000 V three-phase scheme similar to the one in Budapest. The three-phase system was thought ideal for working services at fixed intervals and to identical timetables on the Circle lines where variation of top speed was not wanted. LVDC with conductor rail distribution was chosen by C. T. Yerkes, the American who controlled the District Railway operations over the Circle lines and who pursued an 'Edison-Sprague' engineering policy.

Ganz developed the three-phase system with frequencies of 15 Hz and 16.66 Hz and it was used in Italy and Switzerland. The system in Italy was selected in 1902 after comparative trials between battery-electric traction, LVDC traction using a conductor rail, and 3400 V AC three-phase traction. The Ganz three-phase system was tested on the Adriatic Valtellina railway which had steep gradients where there was a need to control speed carefully. Two locomotives rated at 440 kW were tested in 1902 and the system judged a success. An extensive three-phase railway network was developed in Italy but not in other countries. Important stages were the electrification in 1910 of the Giovi line of the Italian State Railways which had gradients of 1 in 29 and six tunnels, one being 2 miles in length. Steam traction was inadequate and goods could not be cleared quickly enough from the Genoa docks. Sometimes a backlog of 575,000 tonnes accumulated. Ten-coupled electric locomotives were supplied by Ganz and Westinghouse with power outputs of 2000 hp and 2600 hp drawing supply at 3000/3500 V AC and 15/16.66 Hz. These Series E30 and E50 locomotives demonstrated the superiority of electric traction over steam traction on mountain railways, and after 1920 considerable three-phase electrification was undertaken using 10 kV, 45 Hz. Development went on until 1933, but following widespread wreckage of the system in the 1939–45 war, it was declared obsolete and replaced with 3 kV DC. This particular system, with its two supply wires above the track, disappeared when the last section closed in 1976. Other examples of this form of railway were less extensive than the Italian network. Brown-Boveri offered to electrify the 23 km Simplon Tunnel as a means of establishing their three-phase traction system. Development work was completed before the tunnel was opened in 1906, and test runs began in April of that year using 1C1 locomotives drawing power at 3000 V AC, 16 Hz. Only two running speeds were possible, achieved by pole changing, but the engines were a success, and were

118 *Electric railways 1880–1990*

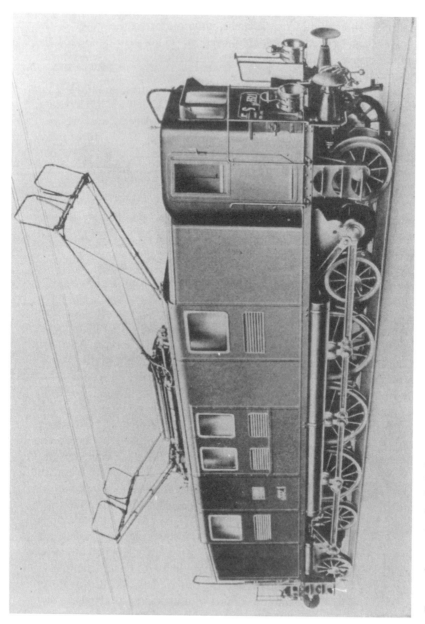

Figure 8.1 Express locomotive for the Italian State Railways' three-phase 10 kV 45 Hz system. Two overhead contact wires were required (one per phase) with the running rails taking the third phase.
Source: Italian State Railways (FS).

followed in 1908 by the D type engine which had four running speeds, and the 1D1 type. The Brown-Boveri three-phase system did not extend throughout Swiss railways, and when the Loetschberg railway opened in 1910 it used the single-phase system with AC commutator motors on the locomotives.

The Brown-Boveri three-phase railway through the Simplon Tunnel was successful enough to remain in operation until 1930, when it was converted to single-phase in common with the main Swiss network. The scheme was applied to one railway in the USA, and to none in Britain. The Great Northern Railway (USA) electrified the original Cascade Tunnel in 1909 using the General Electric 6600 V, 25 Hz three-phase system, with two trolley wires. This was a short line of 10 km length including the 4.2 km Cascade Tunnel. It was worked by B-B locomotives weighing 104.5 tonnes in running order, which were rated at 1500 hp for one hour, and when used in multiple were capable of hauling trains of 2500 tons up gradients of 1 in 45. Because of the short journey and the nature of their duty, which was to take trains through the tunnel, the locomotives had only one running speed, set by the supply frequency. The original scheme was replaced in 1928 when a new Cascade tunnel was bored, 12.5 km long, and single-phase AC electrification was installed over 115 km, using motor-generator locomotives taking supply from the catenary at 11.5 kV, 25 Hz for conversion to DC (CERA, 118, 1979). This line was worked by extremely large motor generator locomotives including single-unit engines delivered in 1948 which weighed 320 tons and were rated at 3730 kW or 5000 hp. The first Spanish railway electrification scheme was three-phase, with two separate trolley wires. This was installed by Brown-Boveri between Gergal and Sante Fe, where steep gradients of 1 in 37 were encountered, and were worked by five B type locomotives fitted with two 160 hp motors drawing from the supply at 5500 V, 25 Hz. These were supplemented in 1923 by two additional locomotives fitted with 360 hp motors.

One noteworthy three-phase scheme was installed in 1912 along the locks of the Panama Canal to enable electric 'mules' to guide the ships through sections where there was little clearance. The three-phase system enabled the speed of four mules working at the four quarters of the ship to be kept constant. Very steep gradients of 1 in 1 were encountered at changes in locks level and combined rack-and-adhesion units were used. The need to throw ropes and cables from ship to shore, and the need for the mules to run close to the canal side, ruled out overhead trolley wires and two separate low level conductor rails were used with the rails taking the third phase. Voltage was only 200 V and frequency 25 Hz. The system proved satisfactory though of limited application.

Another notable three-phase system was the temporary installation used to test high-speed railcars, constructed by Siemens and Halske, and AEG, between 1899 and 1903 (Lasche, 1902). Some engineers dismissed these tests as having little significance, but they demonstrated the potential in electric traction for very high speeds and the various plans for high-speed electric railways considered in the 1890s found practical if temporary realisation in these experiments. The main trials were carried out between 1900 and 1903 by AEG and

Siemens & Halske, with state backing, on a military railway between Marienfeld and Zossen near Berlin. The two high-speed cars and a lower-speed locomotive used three-phase traction motors and precise speed control was exercised by signalling to the lineside power house via an electric telegraph. The speed of the steam engine was varied, which in turn changed the frequency of the three-phase supply and the speed of the motors. This method served well enough with a reciprocating engine and it had been considered by Heilmann in his investigations into the basic forms of electric railway, but it could never be used with a turbine and was only suited to an experimental installation. Trouble was experienced with bogie hunting but the electric trains were run at 130 mph. There were two high-speed vehicles taking the form of large six-wheeled passenger coaches, each of which carried two sets of three bow-collectors, one pair per phase. The running rails were not used for one of the phases, as they were in the Italian and Swiss three-phase installations. Power was supplied at 10 kV, and frequency was either 25 Hz or 50 Hz, depending on the speed required. Current was transformed on the Siemens car to 1150 V, and on the AEG car to 435 V. In high-speed running in 1903 the Siemens car reached 206.7 km/h (128.5 mph) and the AEG car reached 210.2 km/h (130 mph) with the power house supply at 45 Hz rather than 50 Hz due to a defect in the driving engine. The Bo-Bo locomotive was not intended for such high speeds. It was Siemens built with four motors of 250 hp each, weighed 52 tonnes and reached 64 km/h. The success of this demonstration of high-speed traction stimulated the builders of steam locomotives to offer a prize for the design of a high-speed engine capable of matching the electric units. The award went to a 4-4-4 locomotive designed by Kuhn and built in 1904 by Henschel which ran at 145 km/h (90 mph) but which could not equal the performance of the electric cars. The speeds reached on this three-phase line were not approached by steam traction until the mid-1930s. Despite the success of the Prussian high-speed electric cars, the railways through the Simplon and the Cascade tunnels, and the considerable network in the hilly districts of North Italy, the pure three-phase railway pioneered by Brown of Brown-Boveri and by Kando at Ganz never became a potential general standard. It was displaced after 1910 by the single-phase AC system, and by the high-voltage DC system. The intrinsic drawbacks to the form were too severe. Not only were two trolley wires needed, which increased maintenance and made junction networks complex, but speed control was limited. Before the solid-state locomotive-mounted inverter became available in the 1980s, the usual method for controlling the speed of the three-phase traction motor was by pole changing or by concatenation. These were used on the phase-splitting locomotives on the Norfolk & Western Railway and the Virginian Railway, and on the locomotives of Italian State Railways, or on the original Simplon Tunnel line. Both techniques could be combined to provide six or seven running speeds. Concatenation or cascade connection was widely used with three-phase motors, and it was common to insert a resistance in the secondary circuits at starting which was reduced with rise in speed. If only one running speed was needed, which was the case on the original Great

Northern electrification in the USA, this was all the control which was necessary, but if several fixed speeds were required, cascade or concatenation enabled them to be obtained. Assuming the traction motors to be of the same kind, they would be grouped normally into pairs with one stator winding connected to the line, and the corresponding rotor winding connected to the winding of the second motor of which the stator winding was closed through a resistance which was cut out as the motor came up to speed. Concatenation of two motors cut the speed to approximately half that of two motors in parallel on the line. Sometimes up to four motors were concatenated, but polyphase motors in the 1900–20 period could only provide a limited number of operational speeds, which was a serious disadvantage to most general-purpose railways.

8.3 Development of the single-phase traction system

Once manufacturers of electric traction equipment produced reliable equipment with sufficient capacity to displace steam traction, the question of general electrification was raised. Two matters dominated the whole issue. The low-voltage DC supply to short run tramways was unsuitable for long-range heavy-duty work because many small power stations would be needed if LVDC generation and distribution was retained. Also, techniques were needed for supplying large amounts of power to a moving train, without excessive voltage drop or over-large currents. A Westinghouse company paper remarks (CERA, 118, 1979; Westinghouse Electric, 1929):

> With these two conditions encountered, heavy railway electrification was practically at a standstill, and, in the minds of those who knew, there seemed but little promise for future railway electrification on a large scale. It is true that a number of attempts were made to build electric locomotives in the early days, but these were mostly by those who were familiar with the traction problem only and who had but little or no experience with the generation and transmission problems.

The advantages of the AC system was recognised, for both distribution and traction, but technical difficulties prevented application, and when solutions were found, they favoured AC distribution in conjunction with DC traction, and made the LVDC railway the most widely used form before 1920. An impasse was escaped by two innovations which turned the LVDC tramway and short length, light-duty system into the heavy-duty electric railway, able to supplant steam traction on urban and suburban networks. These were the rotary converter and the third-rail current distribution (Westinghouse Electric, 1929).

> ... the coming of rotary converters in large capacities was recognized as a great boon to the heavy electric railway. Here at least was a solution of one of the difficulties, for by means of polyphase alternating-current generation and distribution at high voltage, and conversion to direct-current through distributed substations, the problem of centralization of the power stations was accomplished.

The other outstanding problem, concerning the collection of high currents, was solved by the third rail:

This permitted relatively high amounts of current to be transmitted from the substation to the moving vehicles, and thus it appeared, for a while, to solve the second part of the heavy railway problem. In consequence, after the development of the rotary converter and the advent of the third rail, the heavier work went forward by leaps and bounds. However, it may be noted that this was only where traffic was very heavy such as for elevated and subway service in the larger cities, for city terminals, suburban railway service, etc.

It was technically feasible to build a lengthy main line railway using rotary converters, and third-rail distribution at 600 V DC, but this was not economic due to the large number of substations required and the losses in the supply rail. Before 1900, the advantage was with the LVDC railway supplied through rotary converters from AC stations, but alternatives were needed which would enable long-distance main line electrification to be undertaken with economy. American engineers recognised the achievements of Ganz, and Brown-Boveri, in Europe, which company had introduced three phase systems, but for the reasons given above, and because these systems were limited to 3000 V or 4000 V, they were not suitable for general use. High-voltage DC systems were under discussion in 1900, but there were doubts concerning the practicality of them. HVDC was established in France in 1903 by the use of 2400 V DC distribution with two trolley wires, though this was never a standard form. The normative HVDC system using a single trolley-wire distribution was introduced by the 2400 V DC Butte, Anaconda & Pacific Railroad, in Montana, USA, in 1913. Before HVDC emerged there were strong reasons for developing some form of AC system for long-range electrification of railways. B. J. Arnold is credited with the production of a single-phase induction motor, with a variable speed gear, which was used to drive a railcar in the first attempt at AC traction according to Westinghouse company history. The honour of introducing a workable, single-phase commutator motor, with characteristics resembling those of the series-type DC traction motor, is usually given to Lamme, who also developed the motor-generator (Westinghouse Electric, 1929).

As early as 1892, a pair of 10 h.p. commutator type railway motors had been built for an experimental car for Mr Westinghouse. The frequency used was too low, however, to handle the car satisfactorily, and the motors used were also too small, so that the tests were soon abandoned. About 1896, the problem was taken up again, but was not carried very far. About 1899, some 60 cycle series motors of about 40 h.p. were built and operated for about six months, and while not entirely satisfactory, yet the evidence was such that it looked very much as if, with frequencies as low as 25 cycles or less, railway motors of the single-phase commutator type could be built for quite large capacities.

Between 1901 and 1902, the outstanding problems were solved, and single-phase AC, using commutator motors became an established form in the USA and Europe. In 1905, an 11,000 V single-phase locomotive, no. 10003 was tested against two 650 V DC locomotives to identify suitable designs for the Pennsylvania

Railroad New York terminal scheme, which was eventually electrified using third-rail DC for reasons of reliability. The St Clair Tunnel line under the Detroit River was electrified between 1906 and 1908 using 3300 V single-phase AC with 66 ton Baldwin-Westinghouse locomotives. The 11,000 V, 25 Hz, single-phase electrification of the New Haven lines into New York was completed after 1905 and involved 88 route miles, 550 track miles, 100 locomotives and 70 motor cars. These established the single-phase system as a major form of electric railway. This scheme, completed in 1907, had 9 miles of low-voltage DC third rail to gain access to the New York Central terminal station, and employed dual system locomotives which could change from conductor rail supply to overhead trolley wire supply at full speed (Gibbs, 1910; Middleton, 1977). In 1911, the line through the Hoosac Tunnel in Maine was electrified with single-phase AC, 11 kV. The same system was installed on the Rock Island & Southern, which had a 50-mile route with no substation, and which handled 500 ton trains, and in 1914 was employed to electrify the Philadelphia Paoli Division of the Pennsylvania Railroad. By then Westinghouse had bought the rights to the Ganz three-phase system in the USA, and was co-operating with Ganz on railway electrification in Italy. Interurban electrification made use of high-voltage AC from 1906. The Spokane & Inland Empire line worked three- and four-car trains under 6600 V single-phase AC in the trolley, and several branch railways and secondary routes on the Erie (1907), and the Long Island Railroad (1905) used single-phase 11 kV AC. Electric shunting on the New York, New Haven & Hartford system, used single-phase 11 kV, and proved to be more economic than steam switching.

In Europe, the first main line single-phase AC electrification schemes were established before the 1914–18 war, and were generally operated with locomotives fitted with commutator motors. There were several short experimental lines worked by converter locomotives, which were followed by electrification of heavy-duty mountain and tunnel lines. These included the Loetschberg line in Switzerland, part of which was working by 1908, taking supply from the contact wire at 15 kV, 16.66 Hz. The whole railway was not opened until 1913 when the tunnels were completed. This was the single-phase AC scheme which established the system in Europe. In 1909 in Britain, the London, Brighton & South Coast Railway used German AEG equipment to electrify using a single-phase, 6600 V, 25 Hz scheme. This system was converted to the London Standard conductor rail DC system after the formation of the Southern Railway in 1922 in the interests of standardisation (Dawson, 1923; Moody, 1961, 1979). Between 1902 and 1908, extensive comparative tests were conducted by the French Midi railway, with six types of locomotive. The trials led to single-phase, 12,000 V, 16.66 Hz being used, though this was later converted to 1500 V DC after the first AC locomotives gave poor performances (Haut, 1969). The Austrian branch railway in Styria opened for steam working in 1907 but converted to electric traction between 1909 and 1912 with a single-phase supply, 6500 V AC, 25 Hz, and the Swiss Rhaetian Railway adopted single-phase 11,000 V AC, 16.66 Hz. The Silesian mountain railways electric system was started in 1914 using single-phase, 15,000 V, 16.66 Hz, but the 1914–18 war delayed completion until 1922.

By 1908, the frequency of 15 Hz and 16.66 Hz were practically standard in Europe, and the 25 Hz scheme was discarded. Eventually 15,000 V, 16.66 Hz became the norm in Austria, Germany, Switzerland, Sweden and Norway. The contributions of Behn-Eschenburg, and Huber-Stockar, of Oerlikon, were fundamental in establishing the single-phase AC railway worked by locomotives with commutator traction motors (Tassin, Nouvion and Woimant, 1980).

The experiments on the Seebach to Wettingen railway between 1904 and 1905 investigated the performance of an AC railway supplied from a single high-voltage contact wire. These tests introduced the AC/DC converter locomotive and proved the single-phase commutator motor in railway service. The venture was another instance of a major electrical company carrying out an electrification programme at its own expense, to demonstrate the advantages of electric traction over steam traction. In 1901 the Oerlikon company, which was based near Zurich, offered to electrify the 23 km of the Swiss Federal Railways between Seebach and Wettingen to demonstrate the advantages of the systems developed by Huber-Stockar and Behn-Eschenburg. The first locomotive contained a converter similar to one proposed by Ward Leonard in 1895 for providing DC traction motors with current drawn from a contact wire at 20,000 kV. The pioneer Seebach-Wettingen engine was completed in 1904 and it drew supply at the industrial frequency of 50 Hz, and 15 kV. It was the first electric locomotive to use industrial frequency supply. The supply on this line was later changed to a frequency of 15 Hz, which was the widespread norm in central Europe before the 16.66 Hz standard was adopted. The first locomotive contained two transformers, 15,000 V/700 V, with a total of 500 kVA, and a constant speed converter made up of an asynchronous monophase motor coupled to a DC generator which supplied 400 kW at 600 V. The direct current voltage could be varied, and it supplied two traction motors of 150 kW each. These were mounted on the two bogies, with each motor driving the four wheels of each bogie direct through coupling rods. The engine functioned well, but the weight of 48 tonnes was considered high for the low traction power of circa 345 kW. The losses in the several stages of conversion, and the losses in the converter when the locomotive was stationary discouraged repetition of the design even though it drew supply at industrial frequency (Haut, 1969; Tassin, Nouvion and Woimant, 1980). The second locomotive made use of single-phase AC commutator motors, supplied from a contact wire at 15,000 V and a frequency of 15 Hz. The engine weighed 42 tonnes and developed 360 kW traction motor power. This second machine, delivered in 1905 used the commutator motor developed by Huber-Stockar and Behn-Eschenburg at the Oerlikon works. It was of the BB type with the two motors mounted on the two bogies, with the wheels driven through side rods like those on the first engine. Current was collected from the contact wire at 15,000 V, 15 Hz, transformed to 700 V and supplied to the variable voltage motors. This supply varied in steps of 35 V to control speed. The normal speed was 650 rpm, but a maximum of 1000 rpm was permitted corresponding to a train speed of 60 km/h. The original converter locomotive was rebuilt to a design similar to the second unit at this time. The

Evolution of the electric railway 1920–40   125

Figure 8.2   Phase converter locomotive built by Oerlikon for the Seebach-Wettingen trials of 1901–05. The supply was single-phase 15 kV 50 Hz. Note the simple overhead system and lightweight support masts.

Source: Oerlikon.

third locomotive of the A1A-A1A type was delivered by Siemens in 1907, and was carried on two six-wheeled bogies, with two motors per bogie, driving through gears rather than rods. These were single-phase commutator motors, drawing supply at 15 Hz, and each of the four motor provided 165 kW. This engine could be fitted with one motor per axle to increase traction power to circa 1000 kW and the engineers were confident that single-phase locomotives with AC commutator motors could be developed to provide 2000 kW in the near future. In 1907 the Swiss Federal Railways rejected a proposal that they purchase the Seebach-Wettingen electrification and the installation was dismantled. Steam traction took over the workings in 1909, though the great potential of monophase AC traction using commutator motors had been demonstrated. These developments were paralleled by work in Germany which defined standards for contact wire frequency and voltage. Comparative trial suggested that low frequency better suited the series compensated commutator motor, which gained favour over the repulsion motor. Low frequency in the contact wire could be got from rotary-converter substations supplied with a higher frequency from the power station.

By 1907 or 1908, the 25 Hz system favoured by AEG in Europe was abandoned in favour of the 15 Hz or 16.66 Hz system. The latter system became the standard. Between 1903 and 1907 there were 32 km out of a total of 56 km electrified lines in Germany which used 25 Hz. The later schemes had lower frequencies. Between 1912 and 1914, 74 km were electrified in the Bavarian Alps; between 1914 and 1920, 95 km in Silesia; in 1913, 46 km in Baden; and after 1909, the 25 km of the Dessau-Bitterfeld main line was electrified. The Dessau-Bitterfeld line became the testing ground for 15 types of locomotive, six 2B1 or 1C1 suitable for express passenger work, and seven D type goods engines. The 1911 Class B52 number 10502 was built to operate under 10 kV to 15 kV, 15 Hz, and was fitted with the Winter-Eichberg modified repulsion motor, which never became a widely used form. The engine had two, each rated at 515 kW at 78 km/h. The Dessau-Bitterfeld scheme helped to establish as the norm the single-phase AC railway, supplied with 15,000 V in the contact wire, and with the locomotives using single-phase AC commutator traction motors at low frequency.

8.4 The emergence of the single-phase archetype

The Liverpool & Manchester Railway established the general purpose public steam railway. In Europe, the Loetschberg railway has a strong claim to be the system which established the main line AC railway. This line was opened throughout in 1913, when the Loetschberg tunnel was completed, though part of the route had carried traffic since 1908. It was a difficult line to work. The 75 km long route included a 14.6 km tunnel, sharp bends and gradients of 1 in 37. Electric traction was essential. The single-phase AC system was chosen, rather than a three-phase scheme like the one installed by Brown-Boveri on the

Simplon route. The first section electrified was the relatively level route between Spiez and Frutigen opened in 1908, and this was used as a testing ground for electric traction in the years before the Loetschberg Tunnel was completed (Dawson, 1909, 1923). The Bern-Loetschberg-Simplon line was built to join the Simplon route and was the first main line planned as an electric railway. The electrical system was 15,000 V, 15 Hz (later 16.66 Hz) single-phase AC. The line opened with locomotives which were among the most powerful of the day. The CC Oerlikon locomotives were developed after comparative trials in 1910 between two designs of 2000 hp, one German, one Swiss, built by Oerlikon. The CC locomotives had two large body-mounted single-phase AC commutator motors, driving the two sets of six wheels through jack-shafts, connecting rods, and coupling rods. These successful machines were followed in 1912 by 13 more powerful units, from Oerlikon. These were of the 1E1 type, driven by two body-mounted 1250 hp (933 kW) motors through two jackshafts and a 'Scotch Yoke'. This Be 5/7 design is one of the great classics of railway engineering history, and the engines could take 330 tonne trains up 1 in 37 gradients at 50 km/h. This new electric main line established the single-phase electric railway in European traction. The electrical system, 15,000 V, 15 Hz (later 16.66 Hz) became the standard in Austria, Germany, Norway, Sweden, and Switzerland, where the three-phase Simplon route was converted to it in 1930.

The single-phase AC system, 6600 V, 25 Hz was chosen by Sir Philip Dawson, and supplied by AEG and BTH to the London, Brighton & South Coast Railway which commenced public electric services in December 1909 on the Victoria-London Bridge section. The LBSC obtained powers to electrify in 1903, and their consultant, Dawson, chose the AEG system which was also used on the Midland Railway Lancaster to Morecambe line in 1908. The Bo-Bo motor driving coaches were fitted with the Winter-Eichberg modified repulsion motor, which was also used in locomotives tested on the Dessau-Bitterfeld line. Power vans rated at 746 kW saw service on the LBSC – the only British examples of AC locomotives before the 1950s. It was unfortunate that the LBSC adopted a system which was rejected throughout Europe as a standard in the year before the British scheme started public operation. The voltage of 6600 V was low, compared to the 15,000 V which became the European norm, and the frequency was high at 25 Hz. The system was an engineering success, and it would have been extended had not the LBSC company been absorbed into the Southern Railway which had a much larger LVDC network inherited from the LSWR. When the AC system was converted to LVDC, the British electrical industry lost the opportunity to develop AC railway engineering equipment and this effected its strategy down to the mid-1950s (Bonavia, 1987; Moody, 1979).

The French Midi Railway comparative trials of 1902–08 have been called the 'Electric Rainhill' (Haut, 1969; Tassin, Nouvion and Woimant, 1980) and likened to the Rainhill trials of 1829, during which the Stephenson 'Rocket' was tried in competition with Hackworth's 'San Pareil', and Braithwaite & Ericsson's 'Novelty' (Ahrons, 1927). Six experimental electric locomotives, each of the 1C1 type, were ordered from different manufacturers and were tested under

Figure 8.3 Single phase AC traction motors and Scotch Yoke mechanical drive on Berne-Loetschberg-Simplon Railway Class Be 5/7 2500 hp 1-E-1 locomotives built by Oerlikon and Brown-Boveri in 1910 to operate on single-phase 15 kV 16 2/3 Hz supply.
Source: Oerlikon.

a contact wire drawing single-phase supply at 12,000 V, 16.66 Hz. The specification was that the locomotives should haul 303 tonnes up 1 in 45 at 35 km/h, and 432 tonnes on 1 in 70 at 58 km/h. Locomotive weight was to be circa 85 tonnes with 54 tonnes available for adhesion. Five of the six units had rod drive. The locomotive supplied by Jeumont had individual axle drive. The designs indicated one disadvantage of monophase and three-phase traction motors, which was their large size compared to DC motors of equal power. Consequently, the common practice at this time was to mount them in the locomotive body, and drive the wheels through some transmission system, often jackshafts and rods. Though some DC motors were mounted in this manner, including those of the DD1 on the Pennsylvania terminal lines, it was much easier to mount the LVDC traction motors used on the DC railway and in the oil-electric locomotive by nose-suspension in the bogie. This was a great advantage despite some bad riding and track damage caused by high unsprung mass. The size of the AC commutator motor was eventually much reduced, and used with transmissions permitting bogie mounting, but the advantage lay with the nose- suspended DC traction motor. One of the Midi locomotives, from Brown-Boveri, had the Deri repulsion type motor, with rod drive, but this motor never became standard in railway service. The tests revealed the advantages and disadvantages of each type, but none was selected to work the line. The company introduced instead eight 2Co2 Westinghouse-Jeumont machines of 95 tonnes driven by three double motors, of 260 kW each. The engines were ordered in 1913, delivered in 1914, but prevented from entering service by the Great War. By the time it was possible to use them after 1918, the decision had been taken to accept 1500 V DC as the French standard for new operations and they were rebuilt as direct current locomotives in 1923, drawing current from an overhead contact wire. The rebuilt locomotives were remarkable in that they carried three pantographs and the motors were mounted with the main axis vertical. They could run at 130 km/h and in 1927 were working the fastest train in the world, averaging 108 km/h between Bordeaux and Bayonne despite a speed limit of 120 km/h. The Midi tests demonstrated the good performance of single-phase AC traction with commutator motors, and after conversion to direct current, the railway helped to establish the HVDC railway as a standard form of electric railway.

One point must be stressed. These early electric main line railways were only economic if they were supplied from hydroelectric stations. In some cases the state funded them on strategic grounds to promote the economy of an area. On lines like the Loetschberg and the Simplon, there was no alternative to electric traction, and hydroelectric power was available. These systems proved the engineering success of electric traction, but they were far from establishing general electric working of main line railways as an economic proposition in areas where electricity was generated from coal.

8.5 Converter locomotives

The most common type of converter locomotive enabled an AC supply in the overhead contact wire to be used with DC traction motors. The motor-generator locomotive and the motor-converter locomotive were closely related, and were substations mounted on a locomotive carriage. Both used rotary electro-mechanical converters, and the terms rotary converter, motor-generator and motor-converter are sometimes used interchangeably. There were attempts to distinguish between them. The term 'motor-generator' was applied by some engineers to a set consisting of a driving motor taking current from the contact wire, and a generator delivering current to the traction motors. The term 'converter' was applied to an integrated assembly with a commutator at one end of the armature and slip-rings at the other. The terms were sometimes used as if they were synonymous. The motor generator locomotive combined the advantages of the AC railway, with monophase supply in the contact wire, and the DC traction motor, which could be more easily bogie mounted than contemporary AC motors. The motor generator locomotive and the motor converter locomotive originated with the electric 'Ward Leonard' locomotives of Heilmann, and the Oerlikon converter locomotive on the Seebach-Wettingen line. On railways where there was dense traffic, so that lineside substations equipped with rotary converters were justified, there was little advantage in using motor-generator locomotives. If a railway crossed uninhabited terrain, and traffic was light, the motor-generator locomotive had advantages by reducing the number of lineside substations. It is not surprising that the motor-generator principle found greatest use in the USA, principally on the Great Northern Railway.

The first motor-generator locomotive in the United States was built in 1925 by Henry Ford and his staff at the Rouge River Plant of the Ford Motor Company, in Detroit. Electrical equipment was supplied by Westinghouse, and the mechanical components were made by Ford (Haut, 1969; Middleton, 1974). Henry Ford worked for Edison as a young engineer. He admired the way Edison faced the problems which arose when the first power and light systems were developed. Edison told him that he had to solve all the difficulties himself, component by component, and then form the parts into a working whole. Ford held an important post in the Detroit Edison Electric power house, but left to pursue his own work in motor car engineering. Edison became one of his closest friends (Ford and Crowther, 1922). In 1923, in pursuit of his goal of optimising the efficiency of the Ford Company and of the organisations on which it depended, Ford bought the steam-operated Detroit, Toledo, & Ironton Railroad, which was in a poor state. He reformed it by applying his industrial management philosophy in every department – technical, managerial, legal and financial. Ford was hostile to lawyers and financiers, whom he believed had ruined the railroad industry, and his treatment of them during his reconstruction of the railway was ruthless. Electrification of the railway was planned, starting with a pilot section of 17 miles, and extending electric traction over the entire route from the Detroit factories to the Ohio River, 300 miles away. Anti-monopoly laws compelled Ford to relinquish ownership of the line in the late 1920s and electrification of the railway was never carried beyond the pilot project. This section of the line was electrified using a single-phase AC supply at 11 kV, 25 Hz in an overhead contact wire which was carried on a system of concrete gantries and posts. These survived when the electrification scheme was scrapped as they proved too resistant to cheap demolition. Supply was taken from the motor car factory's own power station, which was enlarged for the purposes, and was in part fuelled by waste heat from the foundries and other installations of the plant. In order to use DC traction motors, a large motor-generator locomotive was designed by Ford engineers, in conjunction with Westinghouse, and erected at the River Rouge motor car plant.

The Ford electrification scheme was unusual and the locomotive and the concrete support arches for the catenary were not repeated despite their success. The locomotive was completed in 1925 and was the first large motor generator unit in the United States. It weighed 393 tons and was rated at 4200 hp. It led to the use of motor-generator locomotives, for shunting purposes, on the New York, New Haven & Hartford Railway in 1926. In 1928, the Great Northern Railway installed a single-phase AC scheme on a new line which replaced the original route through the first Cascade Tunnel opened in 1909 and worked by a three-phase system. This was worked by motor-generator locomotives, inspired by the Ford machine. The Ford locomotive was a two-unit Do-Do + Do-Do. There was a motor generator set in each unit supplying DC motors which were mounted on each of the sixteen axles of the whole machine. Each half of the double-ensemble was divided into two sections, flexibly connected, with each section mounted on the four-axled carriage below. One section carried the circuit breaker, the transformer, the switchgear and the motor blowers, and the other carried the motor-generator, which weighed 29 tonnes. Each 186 tonne unit had two pantographs, and eight nose-suspended traction motors, which provided 1550 kW for traction power, and during trials – some of which were conducted on the New Haven railway – each delivered 2000 kW at 27.5 km/h, and 1400 kW at 40 km/h. Working coupled together, they formed an extremely powerful locomotive, which was successful.

This Ford-Westinghouse locomotive influenced the Baldwin-Westinghouse motor-generator locomotives built in 1927 for the newly constructed second route over the Cascade mountains, which bypassed the old Cascade Tunnel and was electrified using single-phase 11 kV, 25 Hz. Each locomotive was of the 1D1 type, with four DC motors permanently connected in parallel across the armature of the main generator. Contact wire current was transformed in an air blast transformer, with the secondary supplying the stator of a synchronous motor which was rated at 2100 hp, 750 rpm. This drove the main generator which supplied 1500 kW at 600 volts to the DC traction motors. Each engine was rated at 2165 hp for one hour, and two were often worked coupled together. In 1927, General Electric and American Locomotive Company delivered much larger motor-generator units to the same railway. These were of the 1CC1 type, which weighed 235 tonnes each, with a rating of 2450 kW, and a maximum speed of

80 km/h. The contact wire voltage, 11 kV was transformed to 2300 V AC on the engine, and supplied to the synchronous motor which drove two DC generators connected in series, which in turn delivered DC current at 1500 V to the traction motors. These engines worked well, and after the Great Northern Railway discontinued electric traction in 1956, they were sold to the Pennsylvania company and used on goods trains until 1970. The Great Northern Railway introduced further examples of this type as late as 1948, in the form of extremely large Bo-Do + Do-Bo machines, which weighed 320 tons, and provided a continuous output of 5000 hp. In the same year the Virginian Railroad took delivery of four locomotives, each made up of two units carried on four two-axled bogies. These motor-converter locomotives weighed 454 tons, and were rated at 7800 hp, and were able to haul 3000 tons trains up gradients of 1 in 77 at 35 mph. They established no numerous class. In 1948, the mercury-arc rectifier locomotive was being tried in the USA, and the use of locomotive mounted rectifiers to convert single-phase AC to DC became more common despite unreliability in some classes. The motor generator locomotive had the advantage of being able to regenerate current when braking or running downhill, and this was regarded as a valuable feature of the system, though it was one shared with other forms of locomotive.

The motor-generator locomotive was never a common type, and was restricted to services where it could be used to advantage. This included heavy duty shunting in electrified marshalling yards. In 1938 the SNCF introduced two Ward Leonard shunting engines, type CC1001, for use on the 1500 V DC system of the former Paris-Orleans-Midi railway. These were large engines carried on two six-wheeled bogies. They weighed 88.5 tonnes and were designed for 'hump' shunting. They were equipped with Ward Leonard control without resistances to avoid overheating when shunting at low speed, and typical duties included moving 2000 tonnes at 10 to 13 km/h, with speeds of 1.6 to 3.2 km/h on the hump. The motor generator consisted of a compound-wound motor and an anti-compound wound generator, supplying four traction motors, with two excitation rheostats, an auxiliary excitation generator and a secondary auxiliary generator. Current from the 1500 V DC contact wire was fed to the compound motor, which drove the compound wound generator, which supplied the armature of the traction motors which were connected permanently in series. The main motor was rated at 400 kW, continuously, and 525 kW for one hour. The engines worked well in the highly specialised circumstances for which they were designed (Haut, 1969; Ransome-Wallis, 1955).

8.6 Motor-converter locomotives

Motor-converter locomotives were a variant of the motor-generator locomotive which differed in the construction of the electromechanical machinery. Like the motor-generator locomotives they were basically carriage-mounted, self-propelling substations, combining the economy of high-voltage single-phase contact wire supply with the characteristics of the DC motor. In 1910–11, Alioth and Munchenstein constructed for the Paris, Lyons and Mediterranean Railway a two-unit locomotive of the 2-Bo + Bo-2 type (Haut, 1969).

Each half of the locomotive carried a driving cab, a transformer with main switch, and a group of two converters. The traction motors were unusual in being arranged vertically, gears and flexible couplings communicating drive to the wheels. The locomotives were quite powerful for the period; they hauled 200 or 300-tonne trains on a test line (7.3 km long) between Grasse and Mouans-Sartoux. On gradients of 1 in 50 they hauled 150 tonnes at 60 km/h . . . The experiments were quite successful and it is not now clear why they were abandoned.

Current was drawn from the single contact wire at 12,000 V AC, 25 Hz, and the two-unit locomotive was rated at 1800 hp for one hour. Haut mentions that a 2000 hp 1D1 engine, of the same principle, was designed but never built. The Austrian Federal Railways tested phase converter locomotives between 1927 and 1928, using both 1D1 and E types of the Kando design, but these were not developed further when Austria standardised on single-phase electrification at 15 kV and 16.66 Hz. There was a solitary rotary converter locomotive, classified as series 1082, which ran between 1931 and 1941. This was of the 1E1 type. It took single-phase AC at 15 kV, 16.66 Hz, which was transformed and changed into three-phase AC in the rotary converter, for supplying a double DC generator driven by the synchronous motor. The DC supply powered three traction motors which worked in series or parallel to drive ten driving wheels linked by side rods. The low frequency of the contact wire supply was a disadvantage, because, as Haut remarks:

. . . such a locomotive would be more effective under a 50 cycle supply system than under normal railway frequency (16.66 Hz), because it would be possible to lay out the converter as a multipole machine of reduced size, working at higher speed.

This locomotive, jointly built by Siemens-Schuckert and the Florisdorf Locomotive Works near Vienna, worked well in ten years service, but it remained an unrepresentative single unit. The prospect of using the locomotive-mounted mercury-arc rectifier was in view in the mid-1930s, and a rectifier locomotive was used in the Hollenthal trials in 1936. This reduced the advantages of using the electromechanical rotary converter to combine monophase AC supply in the trolley wire with the DC traction motor.

8.7 Phase-splitting locomotives

Phase splitting means using a rotary converter to change a single-phase supply into a polyphase phase supply for induction motors. The technique was pioneered in the USA by E. F. Alexanderson, who used two-phase motors in what he called the 'split-phase' system (Carter, 1922). Single-phase AC supply in the contact wire and three-phase traction motors were combined successfully on the

Figure 8.4 Austrian Federal Railways 1-E-1 rotary-converter locomotive built in 1931 by Siemens-Schuckert and Vienna Locomotive Works (Floisdorf). The contact wire supply of single-phase AC 15 kV 16 2/3 Hz was converted to DC.
Source: Siemens/Austrian Federal Railways.

Norfolk & Western and the Virginian railways in the United States. The phase-splitter was mounted on the locomotives. It was employed with three-phase asynchronous motors on the Norfolk & Western, and the Virginian railways. Both railways were suited to the fixed speeds of three-phase traction because they ran very heavy coal trains over steeply graded, sharply curved lines. During this period, 1910–30, railway economists advised that costs were reduced by working very heavy trains slowly, rather than by moving lighter trains quickly. This philosophy required slow locomotives with high tractive efforts, represented by the first generation of electric main line locomotives, and the compound-expansion Mallet steam engines of that era. In 1915, the Norfolk & Western railway electrified the 48 km Bluefield-Vivian section over which trains of 2800 tonnes were moved along gradients of 1 in 50, round sharp bends, and through a long tunnel (Middleton, 1974). Electric traction eliminated the severe smoke nuisance in the tunnel and the three-phase system provided reliable, fixed speed working, regenerative braking, and greatly improved line capacity. The first locomotives were delivered in 1915 with Westinghouse equipment and were large two-unit machines. Each consisted of two 1BB1 locomotives, with four powered axles driven by three-phase motors through side rods. Supply was single-phase 11 kV, 25 Hz. The double unit was rated at 4100 kW. The first locomotives were joined in 1924 by more powerful machines of a different configuration but with the same principles. These engines successfully worked the section until electric traction was ended in the early 1950s when the route was re-aligned and operated by steam locomotives. The Virginian Railway electrified lines were worked by similar machines.

A single, very large 1CC1 goods engine was built in 1917 for the Pennsylvania Railroad Altoona-Johnstown electrified section which used single-phase AC, 11 kV, 25 Hz in the contact wire (Bezilla, 1980). It was fitted with four induction motors of 1200 hp each, driving through gears, jack shaft and coupling rods. It remained the sole example of phase splitting on the Pennsylvania Railroad which used the single-phase commutator motor on its AC rolling stock until the mercury arc rectifier became available for locomotive mounting in the 1940s when DC motors were adopted. The basic principle of converting single-phase to three-phase on the engine was a sound one, but it required heavy rotating equipment, and this limited its application. In 1924, Krupp built industrial locomotives to work with a contact wire supply at the European industrial frequency of 50 Hz. Locomotive mounted phase splitters converted this to three phase for the asynchronous traction motors. This sufficed for hauling coal and ore trains on industrial railways but the system was not suitable for the general electrification of railways. Most public railways operated different kinds of trains in a variety of services. There were stopping local passenger and pick-up goods trains; long-distance semi-fast passenger trains; and express goods and passenger trains, which ran over a wide range of speeds. This did not suit three-phase motors. Phase-splitting did not overcome the problem of speed control of induction motors which relied on concatenation and pole-changing to get fixed running speeds which were too few in number for general-purpose railway

Figure 8.5 Phase-splitting locomotive of Norfolk & Western Railway built by Alco.-Westinghouse in 1924 to supplement earlier type built by Baldwin-Westinghouse in 1914. Contact wire supply was single-phase AC at 11 kV 25 Hz. The rotary phase-splitter supplied three-phase power at 750 V to the asynchronous traction motors.

Source: Norfolk Southern Corporation.

working. The phase-splitting system did not spread beyond the few railways which adopted it during the First World War.

General electrification of railways after 1920 used the high-voltage direct current system or the single-phase AC system, generally with commutator motors. The DC motor, and the AC commutator motor which was much improved between 1910 and 1925, were more suitable for general electrification than the induction motor because they provided the wide range of speeds required by the complex operations of the general-purpose railway. The AC system generally did not utilise a supply in the contact wire at industrial frequency. This was a disadvantage because frequency changers had to be installed to convert standard supply to match the system supplied to the locomotives: otherwise the railway had to build its own power stations. As national supply grids were constructed in the 1920s and 1930s, ways of using industrial frequency supply in the contact wire were sought. These offered a reduction in lineside conversion equipment and light, simplified fixed works.

8.8 The Kando system

Locomotives with three-phase traction motors were built between the two world wars, despite the problem of speed control. Some were replacements for locomotives withdrawn on the Italian three-phase network, others were additions to the American 'phase-splitting' fleet, and some were experimental machines. Of special importance were the experimental locomotives built by Kando, in Hungary, to investigate railway traction using industrial frequency supply and induction motors. The work done after 1918 by Kando at Ganz resembled the work done on 'phase-splitting' by the Westinghouse company the United States. There were close links between Ganz, of Budapest, and Westinghouse. In 1905, the Westinghouse company built a plant at Vado Ligure in Italy, to make electrical equipment, including locomotives. A design team from Ganz was seconded to this plant for lengthy periods between 1905 and 1915, when Italy entered the Great War on the side of the Allies and against Austria-Hungary. This ended Ganz co-operation, because it was an Hungarian concern. Before the entry of Italy into the war, the Ganz team worked on the Italian three-phase system with two contact wires per track, and after the war Kando sought to simplify this system by drawing supply from a single overhead contact wire at industrial frequency and converting this on the engine to feed induction motors. In 1925, Ganz constructed the prototype locomotive for the Kando system, as it became known. This was a ten-coupled machine, or E type, taking supply from a single contact wire at 16 kV, 50 Hz. The disposition of the electromechanical components owed much to the earlier Ganz-Westinghouse ten-coupled locomotives developed for the three-phase Giovi electrification in Italy in 1910. The Hungarian engine of 1925 used a synchronous converter, with a direct current exciter, which supplied the three-phase induction motor which drove the wheels through coupling rods. Haut (1969) describes the basic principles of this engines as follows:

The 50-cycle 16,000 volt 1-phase line current passed to the primary winding of the four-pole synchronous phase- converter and from the secondary winding of this converter polyphase current of about 1,000 volts was taken off for the traction motor. To facilitate the changing of the number of poles of the traction motor, the phase-converter secondary winding was provided with three, four, and six-phase taps. Built into the locomotive was a single traction motor, the winding of which could be changed over for 72, 36, 24, or 18 poles ... the corresponding economic running speeds were 25, 50, 75 and 100 km/h respectively.

This basic principle was tried in several other engines, including larger ones used in Austria, with single-phase supply at 16.66 Hz. It was used in the Hungarian State Railways V40 class of 29 locomotives built between 1932 and 1934 for the 15 kV, 50 Hz line from Budapest East station to Hegyeshalom on the Austrian border. This line attracted attention because it used contact wire supply at industrial frequency. Speed variation of the induction motors was by changing the tappings of the converter output, and by varying the pole groupings of the single traction motor. Running speeds of 25, 50, 75 and 100 km/h were obtained. The engines were of the 1E1 type, with side rod drive, and they worked well. The Kando system was reputed to be cheaper than the single-phase AC system using 16.66 Hz provided it used 50 Hz in the trolley wire, but it was more expensive that the HVDC railway which became a world standard in the 1920s. Kando influenced research elsewhere into the use of industrial frequency in railway traction, and a German line, electrified on 20 kV and 50 Hz, operated into the 1950s. The Kando experiment preceded the important Hollenthal electrification in Germany, 1936–39, which used industrial frequency supply at 100 kV, which was transformed to 20 kV, and supplied to the trolley wire in single-phase to be rectified or converted using four different methods. This comparative trial helped to decide electrification policy in Europe after World War Two in favour of single-phase AC at industrial frequency. One alternative explored in the Hollenthal trials was the Kando rotary phase converter. The Kando system worked well, but engineers disliked the locomotives, with their very large body-mounted motors with limited running speeds, driving wheels in rigid frames and connecting rod drive. A preference was emerging for bogie locomotives, with DC motors.

By 1940, the Kando system was obsolete, but there were attempts to modernise it. Two 2D2 machines were built in 1940 for the Budapest-Hegyeshalom line which were amongst the most advanced electric locomotives then running. The war prevented them from being properly tested, and at least one was destroyed in an air raid. These were large engines (3200 kW, 5000 kVA) and they were the first in the world to be fitted with apparatus for frequency control of the traction motors, carrying both a phase-converter and a frequency-converter of electromechanical design. According to Haut, a later design, Series V55, was produced after the war with the Bo-Co wheel arrangement containing a frequency converter made up of a normal three-phase slip-ring induction machine with the stator winding fed from the phase converter at 50 Hz, whilst the secondary winding supplied the traction motor current. Five speeds were obtained, 25, 50, 75, 100 and 125 km/h, and the engines were capable of hauling 1500-tonne goods trains at 75 km/h, and 750-tonne passenger trains at 125 km/h. Haut states that in 1962, Krupp and AEG developed the series V43 for use on the Hungarian system. This was a dual-voltage Bo-Bo machine, capable of running with single-phase supply at 50 Hz and either 16 kV or 25 kV. This type used silicon rectifiers, and DC traction motors, and reflected the trend of the time towards using single-phase industrial frequency supply with rectified DC motor current. At that time, there seemed to be no future for the three-phase motor in railway traction.

8.9 The high-voltage DC railway

The first high-voltage direct-current railway did not establish a fruitful line of development. The equipment was built in 1903 by the Swiss company, Secheron of Geneva (originally the Thury company), for the French State Railways line between St George de Commiers and La Mure which was electrified on the two-wire system using 2400 V DC. Two trolley wires were used per track. These were insulated from each other and the locomotives had two bow collectors per wire, giving four in all. The Bo-Bo locomotives weighed 50 tonnes and gave an output of 500 hp. They were followed in 1931 by similar machines rated at 920 hp. These engines worked well over gradients of 1 in 38, but the system never became a standard.

The HVDC railway evolved out of the conductor rail rapid-transit DC railway fed from DC generating stations or from AC stations through rotary converter substations (Agnew, 1937; Barbillon, 1923; Sprague, 1888). These systems used DC voltages in the conductor rail, which were determined by the rolling stock motor voltage, which were of the order 600, or 750, or (circa 1910) 1500 volts. By coupling motors in series, low voltage traction motors could be used on vehicles supplied from a conductor rail or contact wire charged to a much higher voltage. The HVDC railway is associated with voltages which are multiples of voltage limits set by widely used motors, that is 600, 1200 and 2400; 750, 1500 and 3000. Conductor rails took supply voltage up to 1500 V DC, though 2000 V and higher were tried unsuccessfully. The usual practice was to use a trolley wire for 1500 V DC and above (Carter, 1922). These higher voltage DC railways could be integrated with existing DC networks, though some railways installed different systems within their networks. The New Haven electric route used an AC supply for most of the journey but switched to DC conductor rail supply in the New York terminal section. Long-distance tramways, some of which carried goods, adopted HVDC early in the century. In 1911, the Southern Pacific railway used 225 hp BB goods locomotives on a 600/1500V DC overhead trolley line. The engines had four 750 V DC motors. In 1912, the Aroostock Valley railway used an engine with four 600 V 100 hp motors and a contact wire supply at 1200 V DC for shunting and goods purposes. The Oregon Electric company operated goods locomotives with four 600 V 200 hp motors on 600 V and 1200 V DC supplies.

The HVDC form was established on the Butte, Anaconda & Pacific Railway in Montana, USA. Electrically hauled copper ore trains began work on 28 May 1913 and electric passenger trains started on 1 October 1913. This was a heavily used railway, with 32 miles of main line and a total electrified trackage of 123 miles. It was electrified using an overhead contact wire charged to 2400 V DC supplied from a hydroelectric plant, owned by the Montana Power Company and located at Great Falls. This hydrostation supplied electric power for smelters and other mine equipment in addition to meeting railway needs. There were six generators nominally rated at 21,000 kW and the output was stepped up to 102,000 V for transmission over 130 miles of mountainous terrain to Butte, and to Anaconda 26 miles beyond Butte. The railway was equipped by General Electric, which dominated DC railway electrification at this period. The Westinghouse company was developing as a major competitor and was using AC technology, initially neglected by Edison, as an instrument for winning markets. The report on the Butte, Anaconda & Pacific Railway electrification, issued by General Electric, opens with the following statement (CERA, 116, 1976):

... it is ... the first steam road, operating both freight and passenger schedules, to electrify its lines purely for reasons of economy. A number of steam railway electrifications have been made because of peremptory factors, such as terminal or tunnel operation or rapid interurban service. This road, however, cannot be classed as 'enforced electrification,' since no such special limitations have been the determining factors.

The scheme was successful and it was worked by 28 locomotives of 82.25 tons each, and three 40 ton tractor trucks for boosting locomotive power. Each locomotive had four motors, of 320 hp each, with a motor voltage of 1200 V DC. The goods engines were rated at a maximum speed of 35 mph, and a solitary passenger locomotive was rated at 55 mph. The economic success of the railway owed much to the provision of cheap hydroelectric power which was generated to supply the mine, the smelters, and the towns of Butte and Anaconda, which had electric street railways. Without this, the electrification would probably not have been justified on economic grounds. The close connection between the mining company, its towns and services; and the Montana Power Company was a crucial factor. These could act as one, with little opposition, and they were a driving force behind the Milwaukee electrification which was the most important electrification scheme of the early 20th century. It was carried out by the Chicago, Milwaukee and St Paul Railway, which electrified its Montana Division (Harlowton, Montana to Avery, Idaho, 438 miles) by December 1915. It electrified the quite separate Coast and Columbia Divisions (Othello, Washington to the Pacific Coast, 208 miles) by 1919, and the ten-mile route between Black River and Seattle in 1927. All sections used overhead contact wire charged to 3000 V DC. Power was generated in hydroelectric stations of the Montana Power Company, the Washington Water Power Company, and the

Puget Sound Power & Light Company, and was transmitted over the railway company's transmission lines at 100,000 V three-phase, 60 Hz and converted in rotary converter substations to 3000 V DC for supply to the contact wire (CERA, 116, 1976; General Electric USA, 1927; Steinheimer, 1980). The 1915 Montana electrification used BB shunting engines, with four motors rated at 135 hp, and a motor voltage of 1500 V DC. Main line work was undertaken by 2BB + BB2 locomotives fitted with six motors of 430 hp each, with a motor voltage of 1500 V DC. The Coast & Columbia Divisions were worked at opening by 2C1 + 1C2 units, fitted with six twin motors of the geared quill type mounted above the axles, each rated at 700 hp. There was a voltage difference of 1500 V across the twin motor, or 750 V per armature. In addition, the passenger trains were worked over the Coast & Columbia Divisions by the EP2 Class, each of which was fitted with 12 individual, gearless bipolar motors rated at 290 hp, with a motor voltage of 1000 V.

This major undertaking, applied to several hundred miles of a first-class main line railway, established the HVDC railway in the rest of the world. It was visited by engineers from Europe, including a group from the Midi railway of France, and from Britain. It influenced the thinking of British engineers including the Merz & McLellan consultants of Newcastle upon Tyne, and Worsdell and Raven of the North Eastern Railway who had electrified some operations (Raven and Watson, 1919). In 1904, Worsdell electrified the steeply graded North Eastern Railway Quayside line in Newcastle, using an overhead contact wire supplied at 550/600 V DC, and bogie locomotives with four 550/600 V 90 hp DC motors (Carter, 1922; Hoole, 1988; RCTS, 10b, 1990). The more extensive Shildon to Newport on Tees electrification of 1915 used 1500 V DC in the overhead contact wire. The BB locomotives were fitted with four 750 V 275 hp motors (Hennessey, 1970; Hoole, 1987, 1988; Raven, 1922). American DC practice influenced the York to Newcastle electrification project of 1919, for which Merz & McLellan were the consultants, and Raven the company engineer (Hennessey, 1970). The system was designed to use 1500 V DC, in conductor rail, and overhead contact wire. The proposals were revised to advocate use of overhead wire wherever possible, though the original suggestion was to use conductor rail for most of the route. Following visits to the GE works in Schenectady, USA, by Raven and Lydall (Raven and Watson, 1919), General Electric prepared designs for the North Eastern Railway project based on their EP2 1-Bo-Do-Do-Bo-1 'bipolar' gearless locomotive used on the Cascade division of the CM&SP railway. GE suggested alternative designs based on the 2Co11Co2 which used geared twin, quill-drive motors mounted above the driving axles, and which was used on the Coast & Columbia Divisions. Merz & McLellan prepared their own designs, including a light, gearless 2Do2 locomotive, a heavy, gearless 2CoCo2 unit, a light quill drive 2-Bo-2, and a heavy quill drive 2-Co-2 which formed the basis for the only engine actually constructed in connection with this project. This was Number 13, a 2-Co-2 type driven by twin motors mounted above each driving axle with geared quill transmission. Constructed in 1922 by the North Eastern Railway at Darlington with electrical components supplied by

142 *Electric railways 1880–1990*

Figure 8.6 EP-3 class quill-drive locomotive built by Baldwin-Westinghouse in 1919–20 for the Chicago, Milwaukee & St Paul Railway. 3000 V DC. One hour rating was 4200 hp.

Source: Chicago, Milwaukee & St Paul RR.

Metropolitan Vickers, Trafford Park, Manchester, this engine was tested in late 1922 on the Shildon to Newport line but remained unused until it was scrapped in 1950. British main line electrification projects at this time indicate a reliance on American and European precedence (O'Brien, 1920; Raven, 1922).

HVDC railways were widely established following the success of the General Electric system on the Chicago, Milwaukee & St Paul Railway. Japan used 600, 1200 and 1500 V DC. France, Netherlands, Britain, India, Australia and New Zealand used 1500 V DC. Belgium, the Netherlands, Spain, Italy, Czechoslovakia, USSR, Mexico, Chile, India and South Africa installed 3000 V DC schemes, and there were strong arguments in favour of HVDC down to the mid-1950s when the single-phase 25 kV, 50 Hz system became standard. British practice is described in the reports issued by consultants and manufacturers (Lydall, 1928; Merz, McLellan, Livesey, Son and Henderson, 1918, 1924, 1926).

By 1920, there were two general electric traction systems of global significance. The single-phase AC railway, generally using AC commutator motors was represented by the Loetschberg Railway, which used 15,000 V, 15 Hz (later 16.66 Hz), and by the New York, New Haven & Hartford Railroad, which between 1905 and 1907 electrified the main line into New York using 11,000 V, 25 Hz apart from the final 9 miles, which was 625 V DC conductor rail laid down by the New York Central Railroad, whose Grand Central Terminus was shared by the New Haven company. Huber-Stockar, and Behn-Eschenburg of the Oerlikon Company; and Lamme and Westinghouse, of the Westinghouse company made outstanding contributions to the development of this form of railway. The other standards were variations of the DC railway. The LVDC conductor rail system, as used in rapid-transit work and terminus railways, was represented by the London & South Western 'London Standard' network, and by the New York Central Railroad and Pennsylvania Railroad terminus sections, where the voltage might be between 500 and 750 V DC (Moody, 1961, 1963, 1979). There were higher voltage DC systems using conductor rails or overhead contact wire at 1200 to 1500 V DC represented by the Bury to Manchester line of the Lancashire & Yorkshire Railway, and the Shildon to Newport mineral line of the North Eastern Railway (Hoole, 1987, 1988; RCTS, 10b, 1990; Vickers, 1986). There were the main lines electrified at voltages between 2400 to 3000 V DC, represented by the Butte, Anaconda & Pacific Railroad, the CM&SP Railroad, and the South African Railways. In developing the DC railway in all its forms, the contribution of Sprague and the General Electric company was outstanding.

Before 1920, most major electrification schemes came into the category of 'enforced electrification' to use the apt phrase of General Electric and were necessary for continued operation. Later schemes were the result of the need to improve operations where hydroelectric power was available. General electrification after 1920 proved less economic than expected. Before 1920, the exemplary electric railways were built when engineers solved engineering problems which had to be solved if the railway was to function and meet its legal obligations. Engineers' best practice shaped the system. After 1920, there were more frequent clashes between 'engineers' values' and 'shareholders' values' in the private railways, and the influence of state officials increased in the nationalised systems. The emergence of electric railways before 1920 affords a good case study of the evolution of a new engineering system through its major phases with the accent on changing technological form. After 1920, financial, political, and social changes in the broader receiving system became more important.

8.10 The LVDC railway and the interurban network

The LVDC street railways gave birth to heavy-duty rapid-transit railways. Light electric railways developed from the same source, many of which were standard gauge and of considerable extent. The most important 'light' system derived from the LVDC street tramway was the 'interurban' network. This was a long-distance, express trolley-car system, which provided the first high-speed, long-range electric operations. They never carried the huge loads moved by contemporary steam-railways, but some lines carried goods, and the interurbans helped to develop the metropolitan and main line passenger systems powered by electricity (*Railway Gazette*, 1935a). In the 1920s, many interurbans introduced heavy rolling stock and electric signalling and were indistinguishable from electric heavy-duty rapid-transit railways. The first interurbans used street tramways engineering and the overhead system of current collection was normal because many routes ran down public roads where conductor rails were unsuitable unless ducted. There were long lengths of segregated track between centres of population running across sparsely populated rural districts. They developed features of the heavy-duty main line railway apart from the regular, frequent operation of goods trains and the frequent use of locomotives, though their chief feature was passenger trains made up of bogie-mounted cars drawing supply from an overhead contact wire (CERA 20–34, 1977). Middleton (1961) states that the word 'interurban' was coined, or at least popularised, by C. L. Henry, an Indiana lawyer, state senator, and congressman who controlled the Union Traction Company which opened the first Indiana interurban line in 1898. As the name implies, these were electric tramways linking towns. Interurban networks grew quickly, and reached maximum extent in Indiana, Ohio, Michigan, Illinois and Wisconsin, which between them contained 40 per cent of the national total (CERA, 91, 1980; CERA, 107, 1963). The system was best suited to fairly flat, open territory, containing separated but important centres between which there was a need for rapid transit on a scale somewhere between the short-run street trolleys, and the main line steam railways. The practice of running interurban cars down ordinary streets, to 'stop outside the drug store' during a journey of 50 to 100 km, was a characteristic feature of the system. By 1917, there were some 10,000 interurban cars running on 18,000 miles of electrified inter-city railway. Some lines ran eight-car express trains of considerable luxury and fame. A brief description of a typical interurban is given by Flanagan (1980), and the extent of the trolley system is given in McGraw's Electric Railway Directory (1924).

Some interurban routes, like the street trolley railways, began as light railways powered by steam locomotives or horses. After electrification, a distinctive form of electric railway emerged with some lines operating long-distance passenger services with express trains of six to eight large cars in formation. Britain lacked the long-distance tramcar passenger operations found in the United States, but the Blackpool to Fleetwood line, the Immingham tramway, or the Isle of Man lines can perhaps be described as British interurbans, though they were shorter than their American counterparts (Jackson-Stevens, 1971; Pearson, 1970). In Continental Europe, interurban systems grew round many cities to link outlying towns and villages with the centre. An good example is still operating in the suburbs of the Norwegian capital, Oslo, between city centre and the heights above the town.

The scale of American electric 'trolley' operations reached by 1914 can be gauged from the schedules of the Pacific Electric Company, which held 10 per cent of the United States interurban investment, (Hilton and Due, 1960; Middleton, 1961, 1967). In 1914, this company, centred on Los Angeles, ran 1626 trains each day made up of 3262 cars over three divisions. These were the Northern, with 400 miles of track and 33 separate routes; the Western, with 260 miles of track and 12 lines; and the Southern, with 400 miles of track and 12 lines (Swett, 1975, 1979). The company was formed in 1911 by a merger and was controlled by the Southern Pacific railway. It was a large, important electric network, built to main line loading gauge, and equal in size to the first long-distance main line electrified sections of the 'Milwaukee Road', the Pennsylvania Railroad, or the Norfolk & Western Railroad. Another system was centred on Chicago and was built up by the London-born magnate S Insull, who was born in 1860, and who served in 1881 as private secretary to Edison. He formed yet another member of that extraordinarily vital 'Edison faction' which included so many of the pioneers of direct-current ('General Electric') railway traction, ranging from street trolleys, through rapid-transit systems to main line electric traction. Insull was responsible for encouraging important electrical engineering improvements. He advocated the use of copper rather than iron wire; he promoted AC systems; he helped to transform generator design and he supported the extension of AC transmission networks to bring electric power to the largest number of users. Insull typified the utilities magnate, and he built up a business empire from gas, power and traction, and developed interurban lines with through runs of up to 60 miles in length, and speeds of 80 miles per hour. Like Tyson Yerkes and Whittaker Wright, who played major parts in building London's electric railways, Insull fell foul of financial laws and his empire collapsed in the 1930s. He showed the same vision, grasp of technical potential, and energy, as did his 19th-century counterpart George Hudson who developed the steam railway system in the period following the Railway Mania. The interurban systems flourished until 1930, when a decline overtook the American companies caused by motor car competition, though some routes survive to the present day and light rail transport has revived since the 1970s largely due to German endeavour and

146   *Electric railways 1880–1990*

engineering achievement based on Germany's long history of electric light rail operation.

8.11 The Presidents' Conference Car

By the end of the 1920s, the first phase in the history of interurban electric traction was over. It was dominated by lines which had grown out of the direct-current trolley systems, which were becoming obsolete and were facing extinction because of the Ford car, the hire-purchase of automobiles, and the extension of the United States road system. As the trolley system moved towards engineering obsolescence and found its role taken over by the motor car, the interurban railway companies tried to check decline by introducing a new electric trolleycar known as the PCC or Presidents' Conference Car (Carlson and Schneider, 1980). The new car was designed to be compatible with the existing track and power supply, but it improved performance and enabled many interurbans to survive the increasing competition from the motor car. The new model interurban car demanded no new worker skills, and it could be run over existing lines without hindering older rolling stock. It appeared on 1 October 1936, after delays caused by the Depression. It was designed following a conference of presidents of interurban railway companies and street trolley systems in the United States, which recognised the need to retain public confidence in the interurbans despite deterioration of rolling stock following lack of investment, and the advent of the motor car. The project for a new car was first mooted in 1929, the year when the Great Depression began. In their history of the PCC, Carlson and Shneider describe much of the light rail network as obsolete. It had enjoyed a monopoly over large areas until cheap cars, like the Ford T and the construction of good roads challenged its services. In the late 1920s, there were 74,000 electric railway passenger cars in the United States. Conway, Interurban president, claimed that 40,000 of these were over 20 years old, and were 'excessively costly to operate, noisy, obsolete in appearance, and not calculated to stimulate riding or to earn a profit.'

There were several attempts to introduce a new model electric street car, to revive the trolleys and interurbans, but of 3130 street cars ordered between 1927 to 1934, only eight were of revolutionary design, and not one resulted in a quantity production run. It is surprising that any new cars were built during the Depression (though increased poverty may have encouraged a return to public transport) and the rate of replacement, 3130 out of some 74,000 over a seven-year period, would seem lower than usual.

The attempts to improve the interurbans with the PCC saved some railways and slowed the decline of others. The interurban engineers insisted on a standard design, and the PCC was planned as a universal railcar which any system could use. It was built by several firms and ran in several countries and some survived to the present day. Large numbers were built and the only serious rival, the 'Brilliner' of 1935, was so similar that it was classified as an imitation

(Carlson and Schneider, 1980; CERA, 107, 1963; Reed, 1978). In fact, the Brill vehicle was in production first, though the PCC was conceived earlier. These attempts to produce a new electric interurban car, and to develop the internal-combustion engined lightweight train so closely related to it, led to an early use of wind-tunnels in the railway industry, as reported by Reed (1978):

Perhaps the first to heed the advice of the wind tunnel engineers were the railcar manufacturers. Streamlined railcars began to appear in 1932 and 1933 laying the groundwork for the multi-car trains that followed somewhat later. It is not unusual that when the 'Zephyr' and 'City of Salina' were inaugurated in 1934 they closely resembled these gasoline railcars, but in greater length of course. The J. G. Brill Co.' (a well established builder of trolleycars and interurban vehicles which constructed the Brilliner in rivalry to the PCC) 'Bullet electric railcar was actually the first design in America to apply the lessons learned from wind tunnel testing to the design of streamlined railroad equipment, four years before the 'Pioneer Zephyr'.

Windtunnel tests were conducted by F. W. Pawlowski, professor of aeronautics at the University of Michigan, and the Brill Company developed a body design for which energy savings of 40 per cent at 60 mph were claimed. These cars were used by several interurban lines, such as the Philadelphia & Western Railway, which was facing competition from the Pennsylvania Railroad, which had electrified the Norristown line. These cars revolutionised the electric interurban lines, which had survived the rise of motor car transport, and without them more of the system would have been dismantled. Reed writes that the Brill cars built in 1931 were still running on Southeastern Philadelphia Transportation Authority tracks in 1975. The Brill 'Bullet' car appeared in 1931, the Brill 'Brilliner' in 1935, and the PCC proper in 1936, though the latter design originated in Conway's complaint to the interurban presidents' conference in 1929. There are reasons for questioning the allegation that the Brilliner was an imitation PCC because revolutionary design work was done by the Brill company on its own initiative. The PCC interurban trolley car was a successful design but it couldn't preserve the trolleys and interurbans after 1945–50 when a new mass transport system of post-war motor cars and turnpike roads spread over the USA. This resulted in the destruction of much of the trolley and interurban networks, which had been integrated and provided with common stations. Lewis Mumford condemned this as short-sighted selfishness, bad planning, ignorance and a rejection of public-spirited judgement, which wasted technical and other resources, and harmed society. Mumford (1971) saw the destruction of the trolley systems as proof that superior systems can be replaced by inferior ones in an irresponsible democracy. Fortunately, the 1980s has seen revival of the trolleys and the construction of new light rail systems in North America, where some PCC cars are still at work.

Chapter 9
Railway electrification and the thermal-electric locomotive

9.1 The thermal-electric locomotive and general railway electrification

Heilmann's 'Electric Rocket' established the thermal electric locomotive as a means of electrifying railways without fixed works, and the advantages of using an oil engine to drive the dynamo rather than a steam engine were clear. In the early 20th century, an oil engine with a power equal to that of the latest express steam locomotive would need to generate 1000 kW at the brake. Such engines were too large and heavy for locomotive in 1910 and alternative prime movers were sought, including the condensing steam turbine. Two thermal-electric projects after 1900 used condensing steam turbines to 'electrify railways cheaply'. These must be distinguished from later steam turbine electric locomotives, built in the USA between 1930 and 1955, as alternatives to the diesel-electric locomotive. The early locomotive-electric systems were British in concept and execution. In the first years of the 20th century, the economic prospects of electrification were improved by raising the overall thermal efficiency of the electricity supply system from coal input to electricity output. The steam turbine replaced the reciprocating engine for all but low powered installations, and vacuum condensers were standard.

By 1910, there was little doubt that electric traction could outperform the best steam locomotives in speed, power and acceleration, but low system efficiency, and lack of standards covering electricity supply, retarded railway electrification. In 1910, the LVDC system was common, and was associated with General Electric and during the next ten years HVDC and single-phase AC emerged as standards. Third-rail LVDC did not become the chief agent for electrifying main lines, but it did show that express steam locomotives could be outperformed by conductor-rail DC locomotives in adverse weather conditions. This was demonstrated during the acceleration and braking trials on the LVDC section of the New York Central Railroad between the Hudson River tunnels, and the end of the electric section near Harmon (CERA, 116, 1976). A marked voltage drop in the conductor rail at start gave the electric locomotive a lower acceleration than the steam engine but the electric train always overtook the steam train before the end of the test section, and electric traction was proved superior given a reliable supply of sufficient capacity (Middleton, 1974).

In 1910, there was no agreement about which system best served the electrification of main lines. The wide range of systems available were reviewed by Westinghouse (1910), O'Brien (1920), and Dawson (1909, 1923). The *Electrician* leader of 7 January 1910, 'Electric Traction in 1909', indicates confusion arising from lack of standardisation and doubts concerning the economics of main line electrification. Coal-fired traction was wanted on economic grounds, and so there were strong arguments in favour of a thermal-electric locomotive which could burn coal. The findings of Heilmann and Brown, which favoured conventional electrification, were judged inapplicable. During the ten years following their work with the 'Electric Rocket', power station performance had been improved by superheating, vacuum condensing, and the steam turbine. These new techniques promised to make the steam thermal-electric locomotive of 1910 more thermally efficient than Heilmann's locomotives, which used reciprocating engines and were without superheaters or condensers. There was a well-founded fear of selecting the wrong system in the haste to electrify, and many railways deferred their decision. Fixed works were expensive, and railway engineers were attracted to coal-fired thermal-electric locomotive which involved no commitment to permanent electrical systems, such as generating stations.

In 1910, the *Electrician* warned that there were no clear guides for determining which electrical system ought to be used in a particular instance, because 'very similar figures can be made to act as partisans for two mutually opposed systems'.

This warning was repeated several times in the 1920s and 1930s. The 7 January leader remarked (*Electrician*, 1910):

the past year has produced a really remarkable crop of figures relating to the working of electric railways operating under a variety of conditions. The data contained therein should form a useful basis in considering the construction of any future line. The engineer of a new scheme, if biased in favour of the single-phase system, would derive much consolation and benefit from . . . Dalziel and Sayers (who give) in full the advantages of the single phase system which are fully borne out by consideration of the results obtained on the Heysham and Morecambe line of the Midland Railway. Partisans of continuous current will recognize the truth of the statements made by J A F Aspinall before the Institution of Civil Engineers. The former gives interesting details of the electric working on the Liverpool-Southport line of the Lancashire & Yorkshire Railway, and the latter similar data regarding the electric working of the Mersey Railway.

The leader continued (*Electrician*, 1910):

If there still exist supporters of the three-phase system they will find a spokesman in Mr de Muralt the threephase system is making some headway in Italy . . . apparently due to its being considered more suitable for working a heavy service. The steam engineer will find consolation in a Paper by H. E. Brien. The author is frankly against electric working or main lines, but under certain conditions he is able to award it the palm for urban or interurban service. ... the Washington, Baltimore & Annapolis Railroad is changing from single-phase to continuous current at 1,200 volts ... The Seebach-Wetting line is giving up electric traction in favour of steam.

Many companies waited before committing themselves to electrification involving fixed works, and the North British Locomotive Company of Glasgow supported a project for a coal-fired steam-electric locomotive. If the project were successful, engineering companies established during the steam age would continue as suppliers of equipment to railways at home and abroad in the era of electrification. The lack of British expertise in designing heavy-duty main line electric locomotives encouraged the steam engine builders to seize the initiative. A coal-fired steam-electric system could accommodate single-phase motors; three-phase motors; phase-splitting; direct-current motors; and dual-mode working. The builders of steam-electric locomotives would be able to incorporate within their product whichever motor emerged as the best for railway traction. The motors would be bought from an electrical supplier, but the steam engine manufacturers would supply boilers, turbines, auxiliary equipment and running gear. The widespread use of the 'simple' electric locomotive, drawing supply current from fixed generators in central stations would, of course, threaten their whole position.

9.2 The 'Electro-Turbo-Loco' of 1909

The first thermal-electric locomotive using a steam turbine was built in 1909 by the North British Locomotive Co., Glasgow, which supplied the Empire and Colonies with many of their steam locomotives. This 'Electro-Turbo-Loco' was a 1000 kW DC generating station on wheels, complete with vacuum condenser, cooling water circuit, air heaters, and turbine generator. It is described by Duffy (1989), Lomonossoff and Lomonossoff (1945), Macleod (1929), *Railway Gazette* (1910), and Stoffels (1976). Apart from general descriptions of the machinery, no detailed account of the project was ever published, perhaps because the first press announcements were condemned by engineers as premature. This engine first ran as turbine-electric locomotive, and then lay unused for 12 years before it was extensively rebuilt to become a turbine-mechanical locomotive driving through gears. The rebuilt engine was exhibited in 1922 and was the subject of a lengthy paper by one of its designers (Macleod, 1929), but details of the 1909 unit which used electric transmission are scarce. The 'Electro-Turbo-Locomotive' was first introduced to the engineering world and the general public by one of the chief designers of the North British Locomotive Company, Mr (later Sir) Hugh Reid. He was President of Glasgow University Engineering Society, and he described the 'Electro-Turbo-Loco' in his opening address to the 1909 session on 28 October. This was reported in the *Glasgow*

*Figure 9.1 The Reid-Ramsay steam-turbine-electric condensing locomotive built by North British Locomotive Works, Glasgow, in 1909.
Source: National Railway Museum (York).*

Herald and then circulated to the engineering press. The accounts published in *Engineering*, *Electrician*, *Railway Gazette* and other papers, are based on this source, slightly edited. This address, delivered before the engine was completed, drew hostile comment from an anonymous writer to *Electrician* who suggested that the publicity was premature as major design issues were still unsettled (*Electrician*, 1910). He referred to disputes concerning which transmission should be used: DC or three-phase AC. The actual machine was built with DC motors, and it is worth noting that Ramsay, named as a partner in the design with Reid, later produced a steam-turbine-electric locomotive of his own which was powered by three-phase motors. The identity of the anonymous critic has not been discovered. The Reid-Ramsay locomotive was described as 'a self-contained electric locomotive, generating, as the Heilmann locomotive was designed to do, its own electricity'.

It was 20.4 m long over the buffers, and was carried on two bogies, each consisting of two unpowered guiding axles in a secondary bogie and two axles driven by direct-current motors. Steam was raised at 12 bar in an orthodox, superheated locomotive-type boiler at one end of the bridge-frame, and this supplied a 750 kW Parsons impulse turbine which in turn drove the dynamo. The exhaust passed to a jet condenser positioned next to a water cooler built on the opposite end of the frame from the boiler. The exhaust was condensed which made a draught fan necessary. A brief analysis of the design is in Lomonossoff and Lomonossoff (1945). Loaded weight was 132,000 kg, and normally the engine ran with the water cooler in front. Coal and water supplies were carried in bunkers and tanks on both sides of the boiler, similar to the Heilmann locomotives. The impulse turbine ran at 3000 revolutions per minute and drove a DC variable voltage dynamo which supplied power at 200 to 600 volts to four series-wound traction motors with the armatures wound on the four main driving axles – as with Heilmann's 'Electric Rocket'. The Reid-Ramsay machine preserved several of the main features of Heilmann's system, to which were added turbine drive; superheating and vacuum condensing. The exhaust from the turbine passed into a jet or ejector condenser, and the condensate and circulating cooling water passed into the hotwell. Because the condensate was relatively free from oil, unlike condensate from a reciprocating engine where some lubricant was always carried over from cylinders into the exhaust, the boiler was supplied from the hotwell by a pump. Cooling was achieved by taking the condensate and cooling water through heat exchangers, which rejected heat to the surrounding air. The water for the cooler was carried in the side tanks, and circulated by means of small centrifugal pumps driven by auxiliary steam-turbines placed alongside the main turbo-generator. This cooling water passed from the storage tanks through the first pump and into the jet condenser where it became heated in condensing the exhaust from the turbine. The heated cooling water, and the condensed steam fell into the hotwell, from which the boiler feed was pumped. The rest of the water was pumped by the second pump into the heat exchangers, placed at the front of the engine where the cooling air caused by the locomotive's motion streamed into the tube banks aided by a fan. The induced draught of the orthodox steam engine was replaced by forced draught provided by a small turbine-driven fan placed within the cooler to deliver hot air to the fire, and to supplement the cooling air flow. The small switchboard and the instruments required; the controller for grouping the four motors in series, series-parallel, and parallel; and the regulator for controlling the voltage in the electrical circuit and thus the speed of the train, were all placed together on the driver's platform behind the water cooling pipes. The fireman stood at the other end of the 'generator room' behind the firebox of the boiler. The chimney was at the rear of the locomotive. Further releases to the engineering press offered little extra information.

In the *Electrician* of 7 January 1910 appeared the note 'Turbo-Electric Locomotives', which predicted trouble for the North British project. The anonymous communication expressed ignorance of technical details, which would seem to discount Ramsay as the author, and the tone suggests that it was written by somebody who was once involved, but who had withdrawn after a dispute over design policy. This is admittedly speculation. The critic praised the turbo-electric concept but condemned direct current transmission and motors. Much of the communication is a plea for a turbine-electric locomotive with AC motors. After briefly describing the North British machine and quoting the power of the single turbo-generator as 1000 kW, rather than the 750 kW of the press releases, the writer identifies the DC plant as a likely source of project failure:

A 1000 kW plant is no toy ... direct current turbo-generators running under the most favourable circumstances are not machines which may be left to take care of themselves, which will certainly be the case on locomotives. It must also be remembered that 1000 kW is the largest size direct-current single turbo-generator that has yet been constructed for any purpose, larger units always being made in tandem, so that maximum difficulties will be encountered.

He predicted that severe sparking would take place when running over rough track, ignoring both the high standard of British permanent way and the experiences with the Heilmann engines which were DC machines and ran without serious commutation trouble over track no better than normal British way. The unknown author then referred to a petrol-electric motor car, possibly that of Durtnall, which failed because roughness of road caused the DC system to break down, and he displayed a marked hostility to DC motors despite their proven reliability and success on street tramways, underground railways, elevated lines, interurban systems and on main lines electrified on the LVDC system. He identified the commutator as the feature of the Reid-Ramsay design which would bring the whole scheme to nothing and proceeded to advocate AC systems. He wrote that frequency variation of the generator would be the best method for controlling the motors, which should be of the commutatorless three-phase type. At that time the usual means for controlling the speed of three-phase motors was by pole-changing or concatenation. It is a pity that the writer furnished no details of the method he would use to vary speed by varying the

AC frequency, whilst keeping the turbine running at speeds compatible for optimum efficiency. He concluded:

The (turbo-electric) system is one which has a great future before it, and if the present pioneer locomotive is at all successful it will only show more definitely the still greater possibilities when alternating current is adopted.

This attempt to transfer power station technology directly to the railways failed due to poor condenser performance, high maintenance costs, unreliability and a poor power to weight ratio. The space for major components was restricted, which limited fuel capacity and prevented an adequate boiler from being fitted. The anticipated increase in thermal efficiency was not obtained, and any improved economic performance did not offset the drawbacks of the too-small boiler. It ran a few test journeys on the Caledonian Railway and the North British Railway in 1910, and it was then withdrawn and stored at the North British Locomotive Company, Glasgow, until it was rebuilt with mechanical transmission in 1922 and tested down to 1924 but with no success.

9.3 The Ramsay-Armstrong Whitworth locomotive of 1922

The second British steam-electric locomotive was built in 1922 by Armstrong Whitworth, of Newcastle on Tyne. It is likely that the project began before 1914 and was delayed by the Great War and its aftermath. This machine was a self-propelled electricity generating station, with three-phase alternator and traction motors. It was radically different from the Heilmann and Reid-Ramsay locomotives, though Ramsay was the driving force behind this second project, carried out with Armstrong Whitworth (Duffy, 1989; Stoffels, 1976). The Ramsay-Armstrong Whitworth machine was anachronistic. By the time it appeared, the standard forms of the electric railway had emerged, being the high-voltage DC railway; the single-phase AC railway, with commutator motors on the locomotives; or the low-voltage DC system improved by high-voltage supply through substations converting AC to DC. It appeared when national, standardised supply grids were being discussed, when the steam locomotive was being improved, and when the oil engine was a candidate for thermal-electric locomotion. Ramsay did not co-operate with Reid or the North British Locomotive Company when working on the three-phase engine, but the design was probably conceived in Glasgow, as the company set up to develop it had its address in that city. The *Engineer* announced in March 1922:

The experimental locomotive . . . has been designed by Mr D. M. Ramsay, managing director of the Ramsay Condensing Locomotive Company, Limited, St. Vincent Street, Glasgow, and built at the Scotswood works of Sir W G Armstrong, Whitworth and Co., Newcastle-upon-Tyne, under his supervision.

However, the *Railway Gazette* of September 1923 gives the company address as Manchester when the completed engine was in Horwich Works, the main locomotive plant of the former Lancashire & Yorkshire Railway, which by then had merged with the London & North Western Railway. The Chief Mechanical Engineer, George Hughes, was a modern-minded engineer who investigated compound-expansion; superheating; and electric traction. He tested the Ramsay-Armstrong Whitworth engine, and the Ljungstrom turbine-mechanical locomotive of 1925 (Hughes, 1910; Mason, 1974).

The *Engineer* reported that the new engine was designed to save fuel and water, and vacuum condensing was used in conjunction with a steam turbine to increase overall efficiency. The locomotive was a three-phase power station on wheels, and was a test-rig rather than an integrated design. Even so, the power output was low for its size and it did not perform well. The engine was in two sections coupled together as a locomotive and tender with a wheel arrangement of 2–6–0 + 0–6–2 (*Engineer*, 1922; *Railway Gazette*, 1923; *Locomotive, Carriage and Waggon Review*, 1922). The leading unit carried a conventional multitubular boiler which raised steam at 14 bar to supply the turbine-alternator, mounted under the boiler and rated at 890 kW at 3600 rev/min. This supplied four 205 kW three-phase slip-ring motors. Two motors drove the six driving wheels of the leading unit, and two drove the six driving wheels under the tender. The tender carried a remarkable condenser in which water tubes in the form of a gigantic armature rotated through a tank of cooling water. The ensemble was big and complex. The final weight in running order was 154 tons for a total traction power of 820 kW. This compared unfavourably with the contemporary Gresley 4–6–2 no. 1470 of the Great Northern Railway which could provide 1000 kW at the driving wheel rims for a total weight of 149 tons, with greater reliability and thermal efficiency. In its trials on the former Lancashire & Yorkshire Railway between Horwich, Bolton and Wigan, it was outperformed by a 67 ton 2–4–2 tank engine, and despite modifications the design failed. The engine was withdrawn and scrapped, though drawings for a larger, better integrated high-speed design were prepared in 1926 but never realised (Stoffels, 1976). There were many features of the design that were interesting, though hardly suitable for locomotive service. Shocks and vibration prevented condenser vacuum from being maintained so that fuel consumptions were higher than anticipated. The four main driving motors were of the AC slip-ring type, ventilated by fan runners on the rotor shafts. Starting up the turbine-electric locomotive was more difficult than starting a conventional steam engine. To start the Ramsay machine, the auxiliary turbine was run up to full speed to energise the electrically powered auxiliaries; the main turbine was then run up to half speed, and the driving motors were connected in cascade, as speed control was by chain connection or concatenation (*Railway Gazette*, 1923):

> ... with the motors in cascade and the turbine running at half speed the torque from rest to quarter speed will be ... three times the normal torque. The motors are now connected in parallel, the turbine still running at half speed, the speed of the locomotive increases from one-quarter to one-half speed, the torque at the latter being 1.5 times the normal. After this the turbine speed is brought up to its maximum and the torque drops from 1.5 times the normal to normal.

Normal meant 60 mph. Efforts were made to simplify the starting and speed control for enginemen used to controlling a locomotive with a regulator valve and a cut-off lever. This was done through a master control wheel which governed the electrical system, together with a steam throttle valve. Some good runs were made as far as Southport and on one trip a load of 275 tons was hauled at 60 mph over 100 miles, but this was normal performance on the Lancashire & Yorkshire Railway for Manchester to Blackpool, Southport or Liverpool trains pulled by engines less than half the weight of the Ramsay machine (Railway Gazette, 1923).

Further runs took place on 13 May 1923 with 170 tons when an average speed of 38 mph was achieved, but the performance was not even as good as a standard Horwich 2-4-2 tank locomotive. The locomotive was then returned to Armstrong Whitworth and written-off. That must have cost the firm around £15-£20,000 which it could ill afford at that stage when it had only just entered the locomotive business.

This brought to an end all attempts to extend electrification from rapid-transit lines to main lines by mounting the coal-fired, central electricity generating station on a locomotive carriage. After the failure of the Ramsay-Armstrong Whitworth project, the only thermal-electric system seriously considered was based on the oil engine, which was simpler and lighter due to the absence of boiler, condenser and solid fuel firebox. However, regular internal-combustion traction for heavy-duty lines was not possible before the late 1920s.

9.4 Internal-combustion locomotives and the transmission question

In the first twenty years of the 20th century, a large number of internal-combustion engined rail vehicles were built in a variety of forms. Because the oil-engine was heavy, the majority of these vehicles had petrol engines. They ranged from Ford 'T' cars with flanged wheels to petrol-electric bogie locomotives, such as the 350 hp Bo-Bo type built in 1913 for the Minneapolis, St Paul, Rochester and Dubuque Electric Traction Company (Brillie, 1923; Locomotive Cyclopedia USA, 1974). This company was one of the first in the world to be operated entirely by internal combustion engined traction. It introduced petrol railcars in 1910 and General Electric locomotives in 1913, but though an advanced design, the GE locomotives were low in power output compared to a typical 2-8-0 goods steam engine of the period which could provide four to six times the power. In the period before the mid-1930s, the railcar was the most widespread application of the internal-combustion engine (Hoole, 1969; Tufnell, 1984). General Electric built the first American example in 1906, and eventually over 700 'gas-electrics' were constructed for US service, mostly in the 1920s (Locomotive Cyclopedia USA, 1974). The weight of the engine restricted the introduction of oil-engined or diesel locomotives, and the first examples were for shunting (Haresnape, 1984; Marsden, 1981).

Between 1925 and 1936, about 190 diesel locomotives were built for light shunting and goods service but these lacked the acceleration for heavy shunting

158 *Electric railways 1880–1990*

Figure 9.2 Petrol-electric railcar built by North Eastern Railway in 1903.
Source: National Railway Museum (York).

and express work. They were expensive, but provided valuable operating experience with oil engines in railway service. British attempts were noteworthy (RCTS, 10b, 1990; Russell, 1985; Tufnell, 1979, 1984; Webb, 1970). In 1928, the first diesel-electric service in the UK was worked between Blackpool and Lytham using converted electric stock with a 500 hp Beardmore engine and DC traction motors (Tufnell, 1984). In 1932, an Armstrong-Whitworth car operated on the LNER, and was run on a non-stop service between Euston and Birmingham during the 1933 British Industry Fair (Hoole, 1969; RCTS, 10b, 1990). In Germany both electric and hydraulic transmissions were used with diesel engines in high-speed lightweight sets like the 'Flying Hamburger'. In 1934, the GWR introduced diesel railcars, built by AEC, with London bus-type engines and transmissions (Russell, 1985). None of these seriously challenged steam traction and were chiefly used on branch lines and in light suburban service. Despite the widespread use of the internal combustion engine on rail, the heavy-duty railway remained a steam railway because the steam locomotive provided several times the power of the largest petrol- or oil-engined rail vehicle. This remained the case until the 1930s, and for the fastest express passenger and mail services the steam locomotive was without rival until the 1940s in the USA, and until the 1950s elsewhere.

Between 1900 and 1940, the power, weight, reliability and speed advantages were with steam traction (Fitt, 1975; Stagner, 1975). The internal combustion engine was often a petrol motor originating in automotive, marine or light stationary engine use and fitted into small tractors with mechanical transmission for use in shunting, in quarries, and on mineral lines, and light military railways. It could be found on branch lines driving lorries and buses converted to rail use (Brillie, 1923). Heavier vehicles were designed for railway service with petrol engines and mechanical transmission. Examples include the light carriages used on the North Eastern Railway for 'rail-motor' service and the railcars built in 1911 by Motor-Rail Ltd of Bedford for Indian railways. The Motor-Rail company was active in Britain, manufacturing the Simplex series of tractors for mineral lines, and for military railways during the 1914–18 war (Davies, 1973). Petrol-electric cars date from the same period and were found on road and rail between 1900 and 1920. They became common on interurbans, light railways, and branch lines in the 1920s, and were manufactured by General Electric, and – after 1924 – by Electro-Motive of Cleveland (Locomotive Cyclopedia USA, 1974). Internal-combustion engines using petrol or oil, with mechanical or electric transmission, were at work in goods or passenger service by 1910, but were mainly used where steam traction was inconvenient or uneconomic. The quick start up and close down of the internal combustion engine and its ability to be worked by one man were its chief attractions. Its general use required an internal-combustion engine which was thermally efficient and powerful, but not excessively heavy or bulky. The heavy-oil engine was more efficient than the petrol engine, but was heavier, and it was not until the 1930s that it was light enough to supplant the petrol engine in railway service. To replace steam traction in general, diesel engines having a power output of 1000 kW each were needed. By 1930, the largest American steam locomotives were capable of sustained cylinder outputs of 3500 to 4000 kW, and European engines were capable of half these figures. Peak performances were considerably in excess of these magnitudes. It was in the late 1920s that heavy oil engines evolved to a stage which convinced engineers like Hamilton that they could displace steam locomotives within a few years. Many design problems needed to be solved, not least how best to couple a powerful oil engine to the driving wheels (Acland, 1930; Camden, 1927; Dunlop, 1925; Hobson, 1925). Railway locomotives operate over a wide range of speeds and power outputs, and different types deliver full power at different speeds. Tractive effort was a maximum at start, and maximum power was delivered at the speed at which the locomotive performed its definitive duties. The steam engine achieved this speed variation by cutting off steam admission at different percentages of piston stroke, and by throttling steam supply between boiler and cylinders. In electric traction the characteristics of the DC traction motor were ideal for railway service. In internal combustion traction, a transmission was needed which linked the engine with the driving wheels in a way which gave the required train speed versus power characteristic. This transmission had to reduce rotational speeds from those of the engine to those of the wheels, without excessive losses (Burn, 1922; Duffy, 1988).

In the 1920s, the internal combustion engine was recognised as a component around which a new locomotive might be devised, but there was no efficient, reliable and flexible transmission for linking the motor to the rail. Most of the available methods were bulky, heavy, unreliable, and wasteful through excessive losses. In the 1920s, electric transmission was not recognised as the best for linking powerful engines to driving wheels in main line traction, despite the excellent tractive effort versus speed characteristic of the DC motor, and its success in electric railways. More than one eminent authority, for instance Geiger, rejected it on the grounds of excessive weight, complexity and low efficiency. Several solutions to the 'transmission problem' were proposed. For example, there was the 'Compound Diesel' in which steam locomotive pistons, cylinders and drive were connected directly to the driving wheels and regulated by conventional valve gear to vary the power with speed. Instead of being driven by steam the pistons received compressed air from a tank charged by the oil engine. Another variant drove the pistons with high pressure exhaust gas generated by an oil engine (Acland, 1930; Camden, 1927; Dunlop, 1925; Fell, 1933; Hobson, 1925; Kitson Clark, 1927). Another group were Still engines. In these, exhaust heat from the oil engine, sometimes supplemented with oil firing, raised steam using the hot jacket water from the cooling system (Rennie, 1922). The steam was admitted to one side of the engine pistons, with internal combustion taking place on the other, and the addition of steam power (usually at starting) onto internal combustion power varied total power and tractive effort with speed. The system used direct mechanical drive to the driving wheels (Acland, 1930; Duffy, 1988; Kitson Clark, 1927).

Several types of transmission were tried with oil engines in railway service. These included compressed air, steam and exhaust gas, and many never survived the first trial. The transmissions which survived were mechanical, hydraulic and electric. Each came in several forms (Lomonossoff, 1933a). By 1925, the comparative trials of transmissions by Lomonossoff in Russia and Germany, and experience with petrol-electric and oil-electric railcars suggested that electric transmission was the best, but there were dissenters from this view. In a general review (*Verein Deutscher Ingenieure*, 9 May 1925), Geiger concluded that much better results were to be got from the Still system, and from vapour and gaseous transmissions. A thorough survey is given by Hobson (1925). Air, exhaust gas, and steam were tried. Gas or vapour was delivered from a compressor driven by a prime mover; expanded in the main power motor, which was usually a reciprocating engine, and then exhausted to the suction side of the compressor. The circuit was usually closed but open systems were used with exhaust gas. Hobson (1925) describes a typical early project by Dunlop, dating from 1911, for a 1000 hp engine with air transmission. The oil engine and the compressor were combined by using a common piston of stepped form, with the smaller diameter piston serving the compressor, and the annular ring acting as power piston for the oil-engine. Compressed air was stored in a high-pressure reservoir from which the reciprocating engine was supplied. The scheme incorporated a heat exchanger for transferring heat from the exhaust gas to the stored high-pressure air before it passed to the driving engine. In 1924, Dunlop modified his design to an open cycle. The compressed air mixed with steam generated in a small boiler heated by the engine exhaust. Hobson (1925) remarks:

The Schwarzkopff locomotive exhibited last year at the Seddin Railway Exhibition, may be quoted as an example of the internal-combustion locomotive in which exhaust gas from the Diesel engine is compressed and superheated by a compressor driven by the prime mover. The compressed gas then operates in the driving cylinders as in steam-engine practice. Similar locomotives are now being considered and tested in which a mixture of compressed air and steam is used as the transmission medium, as in the case of the Zarlatti system, and compressed steam, as in the Cristiani system, both now undergoing tests in Italy.

One of the respondents to Hobson's paper describes his experiences with the Zarlatti engine, which was tested near Rome (Hobson, 1925):

... the engine was a small six-wheel-coupled side-tank steam locomotive coupled to a four wheel wagon behind, on which was mounted the generating plant consisting of a 70 h.p. Diesel engine driving an air compressor which delivered air under pressure to the boiler of the locomotive, the boiler acting merely as a reservoir with no fire in the firebox. A spray of hot water was also injected with the compressed air into the boiler, resulting in a charge of saturated air in the boiler. The engine was then operated in the same manner as a steam locomotive. ... taking into consideration the high price of coal in Italy, there was a decided economy as compared with the ordinary coal fired locomotive.

Brown, of the Swiss Locomotive & Machine Works, anticipated that the Zarlatti system would be developed for much higher powers, following tests with a 1000 hp machine with pneumatic transmission. This was the 4–6–4 machine, built at Esslingen Works for the German State Railways. It had a 1200 hp MAN oil-engine which drove a compressor discharging air into a reservoir heated by exhaust gas, from which hot air drove a reciprocating engine coupled to the wheels. Masing quoted low figures for the transmission efficiency of this system: 50 per cent without exhaust heating; 56 per cent with it. An overall locomotive efficiency of 16.5 per cent and 18.4 per cent respectively was quoted. The figures of 50 per cent and 56 per cent are much lower than the 80 per cent transmission efficiency claimed elsewhere for this system.

Brown claimed that only electric and hydraulic transmissions were likely to be successful with high-pressure internal-combustion engines. Steam transmission was used in the Cristiani system in Italy, which was subjected to early trials carried out by R & W Hawthorn Leslie & Co. Ltd on the firm's private railway in Newcastle. A six-cylinder V-type steam compressor was driven by two reconditioned marine petrol engines, which had been used earlier in Durtnall's petrol-electric locomotive, tested by the same company (Durtnall, 1925). A small oil-fired vertical boiler provided starting steam which filled the transmission circuit with low-pressure vapour. This was sucked into the compressor, compressed and stored in a high-pressure reservoir which supplied an ordinary steam engine. The petrol engines drove the compressor. The petrol engine exhaust could be used to heat the steam before entry to the steam engine. The whole apparatus was carried on the frame of a six-coupled shunting steam engine, with two cylinders of 14 in diameter by 22 in stroke, and 3 ft 6 in wheels. The tests were sufficiently successful for the trials in Italy to go ahead. The petrol engines from the Cristiani trials were previously used at Forth Banks Works, Newcastle, in tests with Durtnall's 'Thermo-Electric-Locomotive'. This was first proposed in 1902 but not constructed until 1920 (Durtnall, 1925). Durtnall worked with Heilmann as a steam engineer responsible for the Willans engine used to drive the dynamo on the steam-electric locomotives. He was an enthusiast for electric transmission and the thermal-electric principle. Like Emmet in the USA, Durtnall advocated electric transmission for large ships, which was tried in the 1920s and 1930s. He experimented with electric drive for both railway and road vehicles before the 1914–18 war, and the Forth Banks Works locomotive was intended to investigate the advantages and disadvantages of the petrol-electric system.

The locomotive was driven by two petrol engines, reconditioned and modified after service in a coastal motor boat. Each petrol motor was a six-cylinder in-line type, giving 160 bhp at 450 rpm, coupled to single six-pole DC dynamo placed between the engines, which generated 220 kW at 440 volts and 1200 rpm, for supplying four 68 bhp motors. The motors were mounted longitudinally at each end of the two four wheeled bogies and they drove through flexible couplings, cardan shafts, and worm gears. The locomotive was a B-B type with 3 ft diameter driving wheels and it weighed 43 tons in running order. Top speed was 25 mph. An overall transmission efficiency of 82 per cent between engine brake shafts and driving wheel rims was claimed, made up of generator efficiency 93 per cent, electric motor efficiency 92 per cent and worm gear efficiency of 96 per cent, which were very good figures for the time. Durtnall intended to replace the petrol engines with a suitable oil-engine and so produce a 'thoroughly reliable, high-class shunting locomotive'. He intended to reduce losses in control resistances by using a transformer and accumulators to store surplus energy which would be tapped when needed. The prime mover would run at optimum speed all the time. The system was similar to the hybrid thermal-electric motor cars of the present time (1999). Durtnall's design was for a heavy locomotive, which he proposed for mountain lines and for short, rapid shunts where substantial economies would be achieved through regeneration. The system was applied to a Hawthorn Leslie battery locomotive exhibited at the British Empire Exhibition, Wembley, and described in *Engineer*, 13 June 1924. Durtnall was a strong advocate of both thermal-electric traction and simple electric traction (Durtnall, 1926), and he proposed electric transmissions for traction on rail, road and sea.

In the 1920s, the private locomotive builders in Britain were aware of the potential of the oil engine, but they failed to win a sizable home market for the new form of locomotive. The Americans won their home market when the diesel locomotive programme was taken up and backed by General Motors which brought automobile industry methods and fresh insight into the railway field, at a time when engineers like Kettering, inventor of the electric self-starter, had transformed GM with progressive, creative management (Leslie, 1983). General Motors enjoyed the resources of an enormous combine and supported research into internal combustion rail traction on a scale no builder of steam locomotives could match.

In the 1920s, hydraulic transmissions were investigated for a range of powers up to 1200 hp, and a small 60 hp oil-engined locomotive was tested on the Great Eastern section of the LNER in July 1924. The performance of these early oil-engined locomotives was inferior to that of a steam locomotive which had a better power to weight ratio; greater maximum power, higher speed and more reliability. The oil engined locomotives were in the test stage and the experiments in the early 1920s identified which components and configurations were viable. Engines for petrol, paraffin, benzol, and alcohol were tried and much attention was paid to the relative costs of different types. If the initial capital cost per ton of a steam engine was taken as unity, the cost per ton for an oil engine with hydraulic transmission was given as 1.36 to 1.48; and for a locomotive with electric transmission, was 1.78 to 1.9, though other values were often quoted. The transmission efficiency was the percentage of engine brake power delivered from the oil-engine main shaft to the driving wheel rims and values were quoted as follows for several systems under trial in the mid-1920s. The figures were questioned by critics. They relate to optimum conditions being the figures commonly quoted by the supporters of the system in question. Despite the encouraging experience of Hawthorn Leslie with the Durtnall Thermo-Electric engine, electric transmission was ranked as the least efficient form of transmission. The Cristiani system was rated at 82 per cent from the oil-engine brake shaft to the transmission output shaft, or 72 per cent overall at the driving wheel rims. The Zarlatti system was rated at 90 per cent between brake and transmission output shafts, or 79 per cent overall between engine brake shaft and rail, though this figure is elsewhere reduced to 70 per cent. Transmission efficiency of a contemporary steam engine was 95 per cent; for Lentz oil and gear transmission was 70 to 75 per cent; and for typical electrical transmission was 70 per cent. This latter figure caused some engineers to condemn electric transmission but by the late 1920s it was recognised as the best for engines of large powers.

In 1922, Lenin provided Lomonossoff with £100,000 to construct three experimental prototype oil-engined locomotives with 895 kW engine power. In collaboration with the German Reichsbahn, Lomonossoff produced four units each with a different transmission: electrical, mechanical, pneumatic, and hydraulic. Comparative tests showed electric and mechanical transmission to be the most promising, and by 1925 or 1926 Lomonossoff's work, supported by investigations by others, showed that electric transmission was the best for high powers, despite the reservations expressed in some quarters. Mechanical transmission was the best for lesser powers (Lomonossoff, 1933b). Hydraulic transmissions as then existing were not satisfactory but modern systems such as those developed by Voith provide an alternative for low to moderate power ranges, though present day trends are towards using electrical transmission in all power ranges. Electronics, inverters, and three-phase motors are making electrical transmission the universal norm. Lomonossoff's oil-electric locomotive of 1924 was rated at 746 kW engine power, and was a 2–10–2 or 1-E-1 type with five driving axles in a rigid frame with a guiding truck at each end. It carried 5 DC nose-suspended Sprague or 'tramway' type motors each driving one axle through gearing. Control was by voltage variation obtained through field excitation which gave the desired flexibility of power and speed. Total weight in running order was 120 tons with weight breakdown being 44 per cent for the prime mover; 30.5 per cent for the electrical gear; and 25.5 per cent for the mechanical structure and running gear. During Summer in hot climates, a 30 ton tender containing a cooling water system for the oil engine was needed. Granted the experimental nature of the design, it was an encouraging success. The weight of 120 tons for 746 kW engine power compared favourably with that of steam goods engines of equal indicated power, which sometimes needed extra water tenders on long journeys in hot districts.

In 1925, a locomotive appeared which combined electric transmission and an oil engine within a body with a driving cab at each end, carried on two bogies (*Engineer*, 1925). This form is found in older oil-engined locomotives, and became archetypal. It originated in electric locomotives on the LVDC lines and is exhibited by the General Electric petrol-electric locomotives built for the Dubuque Electric Traction Company in 1913 (Locomotive Cyclopedia USA, 1974). The 1925 engine attracted attention in Europe. It was built for the Calabro-Lucane section of the Italian Mediterranean Company 950 mm gauge line in Calabria. Its layout was modern with an oil-engine coupled to a DC dynamo supplying four motors, each one driving an axle (*Engineer*, 1925). Electric transmission was used to provide heavy trains with extra adhesion on long

Railway electrification and the thermal-electric locomotive 165

Figure 9.3 Russian 180 tons 1C-D-C1 main line diesel-electric locomotive built in 1924 to designs of Y. Gakkel, using 1000 hp submarine engine. Tested 1924–26 then used as mobile electricity generating station. (Photographed in 1980.)
Source: Sergey Dovgvillo.

gradients by supplying power to motors on an auxiliary vehicle. It is not clear if this was ever tried. A similar suggestion had been made by Heilmann, whose thermal-electric locomotive was designed to power axles distributed through a train, though this was discovered to be unnecessary as the engine itself had sufficient adhesive weight. Power to weight ratio of the Italian oil-electric locomotive was only 7.46 kW per ton of locomotive weight based on brake power, compared with 21 kW per ton for the power at rail per ton of the best Italian compound-expansion steam locomotive exclusive of tender. The latter figure is a peak value and a mean output over a long period would be lower, of the order 14 kW per ton. Inclusive of tender weight this would be reduced still further to about 9 or 10 kW per ton. The oil engine was a six-cylinder two-stroke motor with cylinders 200 mm diameter × 300 mm stroke giving 328 brake kW at 500 revs per min. It drove a dynamo rated at 250 kW at 500 revs per min with voltage ranging from 300 V to 500 V. The four traction motors were each geared to one of the bogie axles and were series wound, each giving 54 kW at 400 V, 800 revs per min. Gear ratio was 1:5.3. The power to weight ratio based on rail power was about 5 kW per ton or about half that of the best steam locomotive with tender. This was in part due to a poor transmission efficiency of about 70 per cent but the basic concept was sound, and was better than the other layouts used in early internal-combustion locomotives. The Italian locomotive was followed by improved designs of oil-electric locomotives of the same form, and comparison with the Sulzer oil-electric designs of 1933 show the successfully integrated form at the start, and finish, of the period which witnessed its introduction (Duffy, 1989). Contemporary North American oil-electric locomotives were of the same archetype (Locomotive Cyclopedia USA, 1974). This form was influenced by the layout of electric locomotives at work in Europe and the USA.

By the early 1930s, the oil-electric locomotive was recognised as a possible alternative to the steam engine on non-electrified lines, and was a rival to the electric railway system in some cases. Between 1925 and 1933, when the search for an improved form of steam locomotive was being pursued in earnest, electric transmission solved the transmission question. It was used in oil-electric locomotives which had left the experimental stage and were a promising new mode of traction with components well-integrated and capable of long-term development. In the United States, electric transmission was identified as the norm for moderate and higher powers much more readily than in Europe, and the Americans did not explore alternatives to the extent that Europeans did (Morgan, 1955). In the USA, pioneer research into petrol-engined, and oil-engined traction was dominated by large electrical combines, such as General Electric, and Westinghouse. In Europe, the work was done by relatively small mechanical engineering concerns which built steam engines, or by internal combustion makers with little electrical expertise. The influence of the contemporary LVDC lines on deciding the transmission question in the USA was considerable. They demonstrated the excellence and suitability of the 'tramway' nose-suspended DC motor in railway service, which most diesel locomotives used. In North America, more than in Europe, the oil-electric locomotive was seen as an agent

Railway electrification and the thermal-electric locomotive   167

Figure 9.4   Diesel-electric shunting locomotive built by English Electric for London Midland & Scottish Railway in 1935. Engine power was 350 hp. The standard British diesel-electric shunting engine developed from this type.
Source: English Electric (later GEC/Alsthom).

for electrifying railways without fixed works. American engineers accepted the oil-electric system as the norm once Lemp showed how components could be subjected to integrated control. In the period 1930 to 1955, the oil-electric locomotive may have discouraged conventional electrification by providing 'cheap electrification' without fixed works but it gave railways a mode of traction for working non-electrified parts of the system, which integrated with electric traction far better than steam traction did. In this sense, oil-electric traction was an essential supplement to electrification, and it is doubtful if electrification would have gone as far as it did without the oil-electric locomotive.

Chapter 10

Converters, the mercury-arc rectifier and supply to electric railways

10.1 Introduction

Rectifiers were not needed when the light street railways of Europe and the USA were electrified in the 1880s using low-voltage DC supplied from batteries or dynamos in small lineside power houses. In the 1890s, the first heavy-duty rapid-transit railways were powered from DC generating stations which supplied power to conductor rails, or overhead contact wires. Rectifiers were not required for small projects such as the electrification of the Mersey Railway, Liverpool, in 1903. The need for rectifiers arose when small DC power houses were replaced by fewer large AC stations, which supplied AC transmission lines. The rectifier linked the AC supply with the low-voltage DC conductor rail. The development of the HVDC railway increased the demand for rectifiers which were generally rotary converters of the same design as those used on LVDC railways. The principle of the static rectifier was known by 1910, but the first experimental 'bulbs' were too expensive, too unreliable and too low in power capacity. The rotary converter was established, reliable, and long-lasting. Before 1930, it was able to convert powers far in excess of any practical battery of mercury-arc rectifiers. Rectifiers were also needed on many AC railways, to reduce the supply frequency to that of the AC motors.

10.2 Rectification of power supply

Rectifiers in railway service needed to be reliable, and capable of rectifying fairly large powers. This eliminated the purely mechanical rectifier, based on the synchronously vibrating polarised armature, and the various bridges, based on electric valves, in the form in which they appeared in the first ten years of the 20th century. These rectifying bridges, such as the Graetz Bridge, were developed for industrial purposes once high-powered 'valves' appeared, and the Graetz

Figure 10.1 Main control panel, Lots Road power station, Chelsea, London, in 1953. This governed the supply to the London Transport substations.
Source: London Transport Museum.

arrangement was used with mercury-arc rectifiers in the post-Great War period. Though the potential of the mercury-arc rectifier was recognised in the period 1910–20, the general practice was to install rotary converters and motor generators in lineside substations. Though there were rare cases where converters were used in the inverted mode, to convert direct current to alternating current, the usual practice was to convert AC to DC, and it was traction requirements which focused attention on this particular electrical component. They were very reliable by 1910. There had been trouble with 'inverted rotaries' which could race or run away when excessive field weakening by lagging currents on the AC side caused dangerously high speeds of the converter armature, but speed controls using a separate exciter coupled to the converter shaft prevented this. Converters operating in the normal or AC to DC mode could race unless fitted with speed control. There was an instance where a synchronous converter was supplied from a 4000 kVA turbine-alternator, and fed a DC street railway, connected on the DC side in parallel with other substations and an older central 550 V station. When a steam valve failed and closed in the turbine-alternator, the converter became an inverted converter drawing power from the DC trolley system. The DC ammeter continued to read positively though the current had reversed direction, and the operator weakened the main field to reduce an apparent overload. The alternator and turbine raced and were wrecked. This accident led to the fitting of centrifugally operated circuit breakers, and reverse current relays on the DC side which broke the circuit if the converter operated in the inverted mode. These made the converter safe and suitable for traction purposes.

Both the LVDC and the HVDC railways required converter substations. One of the first DC railways to attract worldwide attention were the electric sections of the Chicago, Milwaukee and St Paul Railroad, which began working in 1915. These electrified sections made extensive use of substations. On the Missoula and Rocky Mountain Divisions there were 14 substations, each equipped with 100,000 V/2300 V transformers and two or three synchronous motor-generator sets feeding the overhead contact wire with 3000 V DC from two 1500 V DC generators connected in series. The distance between substations was 32 miles. The substations were of 4000 kW, 4500 kW and 6000 kW capacity, the total being 59,500 kW. Later, eight other substations were added with a total capacity of 28,000 kW and an ability for this to be increased to 40,000 kW. General Electric and Westinghouse supplied the equipment. The substations were manned, and photographs show substantial buildings with nearby operators' bungalows of a fair size. These CM & StP converters had a long life, and as late as the 1950s, the railway – facing a most uncertain future – installed second-hand electromechanical converters purchased from US electric railways dismantling their systems. General Electric rotary converters installed in Grand Central Terminal New York in 1916, to convert 11,000 V, 25 Hz AC to LVDC DC were still running in 1989 when the last four of the original ten were replaced by solid state rectifiers. This New York Central Railroad system LVDC electrification had pioneered automatic substations which needed no manning,

Figure 10.2 Goldhawk Road substation London underground system showing electro-mechanical converter No. 2 with control and switch panels typical of the pre-1920 period.
Source: London Transport Museum.

Converters, the mercury-arc rectifier and supply to electric railways 173

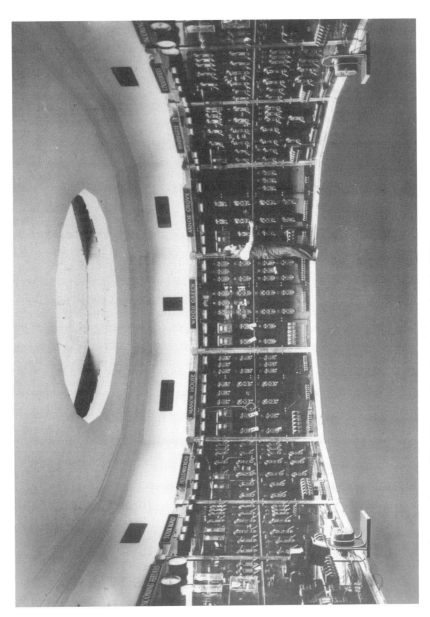

Figure 10.3 Power supply control room Wood Green substation, London, in July 1932.
Source: London Transport Museum.

and by the late 1920s, there were nine manned substations with a capacity of 66,500 kW and six automatic substations with a capacity of 18,500 kW. All used rotating electromechanical converters which proved eminently satisfactory as automatic control was extended from fewer centres and manning levels reduced. In Britain, the first unmanned rotary converter substation was installed by the Underground Company in 1924.

The mercury-arc rectifier was first used as a means of rectifying part of the power supplied, sharing substations with the older, larger rotating devices. The first extensive use of mercury-arc rectifiers was in the 1920s and 1930s, in the USA, when several of the interurban trolley systems were converted to HVDC, from LVDC or AC, and they were generally used to support rotary converters which dealt with the greater share of the power.

10.3 Development of the industrial mercury-arc rectifier

Twenty years of experimental work with electric arcs, mercury vapour, and evacuated vessels preceded the first attempts to make a rectifier based on the arc in mercury vapour. This type was called the static rectifier to distinguish it from the rotary or electromechanical type. Unidirectional current flow by means of an arc struck between a mercury pool and carbon electrodes was discovered by Jemin and others in 1882. In 1889, Fleming investigated unilateral conduction via an arc in air, and in 1892 Aron discovered that losses can be greatly reduced by causing the arc to be struck inside an evacuated vessel. Between 1894 and 1898, Sahulka studied the behaviour of atmospheric arcs between mercury and electrodes of iron or carbon. These were important experiments, but they were not carried out with rectification in mind.

Aron produced mercury vapour lamps between 1890 and 1892, but the successful commercial production of these devices is associated with Peter Cooper-Hewitt, who began commercial manufacture in 1900. To Cooper-Hewitt belongs the credit of making many of the fundamental innovations necessary for a successful commercial mercury-arc vessel, and for integrating them into a reliable, industrial device. His lamp of 1900 was based on the principle of using rectification via a mercury arc, and it formed the basis of the mercury-arc rectifier proper, which developed between 1905 and 1910 as the glass bulb rectifier. These rectifiers could handle only low powers, but were used to provide DC for the electrostatic precipitators employed to remove dirt from smokestack gas. By 1905, the principles of static rectification from a three-phase supply were well understood, and O J Lodge used four Cooper-Hewitt-type mercury 'lamps' or 'bulbs' arranged in bridge to supply the plates of his patent industrial smoke cleaner.

Cooper-Hewitt took out patents for the steel tank rectifier in 1908, and by 1910, General Electric and Westinghouse had constructed the first experimental types, though the glass bulb form remained the norm into the 1930s. Cooper-Hewitt also contributed greatly to the technique of using a controlled grid to regulate voltage in the rectifier, though glass-bulb grid-controlled rectifiers were not practical until developed by Langmuir and Prince and introduced in 1928, with grid-controlled steel tank rectifiers being introduced soon afterwards.

Glass bulb rectifiers were limited in capacity compared to steel-tank types, and it was limited capacity which determined the pace at which the static rectifier replaced the rotary type. Between 1912 and 1914, glass bulb rectifiers could carry more than 100 amperes, and by 1915 the first steel-tank rectifiers could rectify currents of circa 750A. Kraemer's type of steel tank rectifier, of 1919, first produced by General Electric in the USA, was the archetype for later development and led to large-scale commercial manufacture. By 1920, water cooled rectifiers were pioneered by Siemens-Schuckert, so that in the 1920s the reliable, high-capacity rectifier, with grid-control, water cooling, and steel-tank construction became available as a standard rectifier, which could convert industrial frequency AC to DC. It began to find use in railways for charging batteries to supply DC systems, but in the mid 1920s and after, it began to supplement rotary converters in the rectification of traction current, and by the 1930s was accepted as an alternative to the electromechanical converters which was much more compact and equally reliable. Its use was accelerated by the introduction of automatically regulated systems.

General Electric, Westinghouse, Brown-Boveri, Siemens-Schuckert, and AEG produced and developed steel-tank rectifiers, water cooling, and grid-control. There is evidence that the Americans generally led the way in arc-rectifier research, development and application though it can be claimed that they lost initiative to Europe between 1910 and 1920. Certainly the pioneer American work before the Great War was of immense significance. Early tests were made between 1908 and 1914 with glass 'Cooper-Hewitt valves' (Hewittic rectifiers), Westinghouse metal rectifiers, and General Electric metal rectifiers.

Cooper-Hewitt; Langmuir and von Issendorf solved the problem of backfiring. Bela Schaefer developed the metal-clad rectifier, a precursor of the metal-tank type. Kraemer, Toulon and Mittag perfected grid-control between 1924 and 1925. Prince and Langmuir introduced the first practical grid-controlled mercury-arc rectifier in 1928. General Electric, Westinghouse, Brown-Boveri, Siemens-Schuckert, and AEG, contributed to the development and production of rectifiers, including the steel-tank, water-cooled type with grid control. Brown-Boveri pioneered the European multiple-anode steel-case rectifier in 1913, which became a widely used form. By 1925, the industrial mercury-arc rectifier was established. In the development phase, backfire was a serious problem, arising from short circuit of the supply transformer, between anodes, caused by two arcs merging. Backfire damaged seals, and could shatter a glass bulb vessel, and precautionary shielding of anodes was required, but Langmuir and von Issendorf were able to prevent it by the mid-1920s. Capacity per unit was raised and groups of arc-rectifiers could handle a significant percentage of substation power flow, and from 1925 they supplemented the rotary converters.

The multi-anode, steel-tank rectifiers of the 1920s had capacities per unit much higher than those of the glass bulb type, though they required vacuum tight seals and automatic auxiliary pumps to keep the tank free of gas. Reliability was improved by water cooling the arc, and by using heating coils to keep anode temperature above the condensation temperature of the mercury vapour at low load. Arc length had to be increased in the multi-anode rectifier to reduce the chance of backfire, and this lowered efficiency by increasing the drop in arc voltage.

In Europe, metal-tank, mercury-arc rectifier substations were used in Switzerland as early as 1915, but they were only of 240 kW capacity. By 1920, glass-bulb rectifiers were used in Britain to charge low-capacity arrays of batteries from AC supply, and after 1923, they found similar use in France. By 1930 the static rectifier converted AC supply to DC, to feed the conductor rail of the LVDC railway, and the overhead contact wire of the HVDC system. In the 1930s it began to supplant rotary sets for frequency changing on AC lines. However, rotary sets worked well and had long working lives, so that many were left in place long after the mercury-arc rectifier became standard.

At this time, larger capacity Brown-Boveri gridless mercury-arc rectifiers were used in France to supply 1500 V DC. These were expensive, but were lighter and smaller than rotary converters. Britain was slow to develop and use the mercury-vapour rectifier, despite the early use of Cooper-Hewitt valves by Lodge for smoke-cleaning before the Great War. The first 'all British' steel tank rectifier was not installed until 1930, when the British Thomson-Houston Company supplied a 1500 kW, 615 V unit to London Underground railways. Bruce Peebles & Co. started the regular production of steel-tank rectifiers in Britain, and English Electric followed in 1932. Glass bulb rectifiers were used in Britain before the steel-tank form, and both kinds were widely employed after 1930. By 1935, the capacity of the glass-bulb type was about 500 A DC, 600 V. British industry was slow to enter mercury-arc rectifier research, design and manufacture, but rapid progress was made during the 1930s, and by 1939 models were on the market equal to any produced in the USA or Continental Europe. However, the Americans and Germans led in research.

By 1930, the mercury-arc rectifier was the best method for rectifying and inverting current passing to and from transmission lines and industrial circuits. It was superior to the rotary converter, and to alternatives which were sometimes suggested. These included the air-blast rectifier of Erwin Marx, which used an ultra rapid series of compressed air jets to blow out the arc of an AC supply when it passed in one direction, but permitted it to flow in the other direction. Several very large examples were built, and found a limited – but unsuccessful – use in Germany in the period 1930–45. The advent of mercury-vapour systems for rectification, inversion, and frequency changing facilitated the construction of power networks, and it was normative in stationary service by 1930.

Attempts to use static rectifiers in locomotives service were less successful and the reliable locomotive-mounted rectifier required the solid-state industrial devices of the 1960s. Until then, the rotary converter was used in various forms of converter locomotives. Locomotive-mounted static rectifiers were tried as early as 1908 in the USA when Westinghouse used two glass bulb mercury vapour valves to rectify 3.3 kV, 25 Hz AC to 600 V DC. In 1913, General Electric mounted rectifiers on a wagon to supply a 600 V DC motor car, but the rectifiers proved too fragile for locomotive use. Success was achieved however with stationary rectifiers in substations, though low capacity limited application. Between 1913 and 1914, Westinghouse rectified 11 kV, 25 Hz AC to 750 V DC on the Pennsylvania Railroad, and at Grass Lake, between 1912 and 1915, experiments showed that a 5000 V DC trolley wire could be supplied through static rectifiers, though this line reverted to moderate tension DC after the trials. Tassin, Nouvion and Woimant refer to a test, on the Kalamazoo-Grand Rapids line in Michigan in which the voltage in an isolated third rail was raised to 5000 V DC. The Europeans were quick to take up research in this field, and for a while led in research and development until American companies regained the initiative in the 1930s and 1940s.

Between 1930 and 1960, many improvements were made to the mercury-arc rectifier, and the market was dominated by a limited number of types associated with the leading electrical manufacturers, made under licence in different countries. The best known was the 'ignitron', called after the trade name of a patented device for striking the arc, which was developed both by Westinghouse Electric and General Electric. It originated in a discovery of 1933 that the passage of a current from a high-resistivity rod to the mercury in which the rod was partially immersed, would create an arc within a few microseconds at the junction of rod and mercury surface. Another well-known type was the 'excitron' associated with the Allis-Chalmers Manufacturing Company. This was a single-anode rectifier, in which the ignition coil was energised, so that the cathode was ready to conduct current when the anode fired. There were many variants, often devised to get round patents.

10.4 The mercury-arc rectifier in railway traction

An early electrification scheme to use mercury-arc rectifiers was on the Midi railway in France, which used 1500 V DC as standard, influenced by studies of American HVDC installations and by a belief that rectifying higher voltages using rotary sets in series, or using mercury-arc rectifiers, would not be successful at that date. In the event, five mercury-arc rectifiers were tried, alongside rotary converters, and they worked well. They did much to encourage wider employment of static rectifiers in Europe and the USA.

Writing in 1922, Carter provides a useful summary of attitudes to the several systems associated with rectification in railway service at that date. He is rather conservative in his attitude to the mercury vapour rectifier:

(The mercury vapour rectifier) has now been developed in form and capacity suitable for the requirements of railway supply, and although it has hardly yet attained the condition of reliability needed in such work, its development has disclosed no insuperable defect. It should therefore be watched, as a development which may prove to have considerable influence on the future of railway electrification.

Carter is apparently referring to the glass-bulb type. He was certainly aware of the great potential of the mercury-arc rectifier, though he remarks that complete substations would be less efficient than those equipped with rotary units because of less efficient transformers, and increased losses in auxiliaries. Overload capacity was less than with a rotary set, and installed capacity needed to be greater on this account. Cost (in 1922) for a complete installation was higher than for an equivalent plant using rotary units. However,

The mercury vapour rectifier does not impose a limit on the frequency of supply, nor on the voltage of the output. It accordingly appears a fitting development to meet the needs of high voltage continuous current railways taking power from industrial supply. Although the single-phase AC commutator motor was established in Europe and the USA by 1910, and the universal series motor, able to run on AC and DC, found some use on railways including the New York, New Haven and Hartford RR, which had DC and AC sections, the favoured motor was the DC type, usually of the Sprague nose-suspended type, ... although further experience is necessary before it (the rectifier) could be recommended without reserve for such work.

Carter looked forward to the time when mercury-arc rectifier substations would be unmanned, but he did not think this was then in sight,

... for the (mercury) rectifier requires more skilled attention than the rotary. A result to be expected from the sudden changes in line voltage, produced when the arc shifts from anode to anode, is that this form of converting plant will cause much greater interference with neighbouring communications circuits than the usual forms with rotating machinery.

However, this automatic regulation of arc-rectifier substations was to be achieved by 1927 as described below. Carter also referred to the increased use, particularly in the USA, of unattended or unmanned substations, using rotary sets, such as became common in the 1920s. He also makes the very important point that the use of regenerative braking, with the return of energy to the contact wire, and through the substation, demanded reversible converters or rectifiers, which ruled out the mercury vapour type at that time and encouraged the retention of already installed (reversible) rotary converters and motor generators on mountain railways like the Chicago, Milwaukee & St Paul Railroad. Carter makes passing reference to the rectifier locomotive, taking single-phase supply from the contact wire, to be rectified for use in DC motors, but he indicates that trials so far carried out with such machines were not successful.

By 1926 and 1927, mercury-arc rectifiers were being installed in the lineside substations of heavy-duty electric railways to supplement the rotary converters, or to rectify current to charge storage batteries which improved load factor or supplied emergency power. One early example is the electrification of the Illinois Central Railroad, reported in the *General Electric Review* of April 1927. One interesting feature was that the power companies owned the substations, and operated them, selling the rectified power to the railway's 1500 V DC contact wire. The Illinois Central ran ten-car multiple-unit trains needing 5250 kW during acceleration, and 1140 kW for average running. The substations used 3000 kW synchronous converters and mercury-arc rectifiers to convert 60 Hz AC supply into 1500 V DC. Provision was made for 'remote supervisory control for indication to, and control by, the railroad's power supervisor of all d-c feeder circuit breakers in substations and at sectionalization points.'

The substations were six miles apart. There were seven in all, with a total installed capacity of 42,000 kW, of which 9000 kW was rectified by four mercury-arc rectifier sets, two of 3000 kW capacity, and two of 1500 kW capacity. The rest of the power was converted by 3000 kW synchronous converter sets. The *General Electric* report of 1927 stated that:

The rectifier sets consist of two 750 kW or two 1500 kW, 1500-volt bowls operating in parallel; and, in all cases but one, are installed in substations with synchronous converters and operate in parallel with the converters in the same substation. The rectifier overload ratings are 150 per cent rated load for 20 minutes, 300 per cent for one minute.

The voltage-regulation requirements of the railroad were that the voltage at the substation bus should not fall below 1400 V during normal load conditions, and both converters and rectifiers worked well. A major step forward in substation design was the use of the first automatically controlled arc rectifiers to supply traction current to a HVDC rapid-transit railway. This was reported in the *General Electric Review* for July 1927, and the editor introduced the paper with the words:

So far as we know, this is the first description of the automatic control of a mercury arc power rectifier. The instructive information given possesses more than the interest of novelty because the control functions strictly in accordance with the high standard previously set by the automatic control of rotary conversion apparatus.

This claim for priority may be for America only, as later the paper remarks that 'this represents the first application of 1500 volt rectifiers and of 1500 volt automatic rectifier control in this country.' The rectifiers, of the steel tank type, were installed with new rotary converters during the reconstruction of the Chicago, South Shore & South Bend Railroad which started in 1925 and which converted the 6600 single-phase AC system to 1500 V DC. Power supply was from a 33,000 V line which fed eight substations, where conversion and rectification to 1500 V DC took place. The arc rectifiers did not share substation space with rotary converters, but were in substations of their own. One substation contained a 1500 kW arc rectifier, and three contained one 750 kW arc rectifier each. The other four substations contained two 750 kW rotary converters each. Two substations with rotary converters were partially automatic,

180   *Electric railways 1880–1990*

and two were automatic. All the arc rectifier substations were automatic. Out of a total substation capacity of 9750 kW, 3750 kW was rectified using automatically controlled mercury-arc rectifiers. Arc-back or backfire was still considered to be a potential hazard, and a high-speed circuit breaker, connected in the positive lead, was arranged to trip on reverse current flow. This opened the DC side, and via an interlock on the high-speed breaker, simultaneously tripped the oil circuit breaker, energising the reclosing relay, which reclosed the breaker after a delay. Normal operation should follow, but if a second arc-back occurred the process was repeated. If a third backfire took place within a set time period, the station was automatically shut down for inspection. The 1927 article claimed that automatically controlled mercury-arc rectifiers were no more complicated than rotary converters, and were equally reliable. With the introduction of these automatic stations, use of the static rectifier spread rapidly, and in the 1930s it became the favoured method for rectifying from AC to DC; for frequency changing; and for inversion in connection with HVDC transmission lines.

10.5   British railway rectifiers

Britain lagged in the development and use of the mercury-vapour rectifier, and the first 'all British' steel tank rectifier was not installed until 1930, on the London Underground system, and English Electric did not begin regular production of the steel tank rectifier until 1932. Both steel tank and glass bulb forms were widely used, and by 1935 the maximum capacity of a glass-bulb type was about 500 A DC with a voltage of about 600 V. British main line railway companies, and London Transport, began to make extensive use of static rectifiers from 1930 onwards, often installing the multiple anode, steel-tank type with automatic control. They were used alongside older, as well as newly installed, rotary converters. The London Midland & Scottish Railway first used one in June 1931, and they were installed in new substations serving the LVDC conductor rail network in the London area, on Tyneside, and on the HVDC lines in the Manchester district. In the latter area, the electrification of the Manchester, South Junction & Altrincham Railway in 1931 used both new rotary sets and mercury-vapour rectifiers. AC supply was at 11 kV, and was converted to 1500 V DC for the overhead contact wire. BTH supplied ten 750 kW, 1500 V DC rotary sets in 1929, plus a steel tank rectifier of 1500 kW capacity in 1930. The Timperley substation contained the BTH 12-anode, steel-tank mercury-arc rectifier, the first time this type was used on a British railway. It was widely installed in the 1930s throughout London Transport. An early British substation, with automatic static rectifiers, was Balham on the London Transport system. Eventually the mercury arc rectifier was used on such historical systems as the Isle of Man electric railways, and Volk's electric railway in Brighton.

In 1932, two BTH rectifiers, of 1200 kW, 630 V, were installed in the Hornchurch substation on the LMS Barking-Upminster line. In Britain, the

Converters, the mercury-arc rectifier and supply to electric railways 181

Figure 10.4 *Arnos Grove substation on the Piccadilly line (London) in 1933 showing new 12-anode mercury-arc rectifier.*
Source: London Transport Museum.

advertisements in the *Railway Gazette* trace the progressive installation of this form of rectifier. The issue for 19 October 1934 announced that BTH rectifiers with a total output of 37,000 kW were working on the London Passenger Transport Board, quoting as example three 1500 kW, 630 V, rectifiers installed in Barons Court substation. The issue for 11 January 1935 announced the installation in 1932 of two 1200 kW, 630 V BTH rectifiers in the Hornchurch substation on the LMS Barking-Upminster line, and claimed that BTH rectifiers with a total output of 42,900 kW had been installed on British railways. *Railway Gazette* for 27 July 1934 carried a notice announcing the anticipated use of the mercury arc rectifier in railway service as a frequency converter for AC railways.

In the case under notice, a converter of 3,600 kVA capacity will change three-phase 45 KV. 50 cycle supply into single-phase 15 kV. 16.66 cycle current for use in the overhead contact lines of the Basle-Schopfheim-Zell-Sackingen division of the German State Railway, and will replace rotary machines and batteries which have been in service since the inception of the electric service in 1913,

It was claimed that large scale tests of the equipment at the Berlin works of AEG showed an efficiency 'a good deal higher' than that of rotary frequency converters, and substantial reductions in working costs. It was concluded that:

... it may be anticipated that mercury vapour frequency converters will have just as wide a field of application in single-phase systems as the mercury rectifier now enjoys in d.c. traction.

A series of articles on mercury-arc rectifiers, published in *Railway Gazette* in 1934, compares the physical construction, and relative merits of both the glass bulb type, and the steel tank type, at a time when both were finding increased use in traction. The increased capacity was the chief feature of the steel tank form:

The greater mechanical strength of steel compared with glass enables direct current outputs up to 5000 kW or even 8000 kW if required to be obtained from a single steel-tank unit, compared with about 250 kW maximum output from a glass bulb unit, at 500 volts in each case.

However, granted that many substations had a capacity of about 2500 kW, this could be provided by two banks of glass-bulb rectifiers, with each bank having six or eight banks in parallel, or by two steel-tank rectifiers, or even by one. The report stated that both solutions were found in practice, each with advantages of its own. By 1934, the steel-tank rectifier was of proven reliability ('risk of breakdown is almost negligible') and efficiency was 95–96 per cent or better at loads from 0.25 to 1.25 rated output. A review of constructional details is given in the *Railway Gazette* paper of July 1927. The article reports on the 23 Bruce Peebles steel tank rectifiers, rated at 2500 kW being built for the Southern Railway electrification extensions between Eastbourne, Lewes and Hastings, and around Sevenoaks. Also shown is one of three 1500 kW GEC steel tank rectifiers installed at Chiswick Park substation on the London Passenger Transport Board railways. The Hammersmith to Hounslow and South Harrow extension of the LPTB Piccadilly line required 13 1500 kW, 630 V rectifiers in five substations. These five substations were normally unmanned, and four were remotely controlled from a room in the Alperton distribution station. By the early 1930s, the outstanding problems of the automatic, mercury-arc rectifier substation were solved, and the arc rectifier station became the type for future development.

Converters, the mercury-arc rectifier and supply to electric railways 183

Figure 10.5 Hendon substation, London, in 1931 showing British Thomson-Houston rotary converter and 1500 kW mercury-rectifier.
Source: London Transport Museum.

The engineers of London Transport have always enjoyed a reputation for leading innovation, and were responsible for the first large-scale use of water-cooled, continuously evacuating, mercury-arc rectifiers in Britain's railways. The multi-anode, cylindrical tank, automatically controlled static rectifier became practically standard throughout London Transport after 1930 though they continued to work alongside rotary sets dating from an earlier era. Hendon substation was one of the first to be equipped with multi-anode steel tank rectifiers as well as rotary sets, and this type was installed at Arnos Grove, 1933, Wood Green, 1935, and Leicester Square, 1935. By 1935, BTH claimed that their rectifiers provided a total output of 42,900 kW for British railways, mostly for London Transport . British manufacturers were energetic in exporting their rectifiers. In 1936, English Electric and Metropolitan Vickers supplied steel-tank rectifiers (2000 kW and 2500 kW) for Polish railways' 3000 V DC electrification programme, and after the war British firms took part in reconstruction work in Poland.

A great improvement was the introduction of polarised grids, which enabled the current from regenerative braking to be returned to the supply, thus removing a major disadvantage of mercury-arc rectifiers. Railway substations with polarised grids were working successfully in South Africa by 1938. Both the glass bulb type and the steel tank type of rectifier were widely used. In fact, the Hewittic glass-bulb type replaced surviving rotary converters on London Transport in the period of refurbishment in the 1950s and early 1960s, when Russell Square, Bond Street, Charing Cross, Baker Street, and other substations were modernised.

Chapter 11

Signalling, communications and control 1920–40

11.1 Introduction

The history of signalling, communications and control records many innovations which were tried long before they became common practice. The City & South London Railway (1890) and the Liverpool Overhead Railway (1893) used automatic signalling years before it became usual on general-purpose railways, and the London Post Office Railway demonstrated automatic working of an electric railway some 40 years before urban rapid-transit systems employed it. The Post Office Railway ran driverless trains carrying mailbags over 23 miles of track along a route 6.5 miles long. Passengers were not carried. It was the first automatic electric railway in the world, and was of great engineering significance. There were automatic and semi-automatic mechanical railways throughout industry. Typically, wagons were hauled by a chain or cable from a quarry or mine to an unloading staithe or furnace. Many of the operations were controlled by trips or governors, and there was human intervention, as necessary, by an overseer. When labour was cheap, fully automatic operation was not economically justified.

The Post Office had worked a railway of its own in the mid-19th century. This was a 2-foot gauge line between Euston Station and the Post Office North Western District Office, worked by pneumatic power. The length of route was 600 yards. The line was planned by the Pneumatic Despatch Company in June 1859, at a time when pneumatic despatch systems were attracting much attention. In these systems, tubes, or wheeled vehicles were driven down pipes or tunnels by evacuating the passage ahead of the vehicle; pressurising the section behind it, or both. Pneumatic systems pressurised the rear of the passage, and were used to send sealed containers through pipes in large buildings, or across towns in buried pipelines. On a much larger scale, air pressure forced railway carriages through tunnels on demonstration lines like the one in the Crystal Palace grounds in Sydenham. In May 1861, an experimental line 452 yards long was laid down in Battersea Fields, later to be the site of the Battersea

Power Station. The test vehicle was between 7 and 8 feet long, and rode on four flanged wheels of 20 inches diameter on a track of 2-feet gauge. The commercial line started work on 15 January 1863, with 30 trains a day and a journey time of about 70 seconds. The vehicles were slightly larger than the test car, and success was great enough to encourage building a new extension from Euston to Holborn (1863), and from Holborn to Hatton Garden (1865), using a track gauge of 3 feet 8.5 inches, and cars of 10 feet 4 inches in length, weighing 22 cwt. In 1869, this line was extended under Holborn Viaduct, and re-aligned to connect with a new station at the General Post Office, in St Martins-Le-Grand. These lines ran under existing thoroughfares: Drummond St; Tottenham Court Road; High Holborn; Newgate St. Proposed lines under Southampton Row, and from the General Post Office under Gresham St to Pickfords, were never built. The Post Office was not pleased with the service. In December 1873, the journey time from Euston to the GPO by pneumatic railway was 17 minutes, compared to 21 minutes for a horse-drawn van. It proved difficult to keep the system airtight, and to supply sufficient air. One suspects that working costs were high, with energy losses in the steam plant, air blowers and tunnels. Horse vans were cheaper, and the pneumatic railway closed down in October 1874. In later years, traffic congestion increased in Central London and the Post Office again considered the benefits of building its own mail-carrying railway.

The success of the City & South London Railway and other underground lines recommended electric traction for a new Post Office railway instead of the pneumatic system. Electric traction was reliable and automatic operation could be used. In 1911 the Post Office decided to construct an electric underground railway, with driverless trains, to carry mail across Central London from Paddington Sorting Office to East London Sorting Office over a distance of 6.5 route miles. The objective was to reduce journey time. The electric railway was built on a much larger scale than the former pneumatic system (Bayliss, 1978; Sprought, 1992). It had double track tunnels of 9 feet diameter between stations, with single track tunnels of 7 feet diameter at station approaches. Most of the route lay about 70 feet (21 m) below the surface. There were seven stations, which served two main line termini (Liverpool Street and Paddington), and six major sorting offices. Work on tunnel construction started in 1914, but was discontinued in 1917, due to wartime shortage of labour and materials. It was restarted in 1923 and finished in December 1925. A contract for electrical equipment was placed with English Electric in 1924, and the railway opened in December 1927. It proved successful from the start, and there was little change until 1981 when new rolling stock was ordered. The average journey time today inclusive of stops is 25 minutes, with an express service time of 13 minutes for non-stop runs between Paddington and Liverpool Street. The line operates 22 hours per day, for five days a week. Gauge is 2 feet, and each car is 27 feet in length; 16 trains of 1930 stock, and 34 of 1981 stock (from Greenbat Ltd of Leeds) maintain the service. Following mechanical failure, the original 1925 stock was replaced by 50 vehicles between 1930 and 1931, and by ten more delivered in 1936 by English Electric.

The control system remained unchanged from opening into the 1990s, when an entirely new system was installed with Central Train Control and fully automatic routing, scheduling and indication of status. The original system influenced major signalling developments on main line railways in the 1920s and 1930s. The basic power circuit on the car bogies included a brake solenoid which held off the brakes against a spring when the conductor rail was energised and the motor supplied with current. Power was originally taken from the GPO supply at Blackfriars, through a 440 V DC cable which ran along the railway and fed motor-generator sets at each station. Station areas received supply at 150 V DC. Operating voltage outside station sections was 440 V DC. Current was drawn from a centre conductor rail and returned through one of the running rails which was earthed. Gaps in the conductor rail divided the route into sections which could be supplied independently, and the trains actuated track circuits which switched the power supply to create a dead section behind each train. Points were worked by electricity. Train operations were controlled from signal cabins on the stations. The trains were crewless, so there was no need for lineside signalling, but the control or switch rooms contained a switchframe with check-lock levers which controlled points and power supply to the individual track sections. An illuminated track diagram showed train movements to the switchman. Conflicting movements were prevented by interlocking points' setting and power supply. Underneath the control room, which was at platform level, was the relay room containing the relays and contactors for switching points and power supply. It originally housed the lead acid cells for feeding the track circuits with a 24 V DC. supply. These were of 300 Ampere-hours capacity, and were later replaced by two voltage converters.

The original frame represented latest practice, despite the safety requirements being much less stringent than those for a railway carrying passengers. Mail was often extremely valuable and damage could be very costly, and so safe operation was essential. The switchframes used the check-lock principle, with interlocked route selection, train position and power supply. The control levers had six positions. To start a train, the lever was held in the normal position (1) until the mechanical locking was freed, and then moved to position (2) when it was check-locked until the route set was proved to be clear. The lever was then moved through (3) into position (4) where it was check-locked again until the route had been correctly set by the electric point mechanisms. The movement of the lever through (5) into (6) completed the circuits for energising the section on which the train was standing. A signal to send away a loaded train was given by platform staff pressing an overhead push button which illuminated a red indicator light warning staff to stand clear, and informing the signalman that departure was due. He controlled the actual start. To return the lever to normal, it was pushed inward. It was held and locked in position (5) until the section from which the train started, and over which it passed was proved clear. It was further pushed in to be held in (3) until the route was restored to recognised normal setting, and finally restored to (1). In 1992, the decision was taken to introduce centralised train control, described in a later section. The Post Office Railway indicated what electric traction, signalling and control could accomplish when integrated, but automatic operations could not be transferred to passenger carrying rapid-transit railways, and general-purpose systems, in the 1920s. There were formidable technical and economic objections, but the promise of automating signalling and control was never forgotten. On electric railways, the automation of traction was long foreseen, but on steam railways the locomotive required two men, and no system was ever devised to permit one-man control, multiple-unit working, or driverless operations of steam engines. This gave electric traction a potential for automatic operations which steam traction could never share.

11.2 Signalling and interlocking

Many of the techniques which were commonly used after 1945 were first tried in the 1920s and 1930s. Many evolved out of innovations made before the 1914–18 war. The Great War intensified research into all branches of electrical science, and advances were made in communications, lighting, signalling, traction and systems for moving large pieces of military equipment under centralised control. During the 1914–18 war, large numbers of artisans were trained in electrical techniques, and in the 1920s skilled workmen were available to install and repair electric railway signals, electric points and electric communications. In the 1920s and 1930s, signalling and control strategy was closely related to the spread of colour-light signals; powered operation of points and signals; and the use of small-lever or miniature frames to control powered systems. The most favoured power system was the electro-pneumatic, though there were alternatives, and the electric system became the norm in British Railways days. Low-voltage electro-pneumatic points operations were tested in the 1920s on industrial sidings in Northumberland with power taken from batteries located in pits next to the track. Power was too low and the switches moved too slowly so these early installations were abandoned in the 1930s.

In underground service, electro-pneumatic operation was favoured by engineers like Robert Dell because it was more fireproof than the high-voltage all-electric systems used on some lines. The Metropolitan Railway used all-electric systems, with high-voltage pole-line supply, but the high-voltage electric points were liable to failure in wet weather. Dell compared the electric system (favoured by the Metropolitan engineer Workhill) with the electro-pneumatic system and found the latter to be the best, though in some locations the multiple air-lines from centralised compressors took up much room. Automatic signalling was well established by 1920, and should be distinguished from controlled signals. Automatic signals normally show clear, but are changed to danger behind a train until it passes beyond the overlap section of the next signal ahead. Controlled signals normally show danger, and are cleared for the passage of trains.

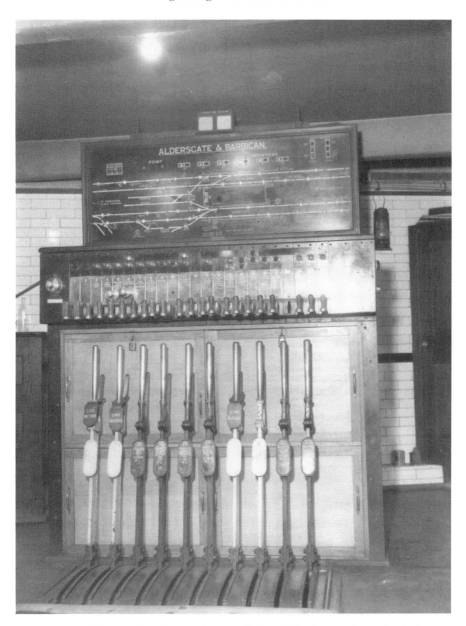

Figure 11.1 Aldersgate Signal Box, photographed in 1953, showing the mechanical points levers, the push-pull route setting handles and the illuminated track diagram. In boxes of this type, signals were usually worked by low-pressure air.
Source: London Transport Museum.

Signals could be arranged to work in both modes. The London & South Western Railway used both kinds between Woking and Basingstoke. In 1902, automatic semaphore signals worked by compressed air and controlled by the track circuits were installed on the main lines, but controlled signals were used at intermediate stations. The automatic signals worked well, and remained in use until 1966 when colour-light signals were installed. Multiple-aspect signalling, using multiple-position semaphores or multi-light signals, increased line capacity because it gave warning two, three or more blocks to the rear of a train, depending on the number of aspects provided for. The more blocks behind for which warning was given, the less necessary was cautionary slowing. Two blocks warning was normal – twice the stopping distance – because a signal could not clear until the train ahead was out of the second block. The standard 'two-block, three-indication' system was established in the USA before it became usual in Britain. It required the length of blocks to be at least the stopping distance of the heaviest, fastest train; but with 'three-block, four-indication' signal spacing could be reduced to half the braking distance. 'Four-block, five-indication' signalling was eventually considered, though 'three-block, four indication' was adequate for many busy routes where block shortening and avoidance of cautionary (unsignalled) slowing was desirable. Multiple-aspect, colour-light signalling, met the requirements of these systems better than semaphores.

In the period 1910 to 1920, signalling policy was reviewed on many railways, and after the 1922 amalgamation, each of the four main British companies had to decide on future standards in this vital matter. United States influence was strong in all aspects of railway engineering: in traction, signalling, control, management and operations. Before the amalgamation, several railways in Britain considered using the three-position semaphore, powered and unpowered, working in the upper quadrant after the American model. These were installed on the Great Western Railway, the South Eastern & Chatham Railway (formed in 1898 by a merger of the South Eastern Railway and the London Chatham & Dover Railway), and the Great Central Railway. Some of the ex-GCR signals survived into the 1960s Their use was criticised by many railway engineers, not least because the 'caution' position of a three-position semaphore was the same as the 'all clear' position for a two-position semaphore acting as a home or starter signal. Some companies reduced the chance of error by working the three-position semaphores in the upper quadrant and the two-position semaphores in the lower. The two types were fixed to the same gantry in some locations. There was another source of confusion: the three-position signal used a yellow light in the spectacle for caution, whereas the two-position distant signal used red for caution. There were some exceptions to this latter rule, and the Metropolitan & District Railway used two-aspect distant signals, with yellow arms, which showed yellow for caution. Three position semaphores had the advantage that every signal could act as a distant and a stop signal and give advance notice about the route ahead. The conventional British system required two arms on the same post to do this, and distant signals could not perform this function. The American use of three position semaphores was due to their policy of 'speed signalling', rarely used in Britain between the wars. Speed signalling was the norm in the United States, and it signalled the speed which a train should take through a particular junction or intersection. In Britain, signalling was based on the philosophy of giving information about the route to be taken and whether or not to proceed. Speeds were decided by the train crew guided by their timetable. The three position signal was a speed signal in the USA. In Britain, it gave an indication about the signal ahead. The number of aspects a signal could show was sometimes increased by fitting two three-position arms on one post.

At this time, electric lamps shone with sufficient intensity to show a clear aspect in bright sunlight, and colour-light signalling was an option. It had many advantages: ease of maintenance, and more complete and convenient integration with electrical interlocking. There was a need for nation-wide simplification and standardisation, and in 1922 the Institution of Railway Signal Engineers set up a committee which published its report in 1924 (Kichenside and Williams, 1963, 1980). The findings were endorsed by the Ministry of Transport which was formed in 1919. This influential document found against three-aspect semaphores, though the ones already installed could be left in place, and those in the Scunthorpe area on the former Great Central Railway lasted into the 1960s. The report found in favour of three-aspect colour-light signals, and suggested that four-aspect signals be adopted, using double-yellow as a standard display for 'attention: proceed at medium speed'. The recommendation not to use three position semaphores opened the way for widespread use of the two-position semaphore working in the upper quadrant, as there could be no confusion over the distant arm aspect. Yellow, not red or white, was the standard colour for the caution signal. Red stood for danger, and green for clear. Lower quadrant signals had a long life, and the Great Western Region of British Railways retained them for its mechanical signalling. Colour-light signals enjoyed many advantages. They had few or no moving parts, and they were operable by trains in automatic systems. Of course, semaphores and disk signals of mechanical type could be worked automatically, but colour-lights were simpler and cheaper to maintain. Electric cables, for power and control circuits, could be run without undue difficulty along tunnels and viaducts, or along main lines for great distances. There were limits to the distance points and signals could be worked from a box by rods, wires, weights and levers. Electric signalling, and electro-pneumatic power operations, extended the range of working from any single box, and enabled many mechanical boxes to be closed. It encouraged centralising of boxes, and the growth of the main box from which a large area was signalled (Wilson, 1900, 1908; Lewis, 1912, 1932; Dutton, 1928; Westinghouse, 1956; Nock, 1962, 1969).

There were two main types of colour-light signal in use from the start, with many variations, major and minor. A good review of British signal types is provided by Kichenside and Williams (1963, 1980). There was the searchlight signal, with a single lamp and lens, with a sliding spectacle of coloured glasses, and there was the multilens signal, with a separate lamp for each aspect, worked by switches. Once switches became reliable, the latter signal was simpler than the searchlight, which needed a mechanism to move the spectacle. The searchlight

192 *Electric railways 1880–1990*

Figure 11.2 Signal cabin and substation, Arnos Grove, London, in December 1932.
Source: London Transport Museum.

signal was used where there was no room for multiple aspect signals, for instance in the tunnels of the underground, but large numbers were installed on certain main lines. The LNER use them extensively. The trend was towards colour light signalling of multilens pattern. For example, on the 'Bakerloo' line the moving spectacle searchlight signals were replaced by two-aspect colour lights by the mid-1920s, which were in turn replaced in 1938 by modernised automatic signals worked from new signal cabins (Horne, 1990). Both types were widely used, but in main line signalling, the multi-lens pattern was much more common. The earliest colour-light signals were two aspect devices, as were the standard semaphores. The 'stop' signals showed red or green, that is stop or clear; and the distant signals showed yellow or green for caution and clear. These two-aspect colour-light signals found use on the high-speed lines of the London Midland & Scottish Railway, and the London & North Eastern Railway, and had the advantage of being clearly visible in bad weather.

The Southern Railway pioneered four-aspect signalling, and installed the first four-aspect signal in 1926. Problems had arisen on the former South Eastern & Chatham Railway lines where the tracks were shared by steam hauled express trains capable of 80 mph, and electric multiple units, drawing current from a lineside conductor rail, which ran at 55 mph. The crews of the steam locomotives needed warning of a 'stop' signal earlier than did the motormen of multiple-unit trains, because the braking distance was longer for the steam-hauled expresses, and so a double-yellow aspect was introduced. The inter-war period saw a steady improvement in the many components needed for automatic signalling circuits, cab signalling, and electrical interlocking.

11.3 Speed signalling

In Britain, signals at junctions indicated the route to be taken by giving permission to proceed, perhaps with caution. Sometimes the route was indicated by an array of white lights separate from the main signal. In the United States and Germany, the signal indicated the speed at which to progress, rather than the route to be taken. Speed signals of the colour-light type used different combinations of colours to indicate speed. Most signals on main line indicated both speed and route, but the prior objective of the system (indicating route or speed at junctions) determined its characteristics. Speed signalling was rarely used in Britain, though experimental sections were set up, for instance at Mirfield in 1932. This scheme was planned by the former Signal and Telegraph Engineer of the Great Central Railway, A. F. Bound who introduced one of the first automatic multiple-aspect colour-light signals on a British main line when he installed three-aspect signals on the steam-worked Neasden-Marylebone section in 1923. In July 1923, Bound became Signal and Telegraph Engineer of the London Midland & Scottish Railway, and in 1932 he installed speed signalling on the Calder Valley main line near Mirfield, between Thornhill LNWR Junction and Heaton Lodge, a distance of 2.75 miles. This was the only full scale

194 *Electric railways 1880–1990*

Figure 11.3   The first type of four-aspect electric colour-light signal used in Great Britain, installed on the Southern Railway electric lines in 1926.
Source: British Railways/National Railway Museum (York).

application of speed signalling in Britain until recent times. All services over the route were steam-hauled. The Mirfield installation used multiple-aspect searchlight signals, with retention of block working and existing signal boxes. As Kichenside and Williams relate (1980, p. 109):

At junction signals, the lights for diverging routes were placed vertically under the main aspect instead of side by side as in current route signalling practice . . . thus the top light or lights indicated the highest speed route and the lower light (or lights) the medium speed route – whether to right or left is not indicated. Beneath this group of lights, lower down the post, was a red marker light to indicate that the signal was in a multiple-aspect area; colour-light signals leading to semaphore signals were not provided with a marker light. ... a fifth aspect was adopted in addition to the standard multiple aspect indications – yellow over green – attention, pass second signal at restricted speed ... The marker light also served as an additional danger signal should the main light fail.

The Mirfield scheme mixed speed signals with route signals. Route signals, fixed side by side to a common post, were used where the speed set for a diverging route varied by less than 20 mph from the speed set for the main route. The Mirfield experiment did not establish speed signalling in Britain but it remained in place until the early 1970s when standard colour-light route signalling replaced it. The rise of train speeds in the 1970s has necessitated advanced notice of route and speed, leading to the introduction of flashing aspects in the mid-1970s when the high-speed trains began to run at speeds up to 125 mph. One of the Mirfield speed signals, made by the British Power Railway Signalling Co. Ltd., London & Slough, is on display in the Oxenhope Museum of the Keighley & Worth Valley Railway. The Mirfield system influenced the signalling system installed on the Camden-Watford electric line, and on the LMS-controlled section of the District Line from Bromley to Upminster which were given colour-light signals and full track circuiting between 1932 and 1933. Stop and repeater signals of the searchlight type were used, worked automatically via track circuits. Junction stop signals were not of the speed type, and they had separate, bracketed signals on the post to show the route to be taken. The Watford line signalling had a calling-on arrangement peculiar to itself which enabled trains halted at an automatic stop signal to proceed with great caution if the overlap track circuit of that signal was clear. Permission to proceed was given by the red marker light changing to a miniature yellow aspect. A repeater was placed to show red behind any train halted at the next stop signal, but this could be passed on red after a halt on one minute though speed had to be dead slow so that the train could be halted immediately, on sight of any obstruction.

11.4 Powered operation and interlocking

Electropneumatic operation predominated in power signalling (Wilson, 1908). Electricity controlled compressed air engines which worked the points and signals. There were all-electric systems from the first days of heavy-duty electric traction, but they did not become the norm until after 1950 when new designs made it easier to install electric point-motors rather than compressors, air lines and compressed air motors. The electric system was more compact than the electro-pneumatic. It did not need bulky air-lines from centralised compressors, and it avoided problems caused by moisture in the air freezing on expansion in cold-weather operation. Interlocking and powered operations could be applied to any kind of signal, including the many types of shunt, repeater and banner signals. Electric semaphores, disks, and moving spectacle signals were extensively used in shunting areas. Before the solid-state era, electrical interlocking was the electrical locking of mechanical bolts, which worked on the same principle as the locking devices of the manual-mechanical signalling and switching system. In pioneer days of powered operations, the interlocking of signals, points and operating levers was mechanical but miniaturised. The power (electrical, electro-pneumatic, or electro-hydraulic) was applied to move the signal or switch. Later, electricity performed the interlocking operation. Electrical operations and powered working of signals and points miniaturised signal cabin apparatus. The locking frame became much smaller than the ones found underneath the signalman's platform in a manual box, and the operating levers were reduced to stubs a few inches long: in later cabins these were replaced by push buttons or tiny switches. Every mechanical locking frame was designed for a particular location and box, and the manual type could take up a great deal of room in big boxes like the ones at Shrewsbury or Preston. Valuable space was occupied where the signal box stood in a station. The miniature, powered frames fitted easily into boxes placed above the running rails on gantries, or set up between converging lines in limited areas.

Powered operation of points and signals was suggested in the mid-19th century, and demonstrated later in the century. It was expensive to install and the first large-scale applications on main lines followed modernisation which increased speed and frequency of service and outstripped the capacity of manual-mechanical signalling. Manual signalling continued in use for low speeds and moderate frequency of service. Electrical signalling and powered operations increased line capacity and lowered staffing levels. In the 1890s and early years of the 20th century, many companies increased line capacity out of London by quadrupling track on main line sections. This was expensive and electrification promised savings by increasing capacity without requiring the construction of extra track. Electric signalling enabled many mechanical boxes to be closed The maximum distance at which points could be worked from signal boxes by hand via a mechanical linkage was about 350 yards, and many boxes were needed to cover a route or a large shunting yard. These could be controlled from far fewer centres using electric signalling and powered points operation. Short blocks, resulting from the need to install a box to operate points for a local yard, could be eliminated by powered working. In these circumstances, the expense of powered operation was justified. Hybrid systems were often found, with mechanical working of signals and electric working of points (or vice versa), and in the absence of a mains supply in isolated locations, battery working of points was used. In some cases, hand-generator sets were installed (Kichenside and Williams, 1980, p. 73). The first big installations were in termini or at busy junctions on intensively used lines, where the advantages of powered operations were evident. Many large boxes were equipped for powered operations in the 1920s and 1930s, though the pioneer ventures were in the 19th century. Mixed boxes were found, as Kichenside and Williams relate (pp. 73–74):

> ... because of the cost, large and complicated centralised signal cabins were often not thought worthwhile in some power signalling schemes, and several existing mechanical signal boxes were retained. Although colour-light signals were installed, they were often

*Figure 11.4 Camden Junction powered-operations signal cabin in 1927 showing the all-electric miniature lever frame installed in 1924.
Source: London Transport Museum.*

controlled by mechanical-type levers, while points within the 350 yards limit remained mechanically operated, and others beyond this limit were converted to power operation.

One of the very first boxes to use powered operations was Gresty Lane No.1 Box at Crewe on the London & North Western Railway, where solenoids to work the semaphore signals and electric motors to work the points were installed in 1898. All train services controlled from the box were steam-hauled. In 1918, the South Eastern & Chatham Railway installed powered signalling at Victoria Station, London, using three-position semaphore signals, working in the upper quadrant, which survived until 1938. The electrified lines of the Southern Railway made widespread use of powered signalling with electro-mechanical interlocking between levers rather than purely mechanical interlocking. On the Southern Railway techniques were developed that became standard practice in the 1950s, including control of signals and points through miniature levers and frames; electromechanical interlocking between individual levers; multiple-aspect colour-light signalling; full track circuiting, and automatic signalling. Other companies were not so progressive. The Great Western Railway never adopted the standard multiple-aspect colour-light signal, but used colour-light signals exactly as if they were electric-lamp versions of the semaphores they replaced. They displayed the same aspects as semaphore stop and distant signals, or semaphore and distant arms on the same post. In the pioneer systems installed before the mid-1920s, each lever controlled one piece of equipment: it performed a distinct function, and was interlocked with other levers also performing different, distinct functions. As remarked by Kichenside and Williams (p. 74): 'power operation provided the opportunity to make one lever control more than one item of equipment – in fact to start a chain reaction through several pieces of equipment.'

This was to lead to the technique of route-setting or route-control using the check-lock lever action employed in the signal boxes of the Post Office Railway in 1927. In the same year, in Newport, Monmouthshire, the Great Western Railway introduced route-controlling in two signal boxes where miniature levers controlled semaphores and points converted to electric working. Each lever in the two signal cabins controlled a complete route from one stop signal to the next. As was common in British route signalling, a separate signal arm was provided for each diverging route, so there was one lever for each stop signal arm. Operation resembled Post Office Railway signal-box practice. When a lever was pulled, it was checked in two intermediate positions. In the first, the equipment checked that the track circuits showed that the route selected was clear. In the second position, a check was made of the points setting over the route. Finally, if these checks positively signalled clear, the lever could be pulled fully over and the signals cleared. In GWR installations interlocking between levers was by rotating cams. The technique was not applied further by the GWR, and when that company introduced powered operations at Paddington, Bristol and Cardiff it used individual control levers for each particular function, without route-setting.

Signalling, communications and control 1920–40   199

Figure 11.5   London Bridge (Southern Railway) signal box in 1928 with the 312-miniature lever frame which worked points and signals electrically, though interlocking was mechanical. Later developments used electrical interlocking. Note rotary clock-type train describers.
Source: British Railways/National Railway Museum (York).

On the London Underground, route-setting dates from September 1932, at Wood Green where the setting was controlled by a train describer located elsewhere, and not from a signal-cabin frame. The system of route-setting or route-control was taken up by the London & North Eastern Railway in the 1930s, and developed for high-speed main lines which were steam-worked at that time. Westinghouse installed for the LNER a miniature signalling frame with thumb switches at Wembley Park in December 1932, and on Goole Swing Bridge in 1933, but the noteworthy achievements were in main line signalling. The main improvement was replacing the mechanical interlocking of operating levers by electrical circuits, giving rise to 'route-relay interlocking'. Levers had moved the mechanical locking mechanisms of the miniature frames, albeit on a much smaller scale than found in the manual-mechanical boxes. Mechanical components moved by hand gave way to electrical circuits which controlled powered operations of signals and points with electrical interlocking. These were activated via thumb switches, which were usually turned rather than pulled out like the switches in the Post Office Railway control cabins.

These thumb switches could be arrayed on a panel, and they took up far less space than did a miniature lever frame. Each switch controlled one route, from one signal to the next. Turning the switch 90 degrees from normal setting, activated circuitry which checked that the route selected was clear, monitored track circuits, set points correctly, and cleared the signal for that switch. Automatic signalling restored the signal to danger when the activating train passed the signal, and the switch could be reset to normal. As Kichenside and Williams remark:

While the train is still on the portion of line controlled by the switch, track circuit occupation holds the route previously set for the train even though the signal behind the train may be at danger and the thumb switch returned to normal.

This system was first used at Thirsk, and later at Northallerton, Leeds, and Hull. The Northallerton box contained an important innovation in the form of an illuminated track display diagram, on which illuminated white lights showed the route set up. At Thirsk, the thumb switches were mounted on a track diagram, so that each represented the signal it controlled, and showed its geographical position. Later policy was to mount the switches, which were designed to be smaller, on control panels, separate from the illuminated track diagram. The largest example of this system was the one designed for York. Installation started in the late 1930s, but the war delayed the project, and it was not completed until 1949. The Northallerton and Thirsk boxes controlled traffic over the fastest sections of the LNER. The usual type of colour-light signal on the LNER was the searchlight type. Despite these considerable improvements to electrical and electro-mechanical route selection and interlocking, there were sites where more primitive apparatus sufficed. Hand-worked ground frames, with levers and rodding, were found in little-used sidings, even on such extensively power-operated systems as London Transport. Some of these Westinghouse ground frames survive to this day at Morden (installed in the 1920s),

Signalling, communications and control 1920–40 201

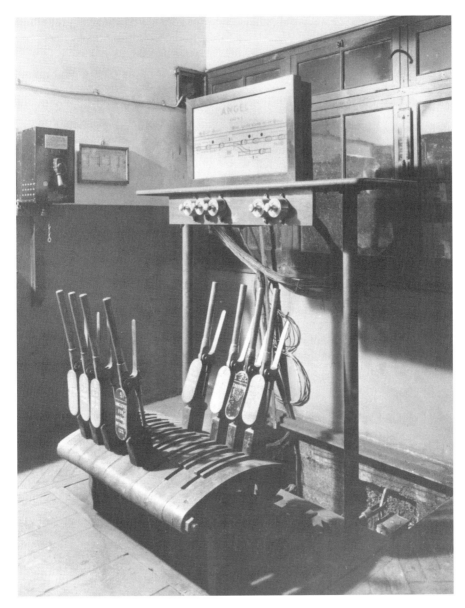

Figure 11.6 Ground lever frame, at Angel, London Transport, in 1959.
Source: London Transport Museum.

Golders Green, Totteridge and Whetstone where they are held ready for emergency use. In each case, they are controlled from the Cobourg Street centre, and are remotely released for local hand-working when needed.

11.5 American practice 1920–40

Developments in Britain, Continental Europe and North America were generally similar. The USA was usually the first to apply an innovative system on the wide scale using equipment which was often manufactured under licence in several countries. For example, the mechanical interlocking used on London Transport throughout much of the 20th century used Hambay's American patent of 1889, and many later, more advanced techniques were first established in the USA. American main line steam railways in the 1920s used cab signalling along the lines of the pre-Great War pioneer trials in the USA and Britain, but with improvements. The American cab-signalling system sent a continuous message through the rails, using slow-pulsed AC, to a receiver-amplifier on the locomotive, where a visual and audible indication of the signal aspect was provided in advance of actual sighting. Ruffell (Signalling Record, 1995) remarks that the original development of the coded track signal and circuit was on the Pennsylvania Railroad, in co-operation with the Union Switch and Signal Co., during tests of cab signalling and Centralised Traffic Control. The cab-signalling was four-aspect, with codes of interrupted 100 Hz AC currents in the rail at a frequency chosen to avoid interference from nearby industrial and domestic supplies. The Pennsylvania Railroad also investigated DC coded track circuitry, at Hulton, Pa., over a length of 3666 yards, which became a standardised length for such installations. Other systems of cab signalling used inductance-contacts, or mechanical strikers. Later developments enabled DC and audio-frequency signals to be sent through the rails to the cab-signals.

Safety devices included Automatic Train Stops for alerting train crew who might pass a danger signal. These were usually a mechanical trip, arranged to operate a valve to apply the brakes. They could be fitted to steam locomotives but worked best on electric railways where they were employed from the earliest years. Many are still working.

By the 1920s, a cab warning signal suitable for main line use on steam railways was available. A magnetic detector on the locomotive was actuated by a lineside inductor, unless neutralised by an electromagnet activated by the signal being set to clear. There were variants, and the simpler devices provided no speed control but merely alerted crew and applied the brakes. Automatic Train Control uses pulses sent through the rails to control speed and bring it into accordance with the signal aspects. Though anticipated in the 1930s, its application was more a matter of the post-1945 era.

In the late 1920s, developments in pulse-coded electronics led to control of all the signals and switches in an area over two line wires only. It became possible and economic for the despatcher to control switching and signalling through an electrical signalling machine, sending out pulse-coded messages from his office. The Western Electric system became a world standard in this field. The Traffic Control System (TCS) and Centralised Traffic Control (CTC), have developed further with the advent of better techniques for sending messages. Both TCS and CTC existed earlier, though not always under these names, and were reliant on the telephone, telegraph or wireless for sending the messages in the 1920s.

In the USA it became possible to control all train movements by signal aspect in a district where block signalling and interlocking was total, using TCS and CTC, with the despatcher controlling signal setting directly or through signal box staff. As always with signalling and control systems, expense delayed widespread application of innovations, and in the 1950s and early 1960s, it was possible to find many British locomotives, operating over main lines on express passenger trains, which had no automatic warning system or cab signalling for indicating a signal setting in advance. Many of the devices installed in the 1960s were based on principles demonstrated in the 1920s, or even earlier. In Britain, coded track circuits after the American example were not applied extensively on main lines until after the 1939–45 war. Ruffell (1995) states that there was only one main line installation in Britain between the wars for sending signals through the rails to the train. This was the experimental continuous cab signalling system between New Barnet and Potters Bar on the LNER main line, with which tests were conducted using a 4–4–2 type steam locomotive. He refers to the first coded track circuits on the GWR as dating from May 1944 at Basildon. It would seem that between the wars the USA led the way in signalling, communications and control as it did in traction.

11.6 British railway communications and control 1920–40

Traffic control was centralised by the telephone, which revolutionised management because it enabled the controller to determine priorities and modify train operations. One of the first examples of centralised control by telephone in Britain was on the Midland Railway at Masborough in July 1907. The system marketed by the American Western Electric Company became a world standard, and was probably the most widely used in the world. In this system, large numbers of telephones (sometimes referred to as 'way stations' in Britain) were connected in parallel across a single pair of wires. A typical number was 241, with each telephone having a selector of the 'ratchet wheel and pins' type, so that selected phones could be called. The signals were pulse-coded by a controller, which – before the mid-1930s – contained a clockwork motor and a cam-operated pulse generator. In the 1930s, these were replaced by dial driven uniselectors to create the pulses. This reduced the size of the controller and saved space, which was always valuable in a control centre. The original Western Electric system was able to work with later equipment and it was used in Britain into the 1980s when the last examples were replaced by solid-state systems. Two-way loud speakers, by Westinghouse, were used widely on London Transport for communication between trainmen and control centres during shunting and operations in depots, and there was steady improvement in trackside communications between the wars.

There was considerable improvement in railway telephony in the 1920s and 1930s. The omnibus circuits continued to be used, but the danger of simultaneous calls and confusion led to the introduction of selective systems employed in conjunction with the signal post telephones needed by train crew to contact a signal box or control centre. Systems were devised with call button identification of the signal from which the call was being made. This gave an audible alarm in the signal box or control centre and locked out other phones on the same line. One widely used design was supplied by STC of New Southgate, London. The omnibus system survived down to the 1950s, but it was condemned as being responsible in part for a major accident and was rapidly replaced. The STC system for signal telephones was electromechanical, like the early pulse-generators for centralised train control. Clockwork motors, cam-controlled, generated the pulses. Each signal post phone had a unique identity code, and other telephones were excluded when one was being activated. In the 1950s, selective systems became usual.

11.7 Train describer-recorders

Some mention should be made of the many kinds of train describer-recorders which were used on steam and electric railways, and which played an important role in route-setting on some underground operations. The describer-recorder or recording train describer originated in the 19th century as a machine for simplifying and speeding the reception and despatch of descriptions of trains to and from signal boxes or control centres, with some means for displaying the status of several trains being monitored. An early example was the Walker Describer which was invented by the Telegraph Superintendent of the South Eastern Railway. This used a clockwork-driven pointer moving over an indicator dial with stops next to the description to be sent. Pulses of current were sent to a receiver elsewhere which needed to be synchronised to work properly. This could only cope with one train at a time, and information received had to be booked manually. In the period 1900–05 describer-recorders were invented which stored information about several trains in the form of dispositions of movable studs on a rotating, electrically driven drum, using the encoding principles developed for printing telegraphs, or for data processing pioneered by Hollerith. These were used on the London electric railways in the first decade of the 20th century. Their capacity to store and handle information, and to pass it on, was increased by techniques borrowed from telephony.

In the 1920s, describer-receivers spread from the electric rapid-transit lines to the electrified main lines. Their advantage lay in their ability to receive, record and transmit information about many trains passing with high frequency down a common route: hence their early use on the District and other London

Underground lines. The introduction of colour-light signalling and continuous track circuiting in the 1920s, not only increased frequency of trains passing through complex routes, but enabled several trains to occupy one line between two signal boxes, using what is today called the Track Circuit Block. The need arose for a form of describer-recorder which could receive, display, store and record information until the signalman dealt with it, and which could pass on data to other boxes. The possibility of automating this sequence and using it to control trains, was recognised early – long before reliable apparatus was available at low enough cost for railways to use. It is worth noting that route-setting on the Underground, by train describer, was demonstrated at Wood Green in 1932.

To meet the demands of the modern electric railway, with colour light signalling and track circuits, several improved types of recording describer were brought out in the 1920s and 1930s and given trial with varying degrees of success, such as the Anderson-Baker Recording Train Describer tried on the Southern Railway in the 1930s, which made a permanent, timed record, and displayed the train descriptions to the signalman. Most of these systems were electro-mechanical, and were often non-standard prototypes offered for trial by their makers. They usually failed to supplant existing apparatus, which worked well enough when supported by abundant, high-grade, low-paid labour – of which there was a ready supply in the 1920s and 1930s. Some devices worked well enough to suggest to far-sighted signalling engineers that describer-recorders, automatic signalling, interlocking and powered operations might one day be integrated into a close-knit network of Centralised Traffic Control and automated operations which could include driverless trains.

Chapter 12
Railway electrification 1920–40

12.1 Introduction

By 1920, electric traction was well established on rapid-transit railways, tunnel lines, mountain railways and sections of main line railway in difficult terrain. These electric railways opened during the period when railway transport enjoyed a near monopoly on land for passenger traffic and long-distance goods movements, though after 1920 competition came from road transport when the state encouraged demobilised soldiers to set up motor bus and road haulage companies to create the motor-vehicle expertise which would be needed in a future war. Air transport using airships or aeroplanes was recognised as a potential threat.

12.2 Electric traction and the efficient use of fuel

In the 1920s, the world was anxious about fuel supply and energy use, and plans were formulated for electrifying whole nations if primary energy was cheap. With so many important engineering matters unresolved, the prospects for general railway electrification in the 1920s were less certain than before the Great War.

There was a 'Coal Crisis' in the 1920s, and general anxiety about energy supplies led to the organisation of regular World Power Congresses to promote efficient energy use, to reduce waste and to tap unexploited resources. The construction of national electricity supply grids was urged as a method for increasing the efficiency of coal, oil and gas use in industry and transport. The plans called for large central generating stations which burned coal in high-pressure boilers for supplying turbine-alternators which provided the high-voltage three-phase supply. Hydropower could meet only a fraction of the projected need, and an increasing proportion of the world's electricity would come from coal. From these 1920s proposals came the national supply grids of several countries, which

208 *Electric railways 1880–1990*

Figure 12.1 The turbine-alternator hall, Lots Road power station, Chelsea, London, in the mid-1930s. Installation of new equipment in the 1930s increased fuel economy.
Source: London Transport Museum.

made railway electrification easier than before the Great War when railways had to build their own power houses. Dawson advocated electrification as a means of introducing a more scientific energy policy throughout Britain. He urged main line railway companies to electrify and reduce fuel costs, but he was opposed by those who regarded electrification as inefficient and costly. There was general agreement in Britain that coal must be used more efficiently. Steam locomotives consumed a significant portion of the national total; and railway companies recognised that fuel costs were crucial in determining the economy of their business – but there was dispute about the best way to lower consumption.

Between 1900 and 1930, innovations improved both the coal-fired electricity generation system and steam locomotion and it was unclear whether electric or steam traction made the best use of coal. There was no indication of how high the thermal efficiency of the electric railway system would rise, measured from coal supply to locomotive drawbar. There was no definite idea what the maximum thermal efficiency of a modern steam locomotive might be, and many improvements were made to the orthodox type in the 1920s. There were suggestions, inspired by contemporary power house practice, that a 'New Form' of steam locomotive was possible, which would give higher fuel economy and equal the electric locomotive in performance (Stoffels, 1976).

In 1922, over 120 private railway companies in Britain were amalgamated into four. These were the London Midland & Scottish Railway (LMS); the London & North Eastern Railway (LNER); the Great Western Railway (GWR); and the Southern Railway (SR). Despite priority being given to reorganisation in the years following the grouping, several schemes for electrifying British main lines were considered, some of them dating from before the amalgamation. These were based on American and Continental main line railways, which used either the high-voltage DC system or the single-phase system worked by locomotives with AC commutator motors. These were considered by mechanical engineers like Sir Vincent Raven, who sought to replace steam traction with electric locomotion (Langdon, 1901; Raven, 1922; Raven and Watson, 1919; O'Brien, 1920). British steam traction system was inefficient in 1920, and many engineers believed that performance and efficiency could not be improved. They thought that impasse had been reached, so that another mode of traction was needed (Fowler, 1922), but in 1922 H. N. Gresley introduced Engine 1470 on the Great Northern railway (Brown, 1961, 1975; Nock, 1945, 1982). This machine played a major role in improving steam traction to such a degree that it was retained and electrification discounted.

Fuel economy and thermal efficiency dominated reviews of traction policy in the 1920s. Grime's 1923 review of motive power and energy costs surveyed orthodox and new forms of steam locomotive, thermal-electric locomotives, oil-engined locomotives; and electric traction. Grime related thermal efficiency to costs and compared the overall systems efficiency of the major modes of traction. He also reviewed the results of efficiency trials carried out in Britain since 1890. The latter showed that there had been little improvement in thermal efficiency between 1890 and 1920, which suggested that British railway engineering was stagnating. In 1890, Acworth expressed concern about the growing ascendancy gained by American and Continental industries over British railway engineering and developments during the next 30 years confirmed his analysis because British railway engineering became relatively backward and remained so until a new British railway was created by the post-l950s introduction of diesel and electric traction, and the extensive reorganisation by Dr R. Beeching (Bonavia, 1971; British Railways Board, 1963).

Thermal efficiency and coal consumption were more serious matters in the 1920s than in later years. In 1931 H. N. Gresley, Chief Mechanical Engineer of the LNER, surveyed new forms of steam locomotive, based on the high-pressure steam generators used in power stations and marine plant, which were being tried to improve thermal efficiency. Gresley set the cost of locomotive fuel at 25 per cent of direct operating expenses. He went on to say (Gresley, 1931):

In Great Britain alone there are 23,000 locomotives and a sum of approximately £45,000,000 per annum is spent on their maintenance, renewal and running. Of this sum nearly £12,000,000 has to be spent on their maintenance and renewal.

Gresley said that maintenance costs were of equal importance to fuel costs, so that new forms of locomotive introduced to cut the latter, should not increase the former. All the proposed experimental types of steam locomotive failed at this point. The attempts to incorporate power station or marine practice into railway traction resulted in complex and unreliable machines which were expensive to maintain, and which were not developed further. Grime argued that the best measure of a traction system was commercial efficiency, which took into account first cost, maintenance and depreciation. He concluded that in the majority of cases there was no reason to depart from the existing form of steam locomotive which he believed could meet foreseeable demands for increased efficiency and improved operation. Many factors effected traction policy, chiefly the financial crises of the 1920s; the low estimated return on capital invested in electrification; the problem of defining the best electric system for general use; and the lack of a successful, powerful main line oil-electric locomotive. At a time when British industry lacked expertise in electric and oil-engined traction, the British steam railway was improved through imported American and French practice. It was therefore inevitable that the steam locomotive be retained down to the late 1930s, and because of the war, down to the 1950s. Grime and other British mechanical engineers cannot be condemned for advocating the retention of steam traction. Sir Henry Fowler, Sir William Stanier, C. B. Collet, and Sir Nigel Gresley had good reasons for continuing with steam locomotives. The major error of the steam locomotive engineers, working on the LMS, LNER and GWR before the nationalisation of railways in 1948 and on British Railways after nationalisation was the failure to develop expertise in oil-engined traction and main line electric traction when by 1930 there were signs that it would be required within 10 or 15 years. Despite the example set by private builders in Britain, and by railway companies and locomotive manufacturers in the United States and Europe, the first prototype, heavy-duty oil-electric and electric main line units did not appear in Britain until after 1948, and these were not used for further development after nationalisation by Riddles and his staff until the Modernisation Plan of 1956 committed British Railways to these newer modes of traction (Haresnape, 1981; Johnson and Long, 1981; Tufnell, 1979). The failure of the private companies (LMS, LNER, GWR) to explore the newer modes after 1930 on their own initiative or in co-operation with the locomotive building industry, preserved obsolete methods of thinking, organising and planning on the railways (Bonavia, 1971; Joy, 1973). The British ignored Lomonossoff's advice that railways should organise small but effective strategic departments for surveying current trends, and for considering long-term engineering policy. There were economic reasons why the three largest companies (LMS, LNER, GWR) sought short-term gain between 1925 and 1950, but the lack of strategic planning and analytical thinking meant that the British Railways senior executive for too long remained unaware of the impending problems occasioned by the retention of steam traction and the failure to cultivate expertise with newer modes of traction. These were recognised by officials outside the traction executive and a radical re-organisation and change of policy were brought about by government pressure.

Grime's analysis was based on American work carried out by Goss (1905, 1907) and Lawford Fry, and by the Pennsylvania Railroad (1905). Grime considered the latest versions of the orthodox steam locomotive represented by the superheated 4-6-2 Pennsylvania K4s and the Gresley A1, with the new forms of steam engine being tried in the 1920s (Macleod, 1929; Stoffels, 1976) . There was much experimentation with new forms of railway locomotive in the 1920s. The turbine-mechanical, vacuum-condensing locomotive, introduced by Escher-Wyss and Zoelly, and based on low-pressure marine turbine developments was attracting much attention (Duffy, 1986; Stoffels, 1976). The Ramsay turbine-mechanical locomotive rebuilt from the 'Electro-Turbo-Loco' of 1909 was under trial in Britain (Macleod, 1929). The Schmidt-Hartmann-Henschel high-pressure locomotive had been constructed in Germany (Duffy, 1992), and other un-orthodox coal-fired traction units were proposed. Grime correctly dismissed these as unpromising. The new forms of steam locomotive were less thermally efficient and less reliable than the orthodox 'Stephenson' type. The turbine-condenser locomotives suffered leakage past the turbines, loss of vacuum, excessive weight, and losses in the geared or electric transmission, which reduced the thermal efficiency expected from 12 per cent to 5 per cent at the drawhook . This could be equalled by an orthodox steam locomotive costing less than half the price.

Grime was unenthusiastic about electric traction. In the mid-1920s the economy of any electric system was dependent on power station load factor. Grime identified this as the crucial parameter for deciding the issue between steam traction and electric traction on main line railways, and he stressed that load factor needed to be improved to make electric traction competitive. The thermal efficiency of a typical 'good' British power station was about 11 per cent, which was not good enough. In 1926, *Electrician* published a survey of British power stations, divided into three categories by power rating: highest, medium, and lowest. The highest thermal efficiency for each category was given. The best of all was from the group made up of the largest stations. This was Barton station with an overall thermal efficiency of 19.85 per cent from coal fuel to bus bar output. Finchley station, in the medium powered group, had a thermal efficiency of 12.72 per cent. Silloth station, in the lowest powered group, was rated at 5.76 per cent. These were the best stations in each group, but as Grime and others like him were using the best steam locomotives to argue their case, it was only fair to compare them with the best power stations, which were capable of thermal efficiencies nearer 20 per cent than 11 per cent. 11 per cent was a reasonable figure for the whole electric power industry based on the sum total of electricity sold and the sum total of coal bought in. Working with 11 per cent gave an overall thermal efficiency of 7.7 per cent between power station boiler and the engine mounted current collector on an electric railway, but the efficiency at the engine drawhook would be less because of losses in motors, auxiliaries and resistors, and because of the power needed to move the engine itself. If it was about 5.5 per cent at the driving wheel rims, it would be 4 per cent or 5 per cent at the drawhook, no better than for a steam locomotive. The advocates of steam traction failed to allow for future improvement to power station efficiency and transmission line performance which increased considerably within the next few years, whereas test data amassed since 1890 indicated that steam locomotive thermal efficiency was reaching a limit. The construction of new, high-thermal efficiency power stations in the 1920s and 1930s, and the modernisation of older ones, greatly increased overall systems efficiency. An example is the improvements carried out to Lots Road Station between 1932–34 when Lord Ashfield became Chairman of the London Passenger Transport Board, formed in 1933. Comparisons in the mid-1920s between a modern superheated steam locomotive, like Great Northern Railway 1470 built in 1922, and a modern electric locomotive showed the steam locomotive to be superior on grounds of overall thermal efficiency, but these comparisons were made between a long-established mode (steam traction) and a new mode full of unrealised potential. In the 1920s steam traction was near a limit beyond which advance within the Stephenson form was difficult to achieve, hence the experiments with new types forms, whereas electric traction was developing rapidly and was proven superior on suburban lines, and mountain routes supplied by hydroelectricity. It was accepted that hydropowered electric railways in the United States, Switzerland and Germany were technically and economically superior to steam-railways, but there was insufficient hydropower to supply railways generally. The comparisons were therefore between coal-fired steam traction, and coal-fuelled electric railways which would be the norm for electrification away from mountains or rivers suitable for hydroelectric plant. Grime concludes:

Sufficient has . . . been said on the subject of thermal efficiency to show that on that score alone there is practically no justification for the electrification of steam railways.

Grime even suggests that comparisons on mechanical grounds indicated the superiority of steam traction:

... neglecting the tender, the steam locomotive is, weight for weight, a considerably more powerful machine than its electrical competitor. When the tender is taken into consideration, however, there is a saving weight of practically 30% in favour of the electric locomotive. In any case, however, since the thermal efficiency at the drawbar is practically the same in each case, the additional weight of the steam locomotive is of no great consequence.

This is a remarkable statement, though factually correct when applied to the latest large British steam engine, no. 1470 of the GNR which could be compared fairly with the North Eastern Railway DC electric locomotive no. 13, designed by Sir Vincent Raven and Merz & McLellan for the proposed York to Newcastle scheme (Hennessey, 1970; Hoole, 1987, 1988; Raven, 1922; *RCTS*, 10b, 1990). Both locomotives were built in 1922. However, the power-to-weight ratio of the electric locomotive was much greater for later designs whereas that for steam locomotives was near its limit. The electric locomotive of today gives at least six times the power per tonne as does a steam locomotive (Harris, 1981).

12.3 The stagnation in British railway engineering

In the 1920s the opponents of electrification failed to recognise the rate at which electric railway engineering was improving. Grime had little faith in main line electrification, and he drew support from reports adversely critical of American lines which had converted from steam traction to electric working. These reports were of questionable accuracy and Grime's conclusions were incompatible with the technical and economic success of the first main line American and European electric railways. The ability of electric locomotives to haul greater tonnage than steam engines in adverse weather conditions with better adhesion and higher tractive effort was demonstrated again and again (Hay, 1982).

On non-electric railways, the oil-engined locomotive was a potential rival to steam locomotives in terms of cost per drawbar horsepower hour but most British railway engineers believed that its many unresolved technical problems placed it beyond immediate consideration. No big British motor car manufacturer or oil engine maker had an adequately backed plan to produce a commercially successful oil-engined railway locomotive. None was in a position to do so. The British motor car industry like the aircraft industry was a collection of many small companies, under-resourced and lacking research and development facilities. The few inter-war British projects for building powerful, main line electric or oil-electric locomotives, equivalent in performance to the largest steam locomotive, were for overseas railways (Webb and Duncan, 1979). A noteworthy example was the two-unit locomotive 9000 built in 1929 as a co-operative venture between the Canadian Locomotive Co., and Westinghouse Electric. This was one of the first large main line oil-electric locomotives in North America. It was two 2-Do-1 machines coupled back to back, each with a

214 *Electric railways 1880–1990*

Figure 12.2 *Electric locomotive designed by Raven for proposed North Eastern Railway 1500 V DC electrification between York and Newcastle. The locomotive was built in Darlington in 1922. Output from six traction motors was 1800 hp.*
Source: GEC/Alsthom (now Alstom).

Beardmore 1330 hp engine. Total weight of the two-unit ensemble was 340 tons. One half of the twin assembly was scrapped in 1939, but the other half worked until 1947 as no. 9001. It is significant that this project to develop 'locomotive power stations', as these engines were called, was supported by Sir Henry Thornton, who had managed the Great Eastern Railway of Britain at the time of the Liverpool Street Station reorganisation of operations (*Railway Gazette*, 1920). Thornton was a strong advocate of 'American methods', and he left Britain for Canada, to help reorganise the Canadian National Railway, formed out of non-remunerative lines in 1923. Thornton had been General Superintendent of the Long Island Railroad before he became General Manager of the GER, and he was one of several Americans who brought new methods into British railways after 1890 (Ellis, 1959). One home-based oil-electric project of the inter-war years was the British-Thomson-Houston-Allen Bo-Bo box-cab shunter supplied to the Ford Motor Company plant at Dagenham in 1932, which was a more advanced design than the first oil-engined shunter to enter service on one of the four big private companies. This was the Paxman engined oil-hydraulic unit of 1932, built on a steam tank engine frame for the LMS railway (Haresnape, 1984; Marsden, 1981).

British electric locomotive design was generally good, and colonial, dominion and foreign practice set the standards. Several successful designs were manufactured in Britain for export to South Africa, India, Spain and Japan (Lydall, 1928; Merz, McLellan, Livesey, Son and Henderson, 1918). Several of these designs originated in Swiss or American practice, which suggests Britain's railway engineering industry was losing initiative in an important field. Evidence of what the British electrical industry could do was the 4–6–2 or 2Co1 express passenger locomotive built in 1925 for the 1500 V DC electrification of the Great Indian Peninsula Railway which used an overhead contact wire. This engine was rated at 2160 hp and was constructed by AEI and Metropolitan Vickers. Another locomotive for the same railway was designed by Merz and Partners, and built by Hawthorn & Leslie, and Brown-Boveri Co. This had the 4–6–4 or 2-C-2 wheel arrangement, like the North Eastern Railway no. 13 built in 1922 at Darlington Works to the order of Sir Vincent Raven. Merz & Partners (now Merz & McLellan) were consultants to Raven for the Newcastle to York project, which would have used 1500 V DC overhead contact wire and third rail, so the similarity between the two designs is not to be wondered at. The various demonstration units, and experimental machines, built during this period disappeared abroad or were briefly tested on British main line railways and were then withdrawn when further funds were needed to keep the projects going. No private builder in Britain could maintain schemes for introducing oil-electric traction or electric traction on the home railways without encouragement from the four major companies, and the certain prospect of a home market for their product (Johnson and Long, 1981; Webb and Duncan, 1979). But the home and colonial railways remained committed to steam traction, and there was a general assumption throughout Britain and the Empire that coal should remain the basic fuel wherever

216 *Electric railways 1880–1990*

Figure 12.3 Locomotive of 2160 hp built by Metropolitan Vickers in 1925 for the 1500 V DC electrification scheme of the Great Indian Peninsula Railway.
Source: GEC/Alsthom (now Alsthom).

possible. This latter assumption grew stronger in the 1930s as the 'coal crisis' of the 1920s receded.

Main line electric traction and oil-electric railway traction remained possibilities in Britain until after 1945 when the pilot projects were introduced, and were then left to stagnate until the government-backed Modernisation Plan of 1956. The new modes of traction were introduced by very large combines, like General Electric, General Motors, English Electric or AEG, which had engineering, manufacturing and organisational expertise over a wide range of mechanical and electrical activities. The steam locomotive makers lacked this broad base from which to start. There is quantified evidence that British expertise in steam locomotive design was lacking after 1890, and there was less improvement between 1890 and 1920 than contemporary mechanical engineers realised (Stanier, 1941). Some engines dating from the early 1920s performed worse than engines running in 1891. The Gresley 1470 in its initial condition showed a drawbar thermal efficiency of 3.93 per cent, which is slightly less than the 3.98 per cent measured on an Adams 4-4-0 on the London & South Western Railway in 1891, during the first competent 'Energy Balances' carried out on a British locomotive (Adams and Pettigrew, 1896; Wilson, 1986). This figure was improved during the 1920s by rationalising engine proportions and valve timing, otherwise Grime could not have sustained the 'thermal efficiency argument' in favour of steam traction The improvements to steam locomotive form came from America (Stanier, 1941). Goss had reformed front-end design by 1908, but the lessons were still being learned on some British railways 20 years later. Woodard's Engine 50,000 which demonstrated 'rational proportions' was built in 1910 and its direct descendent, the Pennsylvania K4s, appeared in 1914. The first British engine to incorporate Woodard's ideas was the Great Northern Railway engine 1470 of 1922 and it took until the 1930s for the ideas of Goss, Chapelon and Lomonossoff to have much impact on the British locomotive (Allen, 1975; *RCTS*, 1986, 1973; Nock, 1982, 1986). Throughout the 1920s and into the 1930s, new designs retained obsolete features from the pre-Goss era, such as badly proportioned draughting and steam passages. The examples of Churchward on the Great Western Railway, and Gresley on the London & North Eastern Railway, were taken up slowly and it was not until the 1930s that rational proportions and greatly improved thermal efficiency became usual in new designs. These improvements were considerable, and were sufficient to dissuade the mechanical engineers of the four big companies from building up expertise in the newer modes of traction as the Russian strategist, Lomonossoff advised.

12.4 Lomonossoff and state ownership of railways

Lomonossoff, was an eminent analyst of railway mechanics who advocated state organisation of research and the funding of engineering development through bodies like the Association of American Railroads. He argued that vital work was not done when left to individual private companies like the British LMS or GWR because these companies severely limited the funding of research. An individual company might spend relatively modest sums to improve locomotive stock, as did the Paris & Orleans Railway under Chapelon's direction, but it would never take up the burden of researching and developing a new mode of traction. A large private builder with the resources of Ford or General Motors might do so if there were big markets to win but most railway companies could not take the risk. Lomonossoff claimed that sustained progress in railway engineering needed direction, regulation and guidance by research institutions, drawing on university expertise, funded by the state. In most instances, electrification was impossible without state support. Sir Philip Dawson criticised the failure to electrify British main lines, and implied that nothing would get done unless the four major home companies were nationalised, and the state funded centralised research. He advocated state-standardised systems of railway electrification. For Dawson and Lomonossoff, electrification was not simply a new technology. It was a new influence, which transformed the relationship between engineering, industry and the government, and necessitated state control. The four companies which dominated British railways were the LNER, LMS, GWR and SR. These did little research into new modes of traction, and British initiative remained with the private builders who were dependent on markets in the colonies and dominions for much of their sales. The privately owned home railways designed and built a large percentage of their rolling stock in their own workshops, and placed orders for builds of their own designs with the private manufacturers only when their own shops could not meet demand. There was a marked split between the private builders with their engineers and consultants who explored electric and oil-electric traction with enthusiasm, and the influential Chief Mechanical Engineers of the four big home companies. These groups had different priorities and values. The latter group had sound but short term economic and technical reasons for ignoring newer technologies and using cheaper methods to improve the existing steam traction system. This divide dominates British railway engineering between 1900 and 1960. Home railways provided no market for electric or oil-electric locomotives and this discouraged British private firms from launching programmes after the example of General Motors, Oerlikon or Brown-Boveri. With no central body to pool the findings of research and development, and to fund experiments, British railway engineering became backward to an ever-increasing extent after 1890 as revealed by reviews of practice by Bell (1946) and the Political and Economic Planning Press (1951). Distinguished consultants such as Dawson, and visitors like Lomonossoff remarked on it, but there was no way of doing anything about it until the railways were nationalised in 1948 and the government funded re-equipment under the 1956 Modernisation Plan (Johnson and Long, 1981). This came too late to save most of the British locomotive building industry, which lacked a home market for electric and oil-electric traction equipment from the 1920s to the mid-1950s, and had therefore been unable to build up the expertise, experience, connections, organisation, patterns of thought, strategy and financial strength to compete with General Motors and (at a later date) with the French and German combines. Much of the assured British market in the colonies (less so in the dominions) was for equipment and methods which were often obsolete. Industrial backwardness and available cheap labour, encouraged the retention of steam traction well into the 1960s and kept long-established firms busy supplying components and new steam locomotives (Political and Economic Planning Press, 1951). Some of the oldest British builders acquired little oil-electric or electric expertise. They underestimated the time needed to build up such expertise and to transform methods of planning, organising and managing every part of the industry to cope with a new kind of technology. They entered the field of diesel and electric traction too late, and went out of business, or were absorbed into big combines when the last markets for steam locomotives in South Africa and India vanished.

12.5 Railway electrification in Britain 1920–40

By 1925, there was disquiet about the underelectrified state of Britain, which was interpreted as a sign of backwardness in engineering (Quigley, 1925). The four big railway companies formed by the amalgamation of 1922 were amongst the largest private commercial enterprises in the world and they dominated railway matters in the British Isles. Engineering was controlled by the Chief Mechanical Engineers despite pressure from the electrical engineers to separate the electrical and the mechanical sections which was done to a limited extent only. Steam traction prevailed, even on electrified networks like the Southern Railway, and the mechanical engineers shaped traction policy. There were good, short-term reasons why they remained conservative during a period of confusion concerning the relative roles and benefits of suburban and main line electrification when the two were not always distinguished in board room argument, parliamentary debate and the engineering press. Throughout the 1920s there were those who regarded all electrification as extensions of suburban networks over longer distances. There was no agreement about the priority of reasons for electrifying railways. Some directors, economists and engineers regarded electric traction as basically a labour saver and a method for maximising engine use. Others saw it as an efficient way of using coal. Another party regarded it as the only method for introducing high-speed, regular services to win back passengers from the roads, and to persuade the public to travel more. Some saw railway electrification as the first move of a grand strategy for expanding the electrical engineering industry and bringing nearer the general electrification of Britain. Sir Ralph Wedgewood, the General Manager of the London & North Eastern Railway, was criticised by the proponents of electrification because he dismissed main line electrification as of no interest to the LNER, and he regarded electric traction on suburban routes as primarily a means of increasing traffic where steam couldn't cope. Those who saw electric traction as an agent for handling existing levels of traffic more efficiently thought this short sighted and feared that without it, the inroads of road transport would continue. Electrification was promoted as a means of reforming a backward British industry, which would introduce new ideas and methods just as the Ford Company's innovations stimulated American industry and the US economy. Engineers were aware of the technical and economic problems delaying railway electrification but they accused the mechanical engineers and senior management of the four big companies of showing a culpable and unjustified disregard for the benefits of main-line electric traction and extended suburban electrification.

1926 was a crucial year. Huber-Stockar reviewed electric railway equipment at the World Power Conference in Basle and compared the main kinds of electric traction system from the technical and economic point of view. Hydroelectric schemes dominated generation in size, capacity and efficiency. The first epoch of the 'electric age' which had been created by heavy current applications such as traction was giving way to the second epoch, dominated by domestic electrification and electronics. In railway electrification, Central European practice exemplified by the single-phase AC railway was gaining in status compared to other forms, including the high-voltage DC railway associated with American long-distance projects. Huber-Stockar claimed that the main line systems built by 1926 had established electric traction as technically superior to steam traction for most kinds of railway operation. The outstanding difficulties concerned economic performance of electric railways and the inability of analysis in the 1920s and 1930s to provide a reliable estimate of likely return on capital invested in them. It was judged likely that electric traction would prove economic, and superior to steam traction where there was dense traffic, where coal was expensive, where hydroelectric schemes provided cheap electricity, where labour costs were high, where gradients were steep, and where capital charges were low. It might take several of these circumstances together to justify electrification on economic grounds, but unfortunately the economic factors could not always be assessed with any accuracy. They varied over time in an unpredictable manner and made estimates of project profitability unreliable. If unions permitted, electric traction led to lower manning levels by reducing the number of locomotives worked and by eliminating the fireman. A Czechoslovakian scheme of 1926 suggested replacing firemen by train guards riding in the cab of the locomotive rather than in the traditional van or compartment, but union co-operation was refused in times when unemployment was widespread. Most Central European railway electrification carried out after the 1914–18 war got state backing, which was essential if the great costs and uncertain economic outcome were to be faced. Where new nations were formed out of the old empires, electrification was an agent in the birth and growth of new, progressive states rising from the chaos of world war, civil war and revolution. In places like Czechoslovakia and the Soviet Union electrification of railways was supported for this reason. The vast Russian network would never have been electrified had not electricity been regarded by Lenin as a vital instrument for forging the Soviet state.

Financial questions concerning electrification bothered railway engineers and managers, but were given no satisfactory answer. How were railway rates to be fixed? How would changes in rates effect the cost of coal shipped? How would rates effect electricity costs? Would the electricity suppliers charge favourable rates to the railways? These questions went unanswered at a time of economic depression. Steam traction was being improved by relatively cheap reorganisation of existing equipment and worker skills, with the result that main lines were not electrified despite anxiety that Britain was deficient in industrial electrification. As Sir Charles Bright remarked in 1927: 'The electrification of Great Britain is overdue.'

The advocates of main line electrification argued their case with persistence and energy in the professional institutions, in the press and in parliament. Rational discussion was not helped by bitter arguments resulting from inaccurate newspaper reports of electrification projects supposed to be under discussion by the big companies. Sir Ralph Wedgewood denied one 1928 newspaper report that the LNER intended to electrify part of its route out of Liverpool Street and he claimed that well organised steam-worked rapid-transit services on surface lines had a carrying capacity in excess of any electric railway then working in Britain. Sir Ralph Wedgewood's attitude is typical of senior managers and engineers on the British home railways between the wars. Many industrialists shared his view, which can be summarised as follows: 'Don't risk capital in radical technological change. Wherever possible, make do with existing, paid-for equipment, which enables short-term profits to be made.'

The latter years of the 1920s and the early 1930s found the advocates of electric traction disappointed. Not only was there a lack of it, compared to other countries, but much of it was extensions of the 'London Standard' Sprague Exemplar system rather than installation of newer forms of electric railway (Moody, 1979). Many writers in the engineering press regarded 'electrification of railways' as meaning 'extension of the London Standard'. A review of home practice, published in 1928 surveyed the British scene, dominated by LVDC lines. Several plans had never been implemented. These included the plan to electrify the ex-South Eastern & Chatham lines, using 1500 V DC on the three-wire system, with 3000 V DC across the outers and protected third rail distribution. As Bonavia remarks (1987):

This had not led to physical works being started, largely because the railway's proposal to build its own power station at Angerstein Wharf on the Thames had been blocked by the Electricity Commissioners working under the Ministry of Transport. The LSWR had built its own power station at Durnsford Road, Wimbledon, but the LBSCR had purchased its current from the London Electric Supply Corporation, and this was more in line with government policy.

The project was never started, and was overtaken by the amalgamation of 1922, when the South Eastern & Chatham lines became part of the Southern Railway which standardised its electric network using the LVDC third-rail system. This involved removing the single-phase, 6600 V AC system with overhead contact wire installed on the London, Brighton & South Coast Railway between 1909 and 1925. This was converted to LVDC following the decision of the SR

Board in August 1926 that all future electrification would use 600 V DC. The most influential system from which the standard developed was the London & South Western Railway LVDC network which first used a voltage of 600 V DC, and later used 750 V DC. This system was called the 'London Standard', and was found throughout the Southern Railway. There were practically identical systems in the London, Midland & Scottish Railway, the Metropolitan Railway and in the London Transport lines where a fourth rail took the return current. The Lancashire & Yorkshire company used 600 V DC in the Liverpool area, and 1200 V DC in the third rail on its Manchester to Bury lines. In the 1920s it looked as if the 'London Standard' might become the norm for British electrified rapid-transit routes. It was suggested that the LVDC systems round London, Manchester and Liverpool could be linked with each other and connected with new LVDC networks to be built in and around Birmingham. This would provide the basis for a national network of LVDC electric railways. The links would be made using low-voltage conductor rail technology as found on the LVDC lines out of New York worked by the NYC and PRR railways.

In 1928, the *Electrician* contained an article which condemned the recently opened Manchester, South Junction & Altrincham electrification of a former section of the Midland Railway, which used the overhead contact wire system at 1500 V DC. The article called this a 'hopeless departure from standard' because it was not only incompatible with the 'London Standard' (from which it was separated by over 150 miles) but used a system different from both the former Lancashire & Yorkshire network and the former London & North Western network in the Liverpool and Manchester areas. The critic thought that one day a single electrified network would cover all North Cheshire and South Lancashire, and this should be planned for by standardising on LVDC electrification, and barring anything different. He suggested the 'London Standard' third-rail DC system using 750V DC was the obvious norm for rapid-transit networks and this remained a common opinion until about 1950, when the single-phase AC system for suburban railways found favour, largely because it could be used in conjunction with single-phase AC main line electrification. So strong was the feeling against the introduction of an 'out of standard' system, like the Manchester, South Junction & Altrincham, that its introduction was likened to Brunel's construction of the broad gauge Great Western railway after the Stephenson 'standard gauge' had been established elsewhere. This attitude was not as unreasonable as it seems at first sight, because the overhead distribution single-phase AC network on the Southern Railway was being dismantled at the time the MSJ & A was opened, simply because it was non-standard. Though main line electric traction on former steam-worked railways was proven in Sweden, Japan, India, New Zealand, Italy, Switzerland, France, Germany and the United States, there was no main line electric railway in Britain, nor was one planned. The nearest equivalent were the fast sections of the 'London Standard' on the Metropolitan Railway or the London to Brighton route of the Southern Railway. Main line electric expertise, Dawson cautioned, was obtained through work carried out in the empire and dominions, and by observing foreign practice. He referred to good results obtained from the third-rail systems used on the Southern Railway, and the four-rail systems used on the Metropolitan railway and the London Transport systems. He thought – or hoped – that main-line electrification was near, because the growth of the national grid would increase overall systems thermal efficiency, would provide a standard supply everywhere and would reduce electricity costs. He hoped that the government would lend the railways money to carry through electrification on a considerable scale which he said would be a safer investment than putting state aid into roads. Dawson had harsh things to say about the consequences of grouping the railways into four big private companies in 1922. This had hindered or stopped several promising schemes, such as the North Eastern Railway York to Newcastle project, and it created a situation in which no national policy for the country's railways as a whole could be pursued unless agreement between the four big railways could be reached. This was unlikely. The Southern Railway was committed to the London Standard and the Great Western Railway showed no interest in electric traction. Dawson was regarded as a biased partisan of electric traction whose judgement was effected by his vision of extensive electrification. His trenchant criticisms, which were widely reported, had little effect on the managers, directors and engineers of the LNER, LMS & GWR, who retorted that the duties they were paid to discharge were to the company and to the shareholders. These duties did not include realising a grand vision of electrification in the supposed interests of the nation and its electrical industry. If the nation and the electrical industry wanted electric railways, they argued, let them guarantee cheap rates of power, biased in favour of the railways and kept low by law.

In 1930, the national road system was being improved, and it was obvious that the inroads being made by road transport into goods and passenger traffic would become much more serious. Much rail passenger traffic within large cities was lost to tramcars, motor buses, motor-taxis and private motor cars. Branch lines were losing traffic to private cars, country bus services and lorries. Much goods traffic was lost following the demonstration during the General Strike of 1926 that well organised road transport could do much that the railways claimed only they could do. Long-distance passenger train services seemed safe, but road improvements, greater car ownership and better bus services were threatening the short-to-medium distance journeys made by commuters and weekend travellers in and out of the larger cities. Electrification of these lines, particularly in the London area, was advocated as a solution to a threat estimated as likely to grow to daunting proportions by the late 1930s. The Southern Railway was electrified out of London, and the LMS worked electric trains over the former L & NWR lines to Watford, but the LNER worked all suburban and commuter traffic out of Kings Cross and Liverpool Street by steam – which is why it became the particular target of the electrification lobby. The Great Western Railway was not in the same category as the other railways as a carrier of suburban traffic.

By 1930, electrification helped railways to compete with road transport by reducing journey time and crowding in busy periods. Conservative engineers thought that the same could be done using steam, petrol or diesel traction. The successes of high-speed steam-worked trains in Britain, and of high-speed petrol- and diesel-engined trains in Germany and the USA convinced them that they were right, and that too much attention should not be paid to the propaganda put out by an 'electrification lobby' of consultants and engineers whose interests were involved. It was difficult for the four private companies to draw up long term strategy in the 1930s, and to agree about the form electrification should take. Leaving the Southern out of it as a special case, would it make sense for the LNER, LMS & GWR to employ different systems if electrification were decided on? The state thought not, and each of the inter-war reports on railway electrification laid down standards to be followed.

Note was taken of the role of the internal combustion engine in railway traction. The diesel locomotive with electric transmission emerged as a 'new form' of tractor in the late 1920s and was regarded as an alternative to electric traction (Seddon, 1957; Webb, 1970). It was called the agent of 'cheap electrification', which provided the traction motor characteristics of the electric locomotive with DC motors, but without the expense of fixed works. At the same time it was recognised as essential if extensive railway electrification was to take place. It was impossible to electrify most of the steam railway as it existed in Britain in 1930. Only part could be electrified – the densely worked main lines and the rapid-transit networks. The rest had to be worked by steam. A range of duties, from shunting, branch line work, much goods traffic, and operations on lightly loaded routes could not be worked by electric traction (Lamb, 1927, 1941). They required another mode. The electrification schemes proposed before 1956 envisaged mixed-working with steam locomotives carrying out many of the duties on non-electrified trackage, and with steam engines working some traffic through electrified routes to and from non-electrified sections. This diminished the economic and technical benefits of electrification in a country like Britain, where electric traction would work in close proximity to steam engines. Steam locomotives and electric traction do not mix well, but the diesel locomotive (Tufnell, 1979; Webb, 1970, 1978), the diesel railcar (Tufnell, 1984), and the dual-mode electro-diesel are complementary. The diesel electric locomotive provided a method for carrying out many of the duties on non-electrified routes, such as passenger work, shunting, and pick-up goods, which was more efficient in fuel and labour than steam traction. The diesel locomotive, the railcar, and the diesel multiple unit set gave the railways motive power for working the wide range of services which lay outside the scope of electric traction and which were an intrinsic part of the railway industry in the 1930s.

Chapter 13

Electro-Motive, General Motors and oil-electric traction

13.1 Introduction

The 1912 trial of an air-started 2-B-2 diesel-mechanical locomotive on the Prussian State Railways was one of the first experiments with a Diesel locomotive intended for main line duties. Present were Rudolf Diesel and H. Lemp, a Swiss engineer who had worked with Edison. Lemp recognised the need to integrate the oil engine, transmission and control if diesel locomotion was to succeed. As Westwood (1979, p. 183) puts it:

Hauling a heavy load at high speed means that the traction motors demand as much current as the generator can provide, but simply increasing the revolutions of the diesel engine is not practical: the magnetic field of the generator has to be strengthened to provide the energy-inducing drag against the revolutions of the engine shaft. On the other hand, when current requirements are low, generation cannot be reduced by slowing down the revolutions because below a certain but quite high critical speed the engine will stall. In such a case the magnetic field of the generator has to be reduced so that with reduced drag it can rotate quite fast without producing a dangerous excess of current.

Early control systems were not well integrated and engines, generators and motors were mismatched. Petrol and oil engines stalled and electrical equipment burned out. This unreliability retarded the introduction of petrol-electric and gas-electric traction between 1900 and the Great War, when Lemp produced an integrated control system, patented in 1914, whilst working for General Electric. In the Lemp self-regulating system, the excitation current was generated from windings driven by the generator, so that excitation current was directly dependent on the generator speed, and the speed of the engine. This enabled oil-electric locomotives to be controlled by a single handle, and obviated manual matching of field control settings. After the 1912 trials in Germany, Lemp returned to the USA and persuaded General Electric to develop a high-speed oil-engine suitable for powering 'portable' electricity generators for use in small towns away from supply grids or hydropower stations. The Junkers engine was considered as a motor, but the power utilities were unresponsive to the idea of using small, local diesel generators. The licences on German engines were allowed to lapse, and were taken up by the Winton Engine Company which later became part of Electro-Motive. Lemp went on to develop the high-speed heavy-oil engine. The GE project for portable generators led to invitations to tender for a high-speed diesel engine of no more than 50 lbs per brake horse power. The light engines designed for such applications were quickly recognised as suitable for locomotive use. The GE invitation was answered by Ingersoll Rand, the makers of industrial equipment which included air compressors and drills, and in 1924 a 60-ton demonstrator engine with an Ingersoll-Rand engine, and electric transmission, was run on lines in the New York and New Jersey area. The demonstration showed that oil-electric shunters worked, and in 1925 a production model was delivered. This was no. 1000 of the Central Railroad of New Jersey, which is accepted as the first commercially produced diesel locomotive, and the first to be sold to a US railway company. By 1926, units of 100 tons and 600 hp were being produced and sold to companies like the Long Island Railroad. These early locomotives had a box-like body with cabs at each end, carried on two bogies. DC dynamo and motors were fitted. This combination was the norm in the USA for ten years (Locomotive Cyclopedia USA, 1974).

In the 1920s, several standard designs were built in the USA as joint efforts between the main electrical manufacturers, the oil-engine makers and the steam locomotive builders. The American steam locomotive builders were well informed about oil engines and their likely use in railway traction, and after 1920, they made great efforts to get internal-combustion engine expertise. They co-operated with General Electric or Westinghouse to build internal-combustion engined units, and they bought oil-engine manufacturing companies and set up their own diesel development divisions. Alco, which co-operated with General Electric, bought McIntosh & Seymour Corporation which became their Diesel Division. In 1931, Baldwin Locomotive Works, which co-operated with Westinghouse, bought the De La Vergne Engine Co. which first built oil engines in 1893 and in 1917 constructed the first solid-injection diesel engine in the USA. In 1930, General Motors bought Electro-Motive and the Winton Company with a programme to dieselise railways which differed profoundly from the plans of the steam-locomotive builders and the strategies of other pioneers of oil-engined traction.

The pioneers were a diverse group. There were diesel engine makers, with stationary engine experience. There were the electrical components manufacturers, usually one of the big combines. There were manufacturers of mechanical components, usually one of the established steam locomotive builders. There was little standardisation at first, and no mass production, except where big electrical combines like General Electric were involved. To displace the steam locomotive, General Motors used a strategy of standardisation of designs, mass production of components, parts exchange, refurbishment of used units, hire-purchase and instruction of customer staff in diesel operations. In 1937 GM

Electro-Motive, General Motors and oil-electric traction 227

Figure 13.1 Central Railroad of New Jersey diesel-electric locomotive No. 1000 built by American Locomotive Company and General Electric in September 1925.
Source: Baltimore & Ohio RR/RL Vickers Collection.

opened their plant at La Grange, near Chicago. This was the first factory devoted exclusively to the production of diesel motive power, in the form of the 1800 hp E1 type. In 1937 and 1938, GM offered to train steam railway staff, free of charge, in the use of diesel locomotives, using facilities in La Grange and a mobile Instruction Car. The steam locomotive builders lacked the resources, the vision, the revolutionary zeal, the wide experience across many fields, and the connections of the great electrical and automobile combines, like GE and GM. They did not direct a comparable operation covering research and development, design, marketing, management, funding and retraining. They did not mount an equivalent project and sustain it for years to preserve their market, using a new form of steam locomotive or a diesel of their own. They were outclassed. By the 1920s, the advantage was with the big combines. The 'Electric Revolution' created industries which made and installed the many kinds of component and system required to generate power and use it, including boilers, fuel handling equipment, pressure vessels, piping, pumps, generators, motors, transmission, switchgear, etc. They also set up specialist departments to pursue advanced analysis, systems planning, research and development. Since the 1940s, electronics has widened the field of activities of the combines which originated in the electrical engineering of pioneer period, and today these companies deal with electronics, weapons systems, nuclear power, chemical engineering, as well as power and traction.

The steam locomotive makers were an older kind of industry which faced a new kind of challenge which they could not meet. They made great efforts, but their history shows a lack of vision, and inevitable failure. They did not produce a Kettering or a Sloan, as did General Motors. Despite this, their contributions to oil-electric traction cannot be ignored. In 1925, Baldwin built a large 1000 hp CC engine, with a Knusden engine which became the works shunter after use in the Richmond area. The co-operation between General Electric, Ingersoll-Rand, and the American Locomotive Company produced several noteworthy types of heavy duty road-switchers and goods engines (CERA, 116, 1976). In 1928, General Electric supplied oil-electric locomotives to the New York Central Railroad . These included the 'combination oil-electric storage-battery switching locomotive' no. 1525. This weighed 128 US tons and could operate as an electric locomotive picking up current at 660 V DC from a third rail or overhead conductor on LVDC section of the NYCRR. It could work as a diesel electric locomotive on non-electrified lines. It had a 218 cell storage battery to help acceleration and to use when the 300 hp Ingersoll-Rand oil engine was shut down. Several manufacturers built multi-mode locomotives of this type. General Electric supplied other locomotives to the NYCRR. For goods service there was no. 1550, a 145 US tons, 750 hp oil-electric locomotive of the 2D2 type, with four driving axles. This was intended for road-freight duty and was expected to haul trains of 550 tons on schedules previously worked by steam traction. The engine was an Ingersoll-Rand solid-injection type. Multiple unit control was fitted to the locomotive which saw service on the Putnam Division of the NYCRR. A passenger locomotive no. 1500 was generally similar to no. 1550 but it had a 900 hp McIntosh & Seymour four-stroke air-injection diesel engine driving the DC dynamo. It was a 2D2 type, with a weight of 175 US tons. These engines, illustrated in Locomotive Cyclopedia USA (1974), marked a great advance towards the general use of oil-electric traction. Number 1500 provided an engine power of 900 hp and a one-hour rated tractive effort of 28,000 foot-pounds for a locomotive weight of 175 tons. The NYC J3a 4–6–4 steam engine built by Alco in 1926 weighed 350 tons in working order. It had a starting tractive effort of 41,860 footpounds, which would be considerably larger than its one-hour rating, and generated 3000 hp in the cylinders. The power to weight ratio of the steam engine was higher than that of no. 1500, but two such oil-electrics run in multiple were able to equal the performance of most steam locomotives by 1930. Double-units driven from one cab played a key role in eliminating steam traction by taking over duties which until then could only be worked by the most powerful steam locomotives.

A noteworthy design was the double-unit no. 9000 delivered to the Canadian National Railways in 1929. This was described by Sir Henry Thornton, the ex-Great Eastern Railway manager, as a self-propelled electric locomotive, equivalent to a CNR Class 6000 4–8–2 steam engine. The remark betrays an attitude common in North America, where diesel-electric locomotives were regarded as agents of cheap electrification because they were self-propelled electricity generating stations with sufficient power to move themselves and their trains. Apart from General Motors, the established suppliers of railway rolling stock concentrated almost entirely on low and moderate powered shunting engines and road-switchers before the mid-1940s (Morgan, 1985; Simmons-Boardman, 1983). Most builders did not plan to displace steam traction totally, as did General Motors, and whilst making 600 hp diesels to replace steam shunting engines they produced nothing to replace the general purpose 2–8–0 or 2–8–4 steam locomotive, let alone the express 4–6–2 or 4–8–4 types. Only General Motors showed how this could be done. In doing so, they built up a lead which was never successfully challenged. The supremacy of the largest, most powerful steam express locomotive was ended when Electro-Motive locomotives worked the 'Super Chief' train, which was introduced on 12 May 1936 to run between Chicago and Los Angeles (Repp, 1980). Two EMD oil-electrics working in multiple reduced the eastbound schedule of the steam hauled 'Chief' by 15 hours. The superiority of steam traction was at an end. World War Two had a decisive effect on strengthening the General Motors position. Because of their expertise in 'Road Diesels', as distinct from shunters or switchers, which all manufacturers made, GM was awarded by government decree all orders for all main line diesels during the war (Stover, 1961). The contracts for switchers went to Alco, Baldwin and General Electric. Electro-Motive built up five years experience in the design of general purpose, express and main line goods locomotive design ahead of the other companies and succeeded in dieselising North American railways with no rivals.

13.2 Development of the Electro-Motive (General Motors) diesel locomotive

The emergence of the oil-electric locomotive in the 1930s owed much to methods developed to improve performance in the engineering, management, organisation and marketing sectors of the American motor car industry. The introduction of diesel-electric locomotives in the USA provides an example of the organised replacement of a system which was long established and at the peak of its effectiveness by one which was both new and superior in every way. A major part was played by Hamilton who founded Electro-Motive and headed its design team both before and after the take over by General Motors. General Motors brought into the railway industry the insights and methods of a modern minded enterprise and great financial strength, at a time when GM had replaced Ford as the leading US automobile maker. General Motors was formed in 1908 by the amalgamation of several motor industries, including Buick, Oldsmobile, Cadillac, Oakland, and eight smaller makers of passenger and goods road vehicles. Chevrolet became part of the corporation in 1918. Staff organisation became company policy and a general team of expert engineers, including production specialists, was formed to advise the managers of individual factories.

An outstanding individual was C. F. Kettering who introduced the electric self-starter in 1912 and thereby made the motor car available to all who could afford it, including women and the elderly. Kettering became the 'scientific mastermind' of GM, in charge of research and engineering programmes (Leslie, 1983). Another genius in management was A P Sloan, who guided GM between 1923 and 1956, first as president, and then as chairman. Sloan introduced assessment of projects by measuring return on investment using accurate cost accounting, in order to make better, cheaper cars than Ford. GM allowed Hamilton and Dilworth considerable autonomy after their take-over of Electro-Motive and the Winton Engine Company in 1930 and the Hamilton-Dilworth team was not broken up, dispersed or disrupted. Electro-Motive continued to function as Electro-Motive Division of General Motors, as it does to this day. Its design strategy received the support of the resources of General Motors, and imaginative methods of marketing and promoting the diesel locomotive were brought in by auto-industry sales directors. No steam engine manufacturer had the resources or the vision to match the GM effort. Only General Electric was in a position to follow suit, which it eventually did.

By 1930, in the USA, electric transmission was standard. No experiments were carried out with compressed steam, pneumatic or gaseous transmission, and trials with hydraulic transmission continued as a relatively minor activity. The new traction system for railways was a LVDC railway with a diesel-powered generating station mounted on a self-propelled carriage. The establishment of the diesel locomotive was achieved using the DC motor and the Sprague bogie, and this should not be forgotten. The American experience with LVDC railways made the diesel-electric form normative for diesel traction and encouraged engineers to regard the diesel locomotive as the agent of cheap electrification and not an alternative to it.

Hamilton identified the main obstacle to the general introduction of the oil-engined locomotive as being the high weight of the power unit. He supervised a research programme involving the Winton Engine Co. to produce a light but powerful engine for use in locomotives which could be used to demonstrate the advantages of internal combustion power to steam operated railway companies. In 1930, General Motors bought Electro-Motive and the Winton Engine Co. and funded research which reduced engine weight from 80 lb/hp to 20 lb/hp in the 201 A engine. This was achieved by using the two-stroke cycle, new alloys, welding and imaginative design. This engine was used in the famous 'Zephyr' trains, one of which, the 'Pioneer Zephyr' ran from Denver to the stage of the Century of Progress Exhibition held in Chicago in 1934. Admirable as these lightweight trains were, total dieselisation was only possible through the agency of a locomotive proper, rather than a multiple unit or a streamlined set. A high-powered general purpose diesel was required which could replace the latest 4–6–4, 2–10–4 or 4–8–4 steam locomotives, each of which was far more powerful than any lightweight petrol or diesel railcar. Only General Motors had a plan to produce such a diesel locomotive in the near future. As Morgan (1955) puts it: 'The development of the obvious, non-articulated, diesel-electric road passenger locomotive fell into the eager hands of EMD.'

Demonstration units were built to show railways that diesel locomotives could match any steam locomotive in performance. The most important were engines 511 and 512 which were introduced in 1935 and ran as a double 3600 hp ensemble and the great exemplar, no. 103 of 1939. At this period, General Motors had not opened the La Grange Plant, and was reliant upon other companies for the manufacture of vital components. The 201 A engines fitted into 511 and 512 were built under licence by General Electric, at Erie, Pa., and General Electric generators and carbodies were used. In the same year, a single unit of 1800 hp, no. 50, practically identical to 511 or 512 was sold to the Baltimore & Ohio Railroad, and an influential double-unit was sold to the Atchison, Topeka & Santa Fe Railroad. This latter locomotive worked the 'Chief' and the 'Super Chief' express trains (Repp, 1980) and proved superior to the 4–8–4 steam locomotives of the 3751 series introduced in 1927 to work the heaviest loads (Stagner, 1975). These pioneer main line diesel locomotives still relied on General Electric and the St Louis Car Co., where many basic components were constructed (Lind, 1978), but in 1937, General Motors opened its own plant at La Grange, Illinois, to become independent of suppliers like GE. The first product from La Grange was the E1, described as a 'semi-standardised line'. It was a 1800 hp A1A-A1A type, with two 900 hp engines, produced in a cab and booster version for multiple-unit operation. This was followed in 1938 by the 2000 hp E6, the first true mass production road diesel, in which the finest traditions of electrical engineering and automotive engineering found expression in the union of several progressive traditions of design, production and operation originated by Edison, Sprague, Ford, Hamilton, Dilworth and Taylor.

Figure 13.2 Baltimore & Ohio Railroad 1800 hp diesel-electric locomotive No. 50 built by St Louis Car Company to Electro-Motive design in August 1935.
Source: Baltimore & Ohio RR/RL Vickers Collection.

It used Electro-Motive generators and motors, and made possible three-unit sets of 6000 hp capable of displacing any steam locomotive, even the most advanced types employed on the prestigious long-distance express trains like the 'Super Chief'.

In 1939, no. 103, the famous demonstration goods engine, was completed. It was a 5400 hp set of four B-B units of 1350 hp each, weighing 402 long tons in working order. This machine is often called 'the diesel which did it', which convinced the steam railway managements that oil-electric traction was efficient and reliable for general-purpose operations. It ran over 20 Class 1 railways in 35 states, and 13 companies ordered similar locomotives. Engine 103 was sold to the Southern Railway after demonstrating that dieselisation of American railways was possible and economic. These EMD units were not the first internal-combustion engined demonstrators on railways but they established the diesel as the future motive power in the USA and Canada, and they had great influence on European and general world practice. The launching of diesel traction in North America was no mere technical exercise, but was part of a co-ordinated plan to displace steam traction, in which General Motors expertise in publicity, use of hire-purchase, and skill at mounting engineering demonstrations was consummately displayed (Morgan, 1955; Stover, 1961). The steam engine manufacturers were dominated by the production of locomotives ordered by individual client companies. They were single-product enterprises, quite different from the international combines which made a wide range of industrial equipment. They could do little but compete in diesel construction at an obvious disadvantage, with various degrees of success. They produced excellent designs of diesel in every power classification, from switchers to four-unit express passenger locomotives (Locomotive Cyclopedia, 1974; Simmons-Boardman, 1983), but they lacked the resources of the General Motors and General Electric combines. In the USA, as in Britain, the former steam locomotive makers were taken over or passed out of business in the 1950s or 1960s, with only General Electric and General Motors left to supply diesel locomotives on a large scale (Pinkepank, 1967, 1973; Simmons-Boardman, 1985). The steam locomotive builders made little effort to change their approach to building and selling their product. They never tried to produce a standard range of steam engines for nation wide use, to be hired out if necessary, despite the government-inspired production of 'standard' designs during the First World War, some of which were developed after federal control over railways was ended. The builders were never able to develop multiple-unit control of steam locomotives from one cab, by one man, a step which would have made a considerable difference to the economics of steam traction. Indeed, nobody seems to have made a serious attempt to devise such a system. The major steam locomotive builders supplied mechanical parts for diesel and electric units and made switchers and demonstrators, but attempted no full-scale response based on a recognition that the future lay with the diesel until General Motors success threatened their existence. Before then, most significant oil-engined locomotive construction was done by General Motors and General Electric. The Ford industry had

Figure 13.3 Twin-unit 2700 hp General Motors diesel-electric locomotive introduced by Dilworth outside the new Electro-Motive factory in La Grange, Illinois. Two such twin-units made up the four-unit 5400 hp ensemble no. 103 introduced in November 1939 which established diesel traction as superior to steam traction within North America.

Source: Electro-Motive Division, General Motors.

constructed a motor-generator locomotive and built aero-engines and equipment for boats and it remains a mystery why Ford never pioneered the internal-combustion locomotive. The most important element in the General Motors' programme was the insistence on standardisation, which enabled locomotives to be sold like motor cars. Motor cars were sold in large numbers because they were a standard design and sold by hire-purchase. General Motors transferred this method to diesel locomotives so that they suited methods for financing rolling stock replacement previously used with carriages and wagons which were standardised products. As Stover (1961) relates:

As the railroads after World War I sought to maintain and expand their traffic by improving and upgrading their equipment, they were increasingly forced to finance the new equipment with a relatively new type of security, the equipment obligation. In this transaction the trustee or legal owner of the new equipment retains legal title until the railroad, with a series of regular instalments, has fully paid for the equipment. Since the equipment is readily movable and can easily be sold to other railroads in the event of default, the equipment obligation features a high degree of financial safety.... Since World War Two the great bulk of all new rolling stock and motive power has been financed in this way.

This system only works if the equipment can be used anywhere and can be transferred from one railway to another. This demands standard design. To function effectively with motive power, equipment obligations required a locomotive capable of operating anywhere in the USA so that it could be transferred, and although the diesel-electric met this requirement (because design was taken out of the clients' hands) steam traction did not. Most of the steam types delivered in the 1930s and 1940s were non-standard, and were designed to meet the requirements of a particular railway. They could not be sold under equipment obligations or hire purchase. The diesel, the electric locomotive, and the multiple-unit railcar created the modern railway after 1945. This could only be done by eliminating steam traction which is why the evolution of the oil-electric unit in addition to the electric locomotive was of first importance. Steam traction could not be eliminated by electric traction alone. The advent of the diesel was essential for it to be done. The Electro-Motive Engine 103 established diesel traction on non-electrified routes in the late 1930s. The electrification of the Pennsylvania main line in 1938 using single-phase AC, 11,000 V, 25 Hz, and operated by the GG1 locomotive showed how electric traction could work the heaviest trains at speeds of 100 mph (Bezilla, 1980; Middleton, 1974). By 1940, there was no duty which steam traction could do better than oil-electric or electric traction. Engines such as the GG1 removed the steam locomotive's last claim to supremacy – to be the fastest mode in express service. By the end of World War Two it was obvious that electric traction and oil-electric traction were the modes of the future, though the rate at which they were introduced in countries outside the USA depended on economic recovery after the conflict, and the availability of aid for industrial re-equipment which was often determined by the state.

13.3 General Electric and the 'Steamotive' project

The General Electric Company was engaged in constructing oil-engined railcars, oil-engined locomotives, dual-mode locomotives, petrol-engined railway vehicles, and electric locomotives from the beginning. The company had a long list of pioneering ventures to its credit, but they were surpassed, as was every other railway engineering industry, by the General Motors campaign of the 1930s to bring about and monopolise the conversion of North American railways to diesels. The General Motors demonstrators established the oil-electric locomotive by the mid-1930s, and in 1936 the company opened the world famous Electro-Motive Division plant at La Grange which produced the first E series locomotives in 1937. General Motors enjoyed all the advantages of being first in the field with a greater fund of expertise than rivals. It improved its position during the period when rivals were trying to acquire diesel expertise themselves. General Electric was the only company with the resources to challenge General Motors, which they did by developing a locomotive intended to rival the GM diesel. This had happened before, albeit with other companies and other technologies. Baldwin Locomotive Works became alarmed at the prospect of losing the goods engine market to the successful Woodard-designed 2-8-2 and 2-8-4 high performance steam engines manufactured by Lima in the mid-1920s (Cook, 1966; Poultney, 1925, 1929). Baldwin launched its own new type of steam locomotive, the high-pressure three-cylinder compound engine number 60,000 in the unrealised hope of winning back much of the market (Fry, 1927). General Electric, realising that they had lost initiative to General Motors and its diesel, launched the 'Steamotive' which was intended to be a replacement for orthodox steam traction and to be a 'successor to diesels'.

It took the form of a steam-turbine-electric locomotive, fitted with vacuum condensers, but it owed nothing to any of the earlier steam-electric experiments of Heilmann, Reid, or Ramsay. It was one of the most radical attempts to introduce a new form of locomotive based on systematic research, careful experiment, and considerable expenditure (Duffy, 1986). One objective was to produce an alternative to the diesel-electric locomotive which would be as closely associated with General Electric technology and expertise, as the diesel-electric was associated with General Motors. Granted the unassailable position of General Motors in the short term, General Electric hoped that by producing a new design of locomotive it could win a large portion of the traction market, especially if low grade oil, or coal was the fuel. At the same time, GE continued to co-operate with Alco to produce a road diesel of their own. The new locomotive was designed to eliminate the undesirable features of the orthodox steam locomotive, such as low thermal efficiency, low adhesion weight, the need to stop to take on water, the inability to be subject to one-man, multiple-unit operation, and the low availability for duty. It was designed to conform as closely as possible to the form of the diesel-electric. Many of the elements such as the DC electric transmission were based on diesel-electric practice, but others were derived from the latest power house technology. The 'Steamotive' was a

Electro-Motive, General Motors and oil-electric traction 237

Figure 13.4 The twin-unit Steamotive built by General Electric in 1938 was made up of two 2500 hp steam-turbine-electric condensing locomotives which could be driven from one cab. Shown in May 1939 undergoing trials on Union Pacific Railroad.

Source: Union Pacific Railroad.

well-integrated system and the designers avoided the 'particularist's error' which arises when a series of technical solutions to a list of problems are not well integrated with each other, so that the ensemble performs badly. The engineering work of the 'Steamotive' was of a high order, and it was probably the most admirably conceived and constructed experimental steam locomotive of the 20th century. The design team did well to introduce this new type within two years (Bailey, Smith and Dickey, 1936; Woodward and Cain, 1939).

The 'Steamotive' was a twin turbine-electric unit, with vacuum condensing and regenerative feedwater heating. It was intended to have a thermal efficiency twice that of a good orthodox steam locomotive, that is about 12 per cent overall. The condenser would eliminate water stops. It was a two-unit ensemble, with multiple unit control. Each unit was a 2-C + C-2 engine of 2500 hp output from the generator, capable of independent operation at 100 mph. The 'Steamotive' was designed to work between − 40 degrees C and + 43 degrees C in the same services as the General Motors main line diesels, with the same degree of standardisation (Lee, 1976). If the key to success of the General Motors diesel was the Winton light-weight oil-engine, the key to the anticipated success of the 'Steamotive' was its quick-start watertube boiler capable of providing steam at 1500 lbs per sq. in (103.5 bar) in 15 minutes from cold (Bailey, Smith and Dickey, 1936; Rudorff, 1938). The fuel was oil, but a pulverised coal version was planned. This boiler unit was evolved from a General Electric design of 1930 for package boilers of moderate output to be used in stationary plant or river craft. Each unit of the twin ensemble was single ended, with one cab, and the boiler, turbine-generator, vacuum condenser, and radiator banks were placed behind the nose. In each unit, the high- and low-pressure turbines drove the DC generator which supplied six axle-hung DC traction motors. A 220 V three-phase alternator was driven from the main shaft to furnish power for train air conditioning and other accessories. The condenser occupied each side of the rear of the body. Fans in the roof drew air through side radiators and discharged it vertically. The locomotive worked well and it attracted attention because of its ability to raise steam from cold in 15 minutes, thus matching the much-publicised capacity of the oil-electric to start and be ready for duty in a short time. After construction in 1938, it was stationary tested at the General Electric Erie plant in Pennsylvania using the water-box rig, and was then set to hauling loaded coal wagons on the test track. Later it was tested in revenue earning passenger service on the New York Central, Chicago & North Western, Northern Pacific and Union Pacific railways (Lee, 1976). Prolonged in-service trials revealed defects Thermal efficiency was less than expected because it was difficult to maintain a high degree of vacuum in the condenser and too much steam leaked past the small radius turbines. Steam consumption was high and the range between refuelling and rewatering stops was less than planned. The 'Steamotive' was difficult and costly to maintain and proved less economic than a modern, orthodox steam engine and a diesel-electric unit. It failed to fulfil its designers' expectations and was withdrawn in 1941 when the tyres needed replacement. It was sent back to Erie and scrapped. No company showed an interest in taking the 'Steamotive' project further, and those who operated it on their trains chose conventional 2-cylinder 4-8-4 steam locomotives (Fitt, 1975) to work services after 1940 until main line diesel-electric power was introduced.

The relative merits of steam, diesel and electric traction in Europe and the USA during this period are summarised by Armand (1947), Bulleid, Ivatt, Hawksworth and Peppercorn (1947), Cambournac (1949), Hurcomb (1948) and Kiefer (1947). Two general points are worth noting. The General Electric design team may have under-rated the excellence of the orthodox steam locomotive, and its ability to outperform their 'new form'. The objections which could be directed against individual features of the practically perfected steam locomotive of 1930, and which were used to guide the design of the 'Steamotive', lost force when it was viewed as a component in a large traction system which had been rationalised. The simple steam engine, represented by the Union Pacific FEF 4-8-4 (Fitt, 1975), proved a more reliable and economic component in the larger traction system than did the 'Steamotive', and was built in its stead to meet Union Pacific needs until diesels were introduced. The 'Steamotive' came too late to displace the Electro-Motive diesel-electric. The 'Steamotive' was outshopped in December 1938 but the General Motors La Grange plant had begun regular production of the standardised E series of main line diesel-electric locomotives in 1937, and was making them available to steam railways through hire-purchase. General Motors' expertise in selling as well as building the new locomotives was a great asset. It reaped the reward of being first with a superior and successful product. General Electric were second with an unproven product and competed at a grave disadvantage. The GM Series E diesel-electric was the result of six or seven years intensive development following twenty years of experiment with oil-engined locomotives in many lands. The 'Steamotive' was a new attempt, designed as a response to the E series. It had no encouraging prehistory and to be successful it had to displace the diesel which was developing fast and which was proving successful. There was no time for lengthy research and development work on the 'Steamotive'. The first demonstration units had to win orders in reasonable numbers, right away, if a market for the turbine-electric was to be won and preserved in the face of General Motors dieselisation. This did not happen and the project was halted when major expense was required to replace worn out components and to cure the engineering defects revealed by the in-service operations. The 'Steamotive' was abandoned. It was a second-comer which failed to gain acceptance as a successful alternative before its rival captured the markets. Though the 'Steamotive' was both an engineering and a commercial failure, it was a remarkable achievement granted the circumstances in which it was launched. Its failure left the oil-electric as the unchallenged, universally recognised future mode of traction for lines which were not candidates for electrification. The ascendancy of the diesel also made the DC traction motor the most common form of motor used on railways, whether dieselised or electrified, and its normative status was unchallenged for some 45 years.

There were two post-war experiments with steam-electric traction, but these were of no significance in the evolution of traction policy (Duffy, 1986).

240 *Electric railways 1880–1990*

They resulted from the desire to develop a coal-burning locomotive for use by railways which owned mines and had access to cheap coal. The companies concerned were the Chesapeake & Ohio and the Norfolk & Western. By the time the C&O M class appeared in 1947 (Atwell and Baston, 1949; Kerr, 1947; Putz and Baston, 1947) and the N&W Engine 2300 appeared in 1954 (*Mechanical Engineering USA*, 1955), the railway engineering industry, from major builders to suppliers of auxiliaries, was dominated by the diesel and there was no place for incompatible forms of traction which needed components which the suppliers of diesel and electric locomotives did not make. General Motors, General Electric, and other makers, supplied equipment for an industry dominated by the diesel. They already supplied electric railways and there was no conflict of interest over dieselisation because the diesel-electric required standard electric traction equipment. But a steam-electric locomotive, whilst using standard electric traction equipment, needed boilers, steam fittings, turbines, and a wide range of minor components which did not come from the major suppliers of electrical and diesel parts. It is significant that the industries supporting the two post-war steam-electric experiments were Westinghouse, the Baldwin-Lima-Hamilton group, Babcock & Wilcox, and the Bailey Meter Company, none of which played a major part in dieselising the railways. The experiments on the C&O and N&W with very large non-condensing locomotives ended in failure, and dieselisation in the USA and Canada went forward unchecked, to be followed by dieselisation of railways world wide.

13.4 Conclusion

Confronted by the success of General Motors, the steam locomotive builders tried to enter the field themselves. In 1940, with the 'Steamotive' a failure, General Electric and Alco co-operated to produced a 2000 hp A1A A1A diesel locomotive, but the entry of the USA into the World War in December 1941 brought government restrictions on further development. General Motors got all contracts for road (main line) diesels, and General Electric and Alco were limited to producing 1000 hp road-switchers. This restriction greatly hindered the efforts of Baldwin, Lima, and Alco to get into the business of diesel locomotive manufacture. Later, General Electric broke its link with Alco in order to manufacture diesels independently. The American Locomotive Company (Alco) did not have the resources of either General Electric or General Motors to continue diesel development and construction on its own, and it stopped building locomotives. It survived as Alco Products, making pressure vessels, pipework, oil rig equipment and domestic equipment like washing machines. Another big manufacturer of steam locomotives, Baldwin-Lima-Hamilton started making diesels late, lacked a standard design, and suffered from the use of an inferior engine. In volume of diesel sales in the USA, the Baldwin group never held better than third place, after General Motors and Alco-General Electric. It met the same fate as Alco. In 1956, it failed in an experiment on the

Norfolk & Western Railway with a steam-turbine-electric locomotive, no. 2300, and with not a single order received for a locomotive of any kind, both Baldwin and its partner Westinghouse Electric left the diesel-electric locomotive building industry. Fairbanks Morse also started late, and was slow in producing a large road unit, though it overtook Baldwin-Lima-Hamilton to occupy third place in the American market before BLH closed down, and for a time it was joint second with Alco-General Electric. General Motors never had less than 65 per cent of all the Class 1 railway orders for diesels in the period when its rivals were in business, and its share was usually greater. Only General Electric survived as a major competitor (Morgan, 1955).

All the above-mentioned companies produced excellent designs which were used on Class 1 American railways. Alco-General Electric introduced its first high-speed passenger diesel-electric locomotive type in 1944, represented by no. 51 of the Atchison, Topeka & Santa Fe Railroad. This was followed by a 2250 hp type capable of multiple unit operation in combinations giving up to 9000 hp output, such as the DPA-4a class built for the New York Central Railroad in 1950. In 1945, Baldwin-Westinghouse introduced a non-standard 2-D + D-2 single unit 3000 hp locomotive, sometimes run in sets of two, which served on the Seaboard Air Line Railroad and the Pennsylvania Railroad. In 1949, BW introduced the RF16 1600 hp goods unit, capable of working in a set of four. Fairbanks-Morse introduced a 2000 hp passenger road diesel locomotive in 1947, capable of multiple-unit operation and the smaller companies, Montreal Locomotive Works and Canadian General Electric, brought out a two-unit 3000 hp goods locomotive in 1950 (Pinkepank, 1967, 1973; Simmons-Boardman, 1983). Yet the advantage remained with General Motors. Nothing could remove the benefit of an expertise in road diesel manufacture going back 6 to 12 years before its rivals, and General Motors reaped the reward not only in the US market but in the world market. General Electric, Alco, and the Baldwin-Lima-Hamilton companies tried to win back shares in the market from General Motors by pioneering new modes of traction which included steam-turbine-electric, oil-fired gas turbine-electric, and coal-fired gas turbine-electric locomotives, but these came to nothing. It is significant that these experimental locomotives of the 1945–60 period all used electric transmission. On Class 1 railways in the USA, the traction issue was settled during the 1941–45 war. Steam traction handled over 80 per cent of US ton-miles in 1946 but the only question worth asking after 1945 concerned finding the best methods to finance and introduce diesels. By the mid 1950s, in a country which had plentiful cheap oil, the diesel was replacing electric traction in the USA, a trend criticised by H F Brown (1961). The Pennsylvania Railroad plans for extending single-phase AC electrification were shelved, and several pioneer electric lines were dieselised including the Detroit River Tunnel, the B&O Belt-line tunnel in Baltimore, and the single-phase system over the Cascades on the Great Northern railway. Eventually, the Milwaukee HVDC systems were given over to diesel working. In 1956, the diesel electric was cheaper in ton-miles per $ than its rivals, and some electric systems were then rated as being more expensive according to this

Figure 13.5 Triple-unit 6000 hp diesel-electric locomotive built by American Locomotive Company–General Electric in 1944 for Atchison, Topeka & Sante Fe railway.

Source: Atchison, Topeka & Sante Fe Railroad (now Burlington Northern & Sante Fe).

criterion than steam traction (Majumdar, 1985). Without the diesel, American railways would have faced far greater crises than the ones they experienced throughout their post-war history.

Other railways in other countries dieselised, at different dates depending on economic circumstances and the rate of recovery from the war. The self-propelled electric power station or thermal-electric system based on the heavy-oil engine became standard in North America and over many of the world's railways (Andrews, 1986; Armstrong, 1978; Majumdar, 1985). It avoided the need for the frequent substations and the high transmission losses of the LVDC system using conductor-rail distribution. The anticipations of Durtnall and other supporters of the thermal-electric system were realised, and in this fashion the majority of the world's railways came to be powered by electric motors, most of which were of the LVDC type.

Chapter 14

The mercury-arc rectifier locomotive

14.1 Introduction

Attempts to produce rectifier locomotives and rectifier motor car units date from the inception of the industrial rectifier itself. General Electric fitted an experimental mercury-arc rectifier to a railway vehicle in New York state in 1908, and Westinghouse tested a locomotive mercury-arc rectifier on the New York, New Haven & Hartford Railroad in 1913, but none of these trials indicated that immediate application was likely. Occasional experiments were made in Europe and the United States, and by the late 1930s small rectifier locomotives were at work on the German Reichsbahn. These were five A1A-A1A shunting engines working under the catenary from a 15 kV, 16.66 Hz supply, with supplementary traction batteries charged through glass-bulb mercury-arc rectifiers rated at 210 kW. One was converted to solid-state diodes in 1957, but the others retained the glass-bulb rectifiers until 1959 and 1960. In the late 1930s, a locomotive-mounted rectifier, for large powers, was expected in about ten years. In 1935, rectifier locomotives were seen as a means of combining the economy of single-phase high-tension AC supply in the contact wire with the admirable traction characteristics of the DC motor. The static rectifier (RG 1935) 'opens new possibilities in electric locomotives with motors using current in a form different from that in which it is delivered at the contact wire or rail'.

Three main classes of 'rectifier locomotive' using mercury-arc rectifiers were envisaged, converting AC to DC; AC to AC; and DC to AC. The term 'rectifier' became associated with the static or mercury-arc equipment, and 'converter' became the label for the rotary system. A contemporary review (*Railway Gazette*, 1935) refers to pioneer designs of 1925:

With single-phase supply to the contact wire and d.c. motors in the locomotive the conversion equipment required is a mercury rectifier, rectifier transformer, and smoothing choke. The first locomotives of this type, built to the designs of Dr Ing. W. Reichel (Z.V.d.I, 1925, p 52) was for three-phase feed with a variable tapping transformer as used on the a.c. locomotives of the German State Railway. These locomotives had not the advantage of grid control which was applied in the later single-phase/d.c. rectifier locomotives.

The rectifier locomotive enjoyed four advantages. Performance was independent of supply frequency. Weight was about 15 per cent less than when supplied with 50 Hz supply compared with a single phase locomotive supplied with 16.66 Hz and using commutator motors. The railway supply could be drawn from the grid at industrial frequency. Motor control was complete and loss free. The single-phase AC/DC type of locomotive was simple in construction. The rectifier was fed by the transformer and the motors were in series with a smoothing choke and received DC voltage varying between zero and the maximum. Reduction in motor weight was an advantage. The single-phase commutator motor was restricted to low frequencies in railway service, and was large and heavy. The DC motor was smaller and lighter. The overhead equipment carrying the single-phase contact wire was lighter than the equipment needed to carry an equivalent DC supply. The rectifier locomotive combined the most advantageous form of fixed works with the favoured traction motor.

The arc-rectifier had a potential role in phase-changing on board locomotives, though the few phase conversion locomotives operated in revenue-earning service used rotary converters or 'phase-splitters' rather than static rectifiers. A review of the 1930s remarked (*RG*, 1935):

> The contact wire or wires carry single-phase or polyphase a.c., as the case may be. The traction motors have no commutators; their general construction is that of synchronous motors, but the utilisation of material is less satisfactory. In addition to the mercury-vapour converter tank, transformers are required for the supply and for the converted output. The latter transformation can be effected by the stator winding of the traction motors.... For individual driving of the axles, each commutatorless motor needs a special converter, an arrangement which is at a disadvantage compared with the parallel operation of d.c. motors in the rectifier (1ph/DC) locomotive.... Phase converter locomotives can be controlled down to zero frequency (d.c.) but they are heavier than rectifier locomotives and ordinary single-phase 16.66 cycle locomotives.

Designs were produced for locomotives taking DC supply from the contact wire or rail which was then inverted using a static rectifier bridge to provide the traction motors with single phase or polyphase AC. The designs generally recommended induction motors. The equipment proposed included an inverted mercury-arc rectifier bridge, transformer, oscillating condenser, and smoothing choke. Frequency variation of the AC was wide, but not down to zero, and voltage was only controllable down to a definite limit (*RG*, 1935): 'In general, the controllability is inferior to that of a rectifier locomotive and the weight is higher owing to the heavy battery of condensers required.'

By 1935, the reliable performance of the grid-controlled mercury-arc rectifier enabled regenerative braking to return power to the catenary. In addition to changes in the grid control circuit, the polarity of the motor circuit needed to be reversed with the direction of current flow determined by the rectifier. The systems first proposed used were for the AC to DC rectifier locomotives. They had an auxiliary rectifier for separate excitation of the motors which operated as generators during braking. The main rectifier operated as a mains-excited inverter, converting the generated DC to single-phase AC which was fed back into the contact wire. The cost of this special equipment was considered justified where substantial energy recovery resulted from regenerative braking, for example on mountain railways. Even then, 'mechanical braking must ... be provided as a stand-by, because regenerative braking is impossible if the a.c. supply voltage fails for any reason'.

One problem experienced with the pioneer mercury arc rectifier locomotives on the 1930s was ripples in the DC output which needed smoothing before being supplied to the DC motors. The term 'waviness' denoted the ratio of the effective value of the superimposed AC voltage to the mean DC voltage obtained with natural commutation. Smoothing chokes were sometimes needed in locomotives with DC series wound motors (*RG*, 1935):

The inductance of the field winding renders a smoothing choke unnecessary if a six-phase rectifier is used, but in the present case of single-phase supply, the waviness of the d.c. voltage is much greater than with a six-phase rectifier, say 47 per cent compared with 4 per cent. . . . the use of a smoothing choke cannot, however, be eliminated when working from a 50-cycle, single phase system, and the magnitude of the choke depends on the waviness permissible from the standpoint of commutation; 10 per cent waviness has no detrimental effects in ordinary series motors, and from 30 to 40 per cent is permissible in specially built machines.

In the late 1930s, the German railways began a series of exploratory experiments to develop locomotives able to run off single-phase 50 Hz supply in the contact wire. These trials of alternative designs of locomotive were as influential as Heilmann's comparisons of different electrical systems in France in the 1890s, and the 'Electric Rainhill' tests conducted on the Midi Railway in France between 1902 and 1908. The German trials were on the Hollenthal line in the Black Forest area of South West Germany where a difficult route with very steep gradients and sharp bends was electrified with AC at industrial frequency in the contact wire. In 1945, this railway lay in the French Zone of Occupied Germany and experience with different types of locomotive persuaded the French to develop the AC railway with high-tension single-phase supply at industrial frequency in the contact wire. This system, with 25 kV, 50 Hz supply, replaced the HVDC standard which had been derived from US practice before the Great War. The electrified line down the Hollenthal was 55 km long, with gradients up to 1 in 18, and curves of 240 m radius. A single substation took current from the grid at 100 kV and transformed it to supply the contact wire with single phase current at 20 kV, 50 Hz.

The four types of locomotive tested were of great interest. All came into operation in 1936. They were of the same general configuration, size and weight, and were carried on two, two-axled bogies with all axles powered. Three were rated at 2000 kW for one hour; the fourth locomotive being rated at 2400 kW. They all took power from the single-phase AC supply at 20 kV, 50 Hz, but each had different conversion, control and traction equipment from the others.

Locomotive E244.21 was a Siemens single-phase AC locomotive, with 8 single-phase commutator motors, mounted in pairs, two per axle, with the motors on any one axle permanently connected in series. It was rated at 2000 kW for one hour. Each motor had 14 poles, and was of the nose suspended type. They were rated at 255 kW (340 hp) at 255 V. Control was by cam-operated switchgear in 14 steps. The motors were heavy: 22 tonnes for the eight, and this severely disadvantaged the type despite good performance. Motor performance was inferior to the normal 16.66 Hz AC commutator motor, and the advantage of operating with industrial frequency supply was offset by weight.

Locomotive E244.0l was built by AEG, and was rated at 2000 kW for one hour. The single-phase supply was rectified in a mercury-arc vessel, regulated by 18 grids which increased or decreased the duration of passage of anode current. To prevent this leading to excessive phase discrepancy and poor power factor, it was necessary to use in addition voltage regulation in the transformer; series-parallel and parallel working of the motors, and field weakening in the motors. This led to a complex control apparatus, and the rectifier regulation grids 'distorted' or interfered with the current supply in the catenary. There were only six running speeds. The problems associated with engine E244.01 required modern solid-state devices for solution.

Locomotive E244.31 was a 2000 kW converter locomotive built by Krupp. It converted single-phase supply to three-phase using the Garbe-Lahmeyer principle, similar to that of Schoen, or Leblanc, in which the single-phase current is converted to three-phase in four traction motors with two concentric rotors, with the induced current conveyed to four separate three-phase motors. Haut reports that the assembly worked to give a shunt-wound characteristic, providing three effective speeds. There were four motor groups of eight motors in all; an oil transformer; cam-operated controllers for all three speeds, and eight water resistances. The weight of the conversion gear and the motors is given as 20.4 tonnes by Tassin, Nouvion and Woimant, with each of the four converters weighing 3.4 tonnes, and each of the four three-phase motors weighing 1.7 tonnes. Though the power factor was unity, the conversion and control gear was too complex, bulky and heavy, and the limited number of running speeds set the Krupp locomotive at a considerable disadvantage.

The fourth locomotive, E244.11 was built by Brown-Boveri, and was rated at 2400 kW for one hour. This was the most successful of the four, and it influenced subsequent practice. It combined the mercury-arc rectifier to convert single-phase supply, after transformation, into direct current, and it controlled speed by means of regulation on the high tension side of the transformer in 28 steps. Output from the rectifier was described as 'wavy', but a smoothing choke reduced the undulations for current supply to the motors. Grid control was not used. The 12-anode rectifier was water cooled, and carried a vacuum pump to maintain low pressure. The four DC traction motors were connected in series in two pairs. Total motor weight was 14.4 tonnes. (Note: There is a difference of opinion between Haut, and Tassin, Nouvion, and Woimant concerning the numbering of two of the above machines. Haut gives the number of the Siemens AC locomotive, with AC motors as E244.11. This text follows Tassin, Nouvion and Woimant and gives it as E244.21. Haut numbers the Brown-Boveri mercury-arc rectifier locomotive as E244.21. This text follows Tassin, Nouvion and Woimant and gives the number as E244.11.)

The Hollenthal line was later converted to a frequency of 16.66 Hz to conform to the German standard. World War Two and its aftermath retarded railway engineering progress in most of Europe apart from neutral Switzerland, and delayed German electrification. The United States, with its immense industrial potential was less effected, and work went ahead during the war so that the USA took the lead in the development of the rectifier locomotive, and produced the first type to operate in regular service. According to Tassin, Nouvion and Woimant, there was an experiment in the USSR in the late 1930s with a rectifier locomotive using a reconstructed VL22 electric locomotive originally built for 3000 kV DC, but fitted with a polyanode mercury-arc rectifier. This trial, possibly following on from the German work of the late 1930s, was halted by the war. During this period, in several countries, there were mobile substations, and power stations, used to bring power to remote areas, or to supplement inadequate local equipment, or to meet military demand. These included mercury arc rectifiers but were not self-propelled and were moved about by a separate locomotive.

14.2 Westinghouse, General Electric and the Pennsylvania Railroad rectifier locomotives

The Pennsylvania Railroad was progressive in civil, mechanical and electrical engineering on railways. In 1910, it brought the LVDC railway to a high level of performance on the railways between Pennsylvania Station and Manhattan Transfer, operated by DD1 electric locomotives. Between 1928 and 1938 it pioneered main line AC electrification in the USA using single-phase AC at 11 kV, 25 Hz, worked by the renowned P5a 2C2 locomotives. In 1935, the PRR introduced the GG1 2CC2 type which was one of the first electric locomotives to outperform the modern steam locomotive in both power and speed. The PRR relied on the single-phase commutator motor, but after World War Two, changes to traction policy were suggested by the major manufacturers. The advent of diesel-electric power made the DC motor the common type and economies would be effected if it were used in Pennsylvania electric locomotives as well as in the diesel fleet. The advantages of the DC motor were admitted, but there could be no question of replacing the single-phase AC supply system to the contact wire at a time when it was fairly new. General Electric suggested motor-generator locomotives for combining the AC supply with the DC motor. GE had supplied these to the Great Northern and the Virginian railways where they worked well. These locomotives were very large and heavy and were incompatible with the PRR post-war policy of operating services with locomotives which ran on four-wheeled bogies, with all axles motored, running them in multiple if need be. American high-powered motor-generator locomotives were of a different form and needed unmotored carrying axles.

In 1950, the PRR needed new electric motive power for heavy duty goods service. General Electric planned a two unit locomotive. Each unit was of the B-B form with one cab. Despite General Electric's expertise with the ignitron and other rectifiers, this goods locomotive, of type E2b had single-phase commutator motors in keeping with traditional PRR AC policy. General Electric wished to employ DC motors of the type used in its diesel locomotives but there was no way of using them because the PRR would not use motor-generator locomotives, and the locomotive rectifier was still unproven in the USA. The E2b of 1951 used single-phase commutator motors similar to those used on the P5a. The two unit ensemble weighed 245 tonnes and developed 3730 kW (5000 hp) continuously.

The Westinghouse Company decided to introduce an equivalent rectifier locomotive and apply the research of their engineer Lloyd J. Hibbard. Hibbard was a great pioneer of the mercury-arc rectifier in railway traction. In 1913 he worked on the Westinghouse experiment to fit a Hewittic rectifier to an electric motor car on the New Haven Railroad. The development of the ignitron rectifier and its success in the metallurgical industry and in railway substations encouraged Westinghouse to fit ignitrons into Pennsylvania Railroad multiple unit motor car no. 4561 in 1949. The vehicle was a standard MP54 type passenger-baggage combination car used in heavy-duty rapid-transit and secondary lines work. It was converted in the PRR Wilmington workshops. The supply from the overhead contact wire (11 kV, 25 Hz) was rectified to LVDC for the traction motors. Engineers from the Westinghouse company, Bell Telephones, and the PRR observed the performance of car 4561 to ensure that interference between rectifer and nearby telephone and signal circuits was not destructive. Tests were successful. There was no undue interference, and the rectifier worked well in motion. There was no excessive mercury pool surge, and vessel vacuum was maintained. The car ran over 2500 miles on trial and then entered regular revenue earning service. The success of car 4561 led Westinghouse to offer the Pennsylvania Railroad a rectifier locomotive with all weight on motored axles able to work heavy goods trains. As Bezilla relates, Westinghouse anticipated that the rectifier locomotive would provide 50 per cent more tractive effort below 17 mph than an AC locomotive with commutator motors. The AC machine would enjoy a slight advantage over 40 mph. Rectifer locomotives had fewer moving parts than diesel electric locomotives or AC locomotives, and the mercury-arc rectifier and its auxiliaries occupied less space than AC control apparatus. They used standard DC motors and ignitrons could be fitted to take current at industrial frequency from the overhead contact wire if the PRR ever decided to work its system at the USA standard frequency of 60 Hz. Rectifier locomotives could be modified from 25 Hz to 60 Hz relatively cheaply, whereas conventional AC locomotives would need extensive reconstruction.

In 1951, Westinghouse and their mechanical engineering collaborator, Baldwin Locomotive Company, constructed an ignitron-fitted locomotive made up of two units, each carried on two six-wheeled bogies. It was classified as E2c. There was a motor on each axle. Each unit had twelve ignitron tubes, two per motor. It was much bigger and more powerful than the General Electric E2b AC machine delivered the same year. It weighed 362 tons, and provided 6000 hp continuously. A second two unit ensemble, the E3b appeared in 1952, and was also fitted with ignitron rectifiers. Each unit rode on three two axle bogies. This type was known as E3b. The DC motors of the E2c and E3b were the same type as those fitted to many of the Baldwin and Fairbanks-Morse diesel-electric locomotives being tried on American railways. At first the rectifier locomotives performed very well. It was claimed that the haulage capacity of one two-unit rectifier locomotive equalled that of three of the P5a or GG1 single-phase AC commutator motor locomotives. This is not a just comparison. The E2c was a goods engine intended to pull trains of 12,000 tons at slow speed. The GG1 could haul 4280 tons in goods service, but it was primarily a high-speed passenger engine, intended to move 1000 ton de-luxe expresses at 100 mph. Likewise the P5a, which did handle goods trains, also worked fast passenger trains, for which the E2b, E2c and E3b were never intended. On goods work the rectifier locomotives did better than the P5a and the GG1. The E2b General Electric AC locomotive hauled less than the rectifier machines which were 117 tons heavier and more powerful by 1000 hp. It pulled 7800 tons compared to 12,000 tons for the E2c and 13,348 tons for the E3b but proved faster. After a promising start to the trials, both the E2c and the E3b began to fail. Breakdowns were so frequent that both types of rectifier locomotive were withdrawn in mid-1954 and stored in Wilmington. The failures and high maintenance costs were blamed on the poor disposition of equipment, because the design allowed for conversion to AC drive if the rectifier experiment came to nought. However, the General Electric non-rectifier locomotive, with AC commutator motors, also failed through over-rating of the motors, and the Chief Mechanical Officer of the PRR, L E Gingerich said that no similar machines would be purchased.

Bezilla blames the failures on the Pennsylvania Railroad company's policy of deferred maintenance in times of financial difficulty. This undermined the quality of engineering and lowered the standards of maintenance for which the railway had once been famous. General Electric specifically criticised the PRR standards for inspecting and repairing electric rolling stock. The PRR moved towards a financial crisis, which eventually led to amalgamation with the New York Central. It sought to reduce operating costs, and rejected the ignitrons as too expensive.

Other railways in the USA experimented with ignitron-fitted rolling stock. In 1952, the New Haven Railroad placed an order for 100 rectifier multiple-unit cars; and in 1954 ordered 10 rectifier locomotives of 4000 hp for passenger train duties. The New Haven and Hartford multiple-unit cars used Westinghouse

Figure 14.1 General Electric ignitron tube for locomotives as used on the Pennsylvania Railroad Class E44 introduced in 1960. Locomotive ignitrons were usually of the order of 12 inches in diameter.
Source: General Electric (USA).

ignitron rectifiers, and were built in 1954 by Pullman-Standard in its plant at Worcester, Ma. The order for 100 vehicles was made up of 89 coaches, seven baggage-coach combination cars, and four club cars. Current collection was via a pantograph from the 11 kV, 25 Hz single-phase AC supply. The order for locomotives did not go to Westinghouse, and the company withdrew from the manufacture of heavy duty traction equipment as it saw no worthwhile future market in the USA for electric locomotives. Unlike GE, it did not make diesel-electric locomotives. The Westinghouse withdrawal left the initiative with General Electric which believed the ignitron needed more development before it could become a reliable motive power component. General Electric enjoyed a monopoly in the construction of electric locomotives in the USA after Westinghouse left the market and in 1953 a committee was set up to develop the rectifier for locomotives. In 1955 the rectifier had been sufficiently improved for General Electric to install them in 10 express passenger locomotives delivered to the New Haven and Hartford railway. These were the 4000hp engines mentioned above, which Westinghouse had hoped to construct. These were the first production rectifier locomotives to operate in the USA. These double-ended locomotives, classified EP5, gave long service.

In 1956 and 1957, GE built twelve 3300 hp C-C units with ignitron rectifiers. They were built to the road-switcher configuration which was to become standard in the USA, having only one cab positioned towards one end of the locomotive and a long, narrow hood covering the main equipment. These were followed by the more powerful E44 rectifier locomotives built by General Electric between 1960 and 1963 for the Pennsylvania Railroad. Sixty-six of these C-C locomotives were built from 1959 onwards and constituted the largest order for electric locomotives placed in the USA since the PRR ordered the GG1 class in the 1930s. Because of the Pennsylvania Railroad's desperate financial position, General Electric leased the locomotives to the railway for 15 years, after which the railway became the owner. The 36 locomotives delivered before July 1962 had the ignitron rectifer, which General Electric had successfully fitted to locomotives supplied to the Virginian; and New Haven railways, but after July 1962, the E44 locomotives were built with the new solid-state silicon-diode rectifier. This was a significant step, and the E44 machine delivered in July 1962 was one of the first locomotives in the world to carry a solid-state rectifier. It is sometimes termed the very first. It was not the first railway vehicle to be so equipped, because tests with the silicon-diode rectifier had been conducted in 1961, on the PRR using a class MP85 multiple-unit car converted in the company workshops in Paoli. Success encouraged General Electric to fit them into the later batch of E44 locomotives. This silicon-diode rectifier was more efficient, more reliable and simpler than the ignitron as it did not need the cooling system and firing circuits of the mercury-arc type rectifier.

Figure 14.2 Virginian Railroad 3300 hp rectifier locomotive built by General Electric in 1956–57 to operate from single-phase AC supply at 11 kV 25 Hz. It had twelve 12-inch diameter ignitron tubes.

Source: Virginian Railroad (now Norfolk Southern Corp.).

14.3 The advent of the solid-state rectifier

Experience with the silicon rectifier E44 locomotives resulted in a complete and rapid move away from the mercury arc rectifier for locomotives, in all forms and derivatives. It was claimed that the failure rate with the early silicon rectifiers was one-sixth that with an ignitron device. The silicon rectifier became the ideal locomotive mounted rectifier for converting AC to DC for traction motors. Eventually, the diesel-electric locomotive builders, including General Electric and General Motors, began testing diesel locomotives with AC alternators, rather than DC dynamos, with rectification to DC via silicon diode rectifiers. The alternator was more compact and reliable than the DC dynamo but the DC traction motor was still wanted in traction. The adaption of the three-phase motor with current-inverter control came after 1980 and required rectification which solid-state devices provided reliably and effectively. In the USA the solid state rectifier became the norm for electric locomotives delivered after the E44 class. All electric locomotives were then being produced by GE, apart from two experimental GM models. Typical is the E50 type, a 204 ton engine, of the CC road-switcher form, built in 1968 for the Muskigum Electric Railroad of the American Electric Power Company which carries coal for power stations. In this machine, an oil-filled main transformer supplied AC power to a group of thyristor controlled silicon diode rectifiers, supplying variable voltage DC through a smoothing reactor (choke) to six GE DC traction motors identical to those used in GE diesel electric locomotives. Continuous rating was 5000 hp. A development is the E60C thyristor-controlled silicon-rectifier locomotive, developed for the single-phase AC, 50 kV, 60 Hz line built to move coal in Arizona on the Black Mesa and Lake Powell Railroad. These engines are rated at 5600 hp for one hour.

14.4 The mercury rectifier in France and Britain

France decided on a standard of 1500 V DC for railway electrification, with permitted development of 3000 V DC, and tolerance of LVDC already laid down. Britain did the same. Both were influenced by the success of the Chicago, Milwaukee & St Paul electrification using HVDC in the Great War period. Experience with the German Hollenthal experiments after 1945 caused French engineers to re-examine the issue of standards for future main line electrification. If HVDC was selected there would be no difficulty in providing locomotives because excellent DC machines were in service on the French railways. If the choice was single-phase HVAC, at industrial frequency, which was being advocated for reasons of economy, then research would be needed to produce an AC locomotive. Various systems would have to be compared to identify suitable forms of motor, and conversion equipment.

The first French experiment with a rectifier locomotive was inspired by the Westinghouse-Pennsylvania Railroad tests with car 4561 in 1949. It took place

256 *Electric railways 1880–1990*

in 1951 and used an old electric motor-car of the kind found in suburban service in and out of St-Lazare on the 600 V DC system. This was fitted with four American ignitrons for service under HVAC single phase supply at industrial frequency, and was then displayed at the railway congress in Annecy. In 1951, the Alsthom company constructed the first SNCF locomotive with mercury-vapour rectifiers, of the polyanode type commonly used to provide 1500 V DC in substations. This locomotive, no. 8051, unfortunately was not able to run under both the 1500 V DC systems and the single phase HVAC supply at industrial frequency. The Congress in Annecy led to the first French system in which an ignitron locomotive worked under the 25 kV, 50 Hz supply. This was in the Collieries of Lorraine where a double-bogie, four axled industrial shunter was fitted with ignitrons. The electrification of the Valenciennes-Thionville section in 1949 afforded an opportunity for trial of single-phase HVAC, at industrial frequency and a comparison of locomotive types. The military withdrew their objections to electric railways, based on the supposition that they would be easily disrupted in war and the electrification went ahead using single-phase AC at 25 kV, 50 Hz – the future international standard. This line, and its extension from Valenciennes to Charleville, did much to establish the 25 kV, 50 Hz AC system. Amongst the types of locomotive tested in 1954 was an AC machine using AC commutator motors of 2000 kW; an AC to DC rotary converter locomotive of 1850 kW; an AC to three-phase AC rotary converter locomotive of 2650 kW; and a rectifier locomotive using ignitrons of 2650 kW. The ignitron locomotive proved superior, and French practice during the era of DC traction motors relied on the ignitron rectifier to convert single-phase 25 kV, 50 Hz to DC on board locomotives, until the thyristor controlled silicon-diode rectifier displaced the ignitron after 1963. The success of engine Bo-Bo 12001 during the comparative trials of 1954 established the ignitron in locomotives as the best form of rectifier then available, but the mercury arc rectifier served rather to introduce locomotive rectifiers which found their practical form in the solid-state silicon-diode. As described in a later chapter, some types of arc rectifiers proved very prone to breakdown and led to the withdrawal of entire classes of locomotive. The experience of British Railways on the West Coast Main Line is one example. The silicon rectifier ended these problems with locomotive-mounted rectifiers. Other solid-state rectifiers were tried, such as the germanium rectifier, used in Britain. Experiments with germanium rectifiers were carried out in converted electric sets on the Heysham branch, during investigations into AC systems between 1955 and 1956, when British Railways was considering changing its electrification policy in the light of French experience, but in the long term the silicon-diode technology became the norm.

14.5 The static rectifier and post-war modernisation

In the early 1950s, the supply systems of European electric railways required modernisation and repair following the damage and deferred maintenance of wartime. The International Railway Congress, held in London in 1954, reported several programmes for improving supply by replacing rotary sets with the latest type of mercury-arc rectifier. At this time, HVDC was the standard system for future development (1500 V and 3000 V), to co-exist with existing LVDC networks, and the static rectifier would serve such works. For example, BTH supplied steel-tank rectifiers for the Euston-Watford route improvements in the late 1940s over the former London & North Western Railway (LMS). The first example of main line electrification in Britain, the Manchester, Sheffield and Wath scheme over the former Great Central Railway (LNER) main line was completed in 1954 after wartime delay, and HVDC (1500 V) was installed between Liverpool Street (London) and Shenfield over the former Great Eastern Railway (LNER). One of the largest schemes was for the Southern Region of British Railways where the 660 V DC third-rail system formed the largest suburban electrification in the world. Within this area, supply was originally at 11,000 V, three-phase, 25 Hz, which was distributed to 49 manually operated rotary convertor substations for conversion to 660 V (originally 600 V) at the conductor rail. The report in *Railway Gazette* announced that :

The whole of the 25-cycle power distribution scheme is being replaced (at a cost of over £11 millions) by a new 50 cycles system, using 71 substations, 70 track paralleling hut equipments, seven switching stations for 33 kV circuits and three additional points of connection to existing 33 kV 50 cycle-fed sub-stations; the whole is remotely controlled from three new control rooms.

This improvements scheme employed the latest type of post-war mercury-arc rectifier, as did the contemporary modernisation plan for the London Transport system, described in a report to the 1954 Congress. Repairing wartime damage, upgrading power stations at Lots Road and Greenwich, and converting supply systems to industrial frequency (50 Hz) were part of the London Transport plan, which included reconstruction of substations. Several innovations were pioneered by LT engineers in the matter of transformers and rectifiers. At Surrey Docks, London Transport engineers installed a compact, totally sealed air-cooled transformer, the first of its kind in the world, and refitting the Bond Street substation marked 'an even greater advance in the compactness and portability of the plant'. The four 900 kW rotary converters, with DC circuit breakers mounted on vertical panels, were replaced by mercury-arc rectifiers totalling 4000 kW capacity with truck-type high-speed circuit breakers, and totally-enclosed air-cooled transformers. No major structural alternations to the building were needed. The Bond Street substation was unmanned, and remote controlled, bringing the total unattended substations to 44 on the London Transport system at that date, compared to one in 1924. London Transport was the first company to use the pumpless, steel, air-cooled mercury-arc rectifier in its substations, and this type was common practice on British railways by the mid-1950s. Postwar reconstruction abroad helped British exports of electrical equipment. In 1949, Metro-Vick. sent equipment for a 2800 kW mercury-arc substation to Poland. In the late 1940s,

English Electric sent steel-tank, pumpless rectifiers to Brazil for the substations of the 3000 V DC railway between Mooca, Sao Paulo and Jundiadi. In the mid-1950s, English Electric supplied to the New South Wales railways, Australia, fifteen 4000 kW, 1500 V rectifier equipments of the air cooled, pumpless type, rated at 650 V, 1500 V, 3000 V DC. By 1955, the single-anode static rectifier, assembled in sets of 6 to 12 units each, was available for outputs of 25,000 V if necessary, though in railway service the normal maximum was 3000 V. Assemblies were available for delivering over 100,000 kW if needed, with capacity related to the number of phases, but by 1965 there was the prospect of replacing them, in the not too distant future, with solid-state devices and their normative role was ended.

It is interesting that the introduction of solid-state rectifiers to British Railways involved experiments on the former Midland Railway (LMS) Lancaster-Morecambe-Heysham electrified section of 1908, which was converted from 6.6 kV, 25 Hz to 6.6 kV, 50 Hz by November 1952 to investigate the advantages of single-phase AC industrial frequency supply in railway traction. The success of the tests led to the replacement of the HVDC standard by the new international standard of 25 kV, 50 Hz. Part of the tests were carried out with carriage-mounted mercury-arc rectifiers, in anticipation of using them on the locomotives for the West Coast Main Line electrification. In 1953, on the Lancaster-Morecambe-Heysham line, British Thomson Houston successfully tested the world's first semiconductor rectifiers, made of germanium, in railway service. As early as 1960, the British Railways Class AL5 locomotives were built with germanium rectifiers and within ten years the solid-state rectifier, using silicon, had largely replaced the mercury-vapour unit in locomotives.

14.6 The first British mercury-arc rectifier locomotive

The first British mercury-arc locomotive was a particularly interesting machine, rebuilt from the former Great Western Railway gas-turbine-electric engine no. 18100, which had been built by Metropolitan Vickers at Trafford Park in 1952 as the second Great Western gas-turbine locomotive. In late 1955 British Railways decided to standardise main line electrification using single-phase 25 kV, 50 Hz rather than the previously recommended HVDC systems. The French electrifications using single-phase AC 25 kV, 50 Hz, particularly the one between Aix-les-Bains and La Roche-sur-Foron, opened in 1950, and demonstrated to visitors including Riddles and Warder from BR, helped bring about this change in policy. Following the Lille Conference of 1955, S B Warder submitted a report comparing HVDC and AC at 25 kV, 50 Hz which resulted in the decision to choose AC for future main line projects at a time when there was a lack of AC locomotive expertise in Britain. Considering the lack of AC expertise, the challenge of AC electrification was well met. The first step was to produce a pioneer main line locomotive to foster design and operating skills, and to serve for crew training. This was done by converting the former gas-turbine unit 18100, which was officially withdrawn in January 1958, and rebuilt in the Stockton-on-Tees works of Metropolitan Vickers and Beyer Peacock, to emerge in October 1958 as Engine E1000, later E2001. The 130 ton gas-turbine-electric unit had been rebuilt as a 105 ton electric unit, with the A1A-A1A wheel arrangement. It was the first electric locomotive in Britain to be fitted with a mercury-arc rectifier.

Four out of the original six MV type 271 DC traction motors were retained. These separately ventilated four pole machines were axle-hung and nose-suspended . . . They took a maximum current of 1100A, and voltage 825. New equipment necessary to its conversion included Stone-Faiveley pantographs, a Brown-Boveri 22.5kV 360A air-blast circuit breaker, and main mercury-arc rectifiers. . . . The Hackbridge & Hewittic mercury arc rectifiers consisted of 16 four-anode glass bulbs connected in bi-phase to provide the supply to the traction motors, the normal load being distributed between the bulbs by anode compensators . . . Continuous rating of the rectifier was 2800A at 975V; each bulb was fan cooled.

The rectifiers were the glass-bulb type. Cooper (1985) remarks:

(These were) of a type widely used in traction substations. Like the steel tank rectifiers which had been developed from industrial designs for the prototype locomotives (for the WCML) they did not need pumps for maintaining the vacuum. The choice of a rectifier primarily intended for stationary duties was dictated by availability and the fact that E2001 was a stopgap until the first AL1 (now Class 81) locomotives were delivered.

The rectifiers were inside a compartment which could be electrically heated in cold weather, though the bulbs were cooled by fans. The auxiliary equipment included an underframe mounted 65 kW transformer with a bridge-connected germanium rectifier for feeding 110 V DC to auxiliary motors, lighting, batteries, control circuits, etc.

This machine symbolised all that was progressive in British railway engineering, and it heralded the establishment of a new kind of railway in post-war Britain. It was used to generate new skills throughout workforce and staff as the single-phase AC railway was introduced. In autumn 1958, it went to the first electrified section of the Manchester to Crewe route where it was used in crew training. Later it went to Liverpool, Crewe and Glasgow. It was withdrawn in 1968, and was sold for scrap in 1972. The classes of locomotive produced to work the West Coast Main Line electric services between Euston, Birmingham, Liverpool, Manchester and Glasgow consisted of rectifier locomotives. The AL1 type of 1959 used multi-anode, air-cooled mercury-arc rectifiers, as did the AL2 of 1960. The AL3 class of 1960 were fitted with liquid-cooled ignitrons, though in 1962, one of the class – Engine E3100 – was equipped with silicon-diode rectifiers. This makes it one of the first in the world to carry solid-state rectification equipment, as the PRR E44 was so fitted in July 1962 to become what Bezilla calls the world's first locomotive to use a silicon-diode rectifier. These early British mercury-arc rectifiers gave trouble in service. So many difficulties were experienced with those on the AL3 class, that there were plans to replace them with roof-mounted, natural-air cooled silicon rectifiers once the solid-state devices were proven in the early 1960s, though the builder, English

Electric, raised objections. The same solution was proposed for the General Electric AL4 class. It was even suggested that these locomotives be scrapped. They were not scrapped, but in 1967–1968, the AL3 class were stored, until fitted with silicon-diode rectifiers in 1971. The AL4 class had originally carried a GEC (GB) single-anode mercury-arc rectifier which was intended to incorporate the best features of the ignitron and the excitron, and to be more resistant to shocks and vibrations. However trouble was experienced to a degree sufficient to result in these machines being stored, in 1967, and like the AL3 type, they received solid-state rectifiers in 1971 and 1972. Unfortunately, repeated traction motor failure led to the entire class of ten units being withdrawn in 1977 and 1978. The AL5 class, dating from 1960, used germanium rectifiers from the start (clearly a use of solid state devices in advance of the PRR E44) but eventually all received silicon-diode rectifiers which generally displaced the germanium type. One can certainly endorse the remark by Haut that 'The number of mercury-arc rectifiers installed in locomotives remained very limited and experimental.' The semi-conductor rectifier was the real solution to the locomotive mounted rectifier, and it remains so to this day. In stationary use, for example in lineside substations, the mercury-arc rectifier enjoyed a successful and economic usage from the 1920s until the 1960s, after which they were generally replaced by solid-state devices when it became economic to do so, though like the older motor-generators and rotary converters, they could survive long after they were technically obsolete, and some are still at work in places where economics does not permit refurbishment.

Chapter 15

Railway electrification in Britain 1920–60

15.1 Introduction

Several schemes for railway electrification were proposed between 1920 and 1956 when the modernisation plan for British Railways was published. There were four reports on railway electrification submitted to the government between 1920 and 1951. Each one built on earlier work and each recommended that national railway electrification should use the HVDC system for main line work, and retain the LVDC 'London Standard' system for extensions to existing rapid-transit networks. The 1951 British Transport Committee report reviewed developments since 1920 and was compiled by a Railway Executive Committee which included a representative of Merz & McLellan (Lyndall). It used the findings of a team chaired by Lord Hurcomb which had examined traction policy in the late 1940s (Hurcomb, 1948; Johnson and Long, 1981). The 1951 report referred to the three major inter-war reports of Sir Arthur Kennedy (1920), Sir John Pringle (1927) and Lord Weir (1931). The 1951 report reflected the times in which it was compiled. In the early 1950s British industry was slowly recovering after the Second World War, material was in short supply, stringent financial restrictions were in force, and the economy was dominated by rearmament. This discouraged innovations in industries which were not directly involved in the defence of Britain. Priority for materials and funds was assigned to atomic power and weaponry, rockets and jet engines, radar, and new equipment for the armed forces. The new technologies, including electro-mechanical and electronic computers and nuclear engineering, attracted the most creative engineering talent. The railway engineering industry had to 'make do' with reliable, orthodox rolling stock until state funds became available for modernisation. This meant continuing with steam traction as the dominant traction mode.

Lord Hurcomb understood the economic situation which constrained traction planners and he expressed disquiet about a well-nigh total neglect of modern modes of motive power and an implicit indefinite commitment to steam traction by the British Railways mechanical engineering staff. Lord Hurcomb was worried by the lack of a programme to gain expertise in diesel traction and by the failure to build on the pioneer work of the pre-nationalisation companies which had begun to investigate oil-electric and electric traction (Johnson and Long, 1981). The 1951 report agreed with the major inter-war reports. It recommended 1500 V DC with overhead contact wire for general use, with the exception of extensions to the 750 V DC third-rail Southern Region system and the fourth-rail London Transport network, both of which were too large and successful to convert at that date. The key factor in determining whether or not a line was electrified was identified as traffic density, with the threshold value being 3 to 4 million trailing ton-miles per single track per annum.

In 1951, the British electric railway industry favoured direct current systems. This attitude reflected its experience with the main line railways abroad which British firms had equipped and with the LVDC lines round London, Manchester, Liverpool and Newcastle (Andrews, 1951). In 1951, the only British mainline electrification system under construction was the ex-LNER Manchester (London Road) to Sheffield (Victoria) route of the former Great Central Railway over Woodhead. This was a mid-1930s plan for a 1500 V DC railway which the war delayed (*Railway Gazette*, 1954). In the 1950s, there was revived interest in Central Europe for electrifying railways using the industrial frequency of 50 Hz, but there were unsolved technical problems. A low voltage 50 Hz AC traction motor was needed or reliable, compact locomotive rectifiers for DC traction motor supply. The influence of French and American practice was crucial and one must pay tribute to S. B. Warder for directing British engineers' attention to Continental developments. Warder was appointed Mechanical and Electrical Engineer to the Southern Region in October 1949, and he took over the SR diesel-electric development programme centred on engine 10201 from the charge of Bulleid and Cock, when Bulleid (former CME of the Southern) moved to Ireland after nationalisation and Cock became the Chief Electrical Engineer to the Railway Executive. Warder publicised the benefits of electrification using single-phase, 50 Hz supply in the contact wire rather than 1500 V DC. Before his strategic recommendation was accepted, British Railways favoured DC, as had the inter-war reports. The report of the committee chaired by Sir Arthur Kennedy, and presented to the Ministry of Transport by the Railway Advisory Committee on Electrification in March 1920, recommended 1500 V DC and 750 V DC, with higher voltages to be multiples of these standards: both overhead and third-rail distribution envisaged. These findings were endorsed by the report of the Railway Electrification Committee chaired by Sir John Pringle and published in November 1927, as a result of which the Ministry of Transport introduced the Railways (Standardising of Electrification) Order in 1932. After this, 1500 V DC became the official norm for all new electrification schemes on main lines. Extension to third-rail networks was to be on the London Standard of 750 V DC (Bonavia, 1987; Moody, 1961, 1963, 1979; Rayner, 1975). Lord Weir's report 'Main Line Electrification' published in 1931 accepted the technical findings of the Pringle 1927 report but touched on a wide range of other matters such as rating of goods, funding, and electricity charges.

It stimulated general discussion of all aspects of railway electrification in Britain.

15.2 The Weir Report

The Weir Report was published on Friday, 24 April 1931. The conclusions were made public by the Lord Privy Seal in the Commons a week earlier (Reader, 1968). The almost-total electrification of the British system was costed between £260 and £320 million, plus £80 million to be spent by the Central Electricity Board on new installations, plus £45 million extra to upgrade existing supply and support systems. The total was between £340 and £445 million in 1931 prices, a very large sum at a difficult time for the national economy which the private companies would not meet without direct and indirect state aid. The lowest sum the railways would have to pay to electrify most of the national system was £260 million. This would give a 7 per cent return on investment without expansion of traffic due to reduced running costs, fuel bills, repairs and wages. The report was limited in scope, and was criticised because it gave no indication of the effect of electrification on the national coal mines, the steam locomotive building industry, and railway employees. Several authorities thought that £500 million would be the minimum total sum needed for complete electrification, which was equal to the sum 'wasted on our roads since 1918'. Other critics indicated – as did the report itself – the potential role of the oil-electric locomotive as an alternative to electrification, as a supplement to electrification, and as a means for getting cheap electrification without the expense of fixed works.

The Weir Report admitted that no accurate prediction of financial risks was possible, which alarmed senior railway executives, major shareholders, and financial advisers at a time when railways faced decline in profits. Offered an immediate, cheap method of improving services by using the latest steam locomotives, and effecting economies through better management of existing resources, the senior staff of the four big companies retained steam traction and were unenthusiastic about the Weir Report. The LNER, LMS, and GWR shunned main line electrification and improved express passenger services with new types like the Gresley A1 and A3 4–6–2; the Fowler-North British 4–6–0; the Collet 60XX 4–6–0, and the Maunsell Class 850 4–6–0 (Ahrons, 1927, 1966; Nock, 1969). There was a clash between two diverse sets of interest and priorities. The electrical industry wanted the home railways to electrify to serve as a proving ground and demonstration system for winning foreign orders. The home companies refused to do this without state aid on a scale they were unlikely to get.

The Weir Committee involved Mr Kennedy, Mr Richardson, Col. Trench, and staff from the private home railways and Merz & McLellan. The technical recommendations were in agreement with the earlier Kennedy and Pringle reports, namely that lines were to be electrified with 1500 V DC or 750 V DC with the Central Electricity Board providing the generating stations, transmission lines, and substations. The prospects of getting cheap electrification through the use of oil-electric locomotives were stated as uncertain because internal-combustion traction was untried, though it was recognised as potentially better than steam traction. The report stressed that electrification of Britain's railways was needed to enable British industry to compete for foreign markets and a sizable scheme was necessary for a proving ground even if lightly loaded lines were omitted from the plan as unsuitable for electric working. Some experts, including the Merz & McLellan consultants, suggested that the former Great Northern Railway between Kings Cross and Doncaster should be electrified though there was debate about including the Cambridge branch in the plan. The journal *Electrician* agreed that electrification of the Great Northern Railway would be a pilot scheme to enable all aspects of main line electrification to be assessed and for developing methods to determine financial risk and technical reliability. The *Electrician* suggested that this GN scheme should be aided by a supply of electricity at a special cheap rate of one-halfpenny per kW hr. This touched on the very controversial suggestion that the railways be supplied with cheap electricity, perhaps below cost price. This was rejected by the electricity commissioners.

Electricity supply cost was perhaps the single most important matter in the whole report. The matter of coal prices, electricity costs, and systems load factors was a complex one and the long-term costs for electricity supplied were uncertain. Coal was supplied to the railways by many individual companies but electrification would make the railways dependent on one supplier, the CEB, which might revise the charges upwards unless some agreement favourable to the railways was reached before electrification was undertaken. Understandably the electricity supply industry was unable to commit itself in advance for many reasons. Not least was the effect such an agreement would have on other customers, like the street tramways, which would resist favourable rates offered to the railways or demand them for themselves. Despite this, those who supported the Weir Report felt that electricity supply costs would need to be limited if electrification of railways was to be carried out. There were other areas of difficulty. Who would control, or own the electrified railways? Were they to be nationalised before electrification? Was nationalisation desirable to create a single company able to carry through large scale electrification and impose a standard system on what in 1931 were four private companies? If the four companies were not grouped into one, would a standard system be imposed by the state? One railway engineer objected to the report because 'there was too much evidence of a desire to help the grid and the general power user'.

Many felt that the railways must judge electrification for themselves without coercion or state intervention. National electrification might be necessary to get British industry under way as an internationally competitive supplier of electric railway equipment, but it was not generally accepted that private companies should act against their own immediate interests to effect this when the Weir Report had nothing definite to say about long-term financial consequences. The statement by the Weir Committee that no accurate prediction of financial risks was possible made the report useless as far as railway directors were concerned, and within a few months of its issue Viscount Churchill of the Great Western Railway Board declared that 'the Great Western Railway will have nothing to do with main line electrification at the present time'.

Individual companies like the Great Western Railway were free to reject the proposal and there were no means for enforcing national railway electrification by the state in the inter-war period. The Southern Railway was committed to extending the 750 V DC third-rail London Standard, and therefore the major question dominating electrification following Viscount Churchill's statement on GWR policy was 'Will the LNER, or the LMS implement aspects of the Weir Report or carry out schemes of their own?'

Both companies had considered electrification based on foreign practice but were reluctant to pursue main line projects in the 1930s. The years following the Weir Report saw prolonged disputes about its findings. The oil-electric locomotive found powerful advocates (Dunlop, 1925; Fell, 1933; Hobson, 1925; Lomonossoff, 1933). It was suggested that an internal-combustion engined locomotive running off oil derived from coal, using German techniques employed in the 1914–18 war, would provide British railways with the coal-fuelled modern traction system it required without expensive fixed works. It was widely admitted that suburban electrification 'might' pay but main line electrification was a more dubious proposition. There were campaigns for a government decision on electrification and for the state to define appropriate spheres of railway and road transport. It was argued that if the electrical industry wanted widespread main line electrification, it should work for a more equal treatment of railways with respect to roads in the matter of rates and taxation. It was admitted that electrification was possible only with state aid so that taxation, rates and electricity charges, rather than engineering matters, would decide the issue. All talk about electrification which left these matters unsettled was futile.

In 1934, when the first successful, light-weight oil-electric express trains were in operation in the USA, the advocates of electrification attacked the steam railway as a producer of filth and a contributor to the formation of fog in large cities. They made the suggestion that the steam locomotive be banished on health grounds from London within 10 to 15 years, just as it had been excluded by law from New York City. Electricity charges continued to dominate the issue of British railway electrification. The private railway companies believed that the Central Electricity Board should be able to sell direct to the railways, with the latter entitled to negotiate with a single authority. It was thought likely that a Bill to authorise the CEB to supply electricity direct to railway companies would reach the Statute book by the mid-1930s, but though the railways wanted electricity from the CEB below cost price, the Electricity Commissioners were quite emphatic: 'No sale to railways below cost'. In the mid-1930s, the advocates of main line electrification and of greatly extended rapid-transit electrification were clearly disappointed and their correspondence betrays signs of frustration. No significant progress had been made since the years immediately before the 1914–18 war and there were no extensive projects under way in 1935. Sir Philip

Dawson, in a letter to the 'Times' described the LNER electrification scheme for the ex-Great Eastern lines out of Liverpool Street station as 'a drop in the ocean' which had been recommended in the 1920s and postponed because of the successful revision of steam-worked operations. It was to be further delayed by the Second World War. Dawson remarked that there had been little improvement on British railway systems in 25 years, and he directed his criticism once more at the LNER and asked why this company could not follow the example of the admirable enterprise of the Southern Railway. The general opinion amongst the advocates of electrification was that apart from the Southern Railway, which was extending its 750V DC third-rail system over previously steam-worked main lines, other companies found excuses for not electrifying their trunk routes. The reluctance of the LNER, LMS and GWR to embark on pilot programmes of electrification puzzled some of the critics, especially as they found no 'coal interest' linking the steam-worked systems with the private collieries supplying engine fuel, unlike the situation in the United States where major railways owned their own mines.

One scheme warrants particular mention, namely the 1938 electrification proposals made public by the Great Western railway, a company which had shown no interest in implementing any part of the Weir Report (Semmens, 1985). A brief account is given by Semmens, who believes it to have been an exercise to compel the colliery owners to reduce coal prices. Semmens refers to the concern expressed by the Great Western Railway chairman, Lord Horne, over the upward trend in coal prices. Every shilling on the price of coal per ton added £100,000 per year to the fuel bill. The average price at the pithead had increased from 12s 11d (64.5p) per ton in 1934 to 17s 1.5d (86.5p) in 1937 with the best steam coals selling at much higher cost. After some opposition from the shareholders, the Great Western Railway commissioned Merz & McLellan to compile a report issued in February 1939 which considered the experimental electrification of the main line west of Taunton, which could be extended if successful. The routes considered were: the main line Taunton to Penzance; the Kingswear Branch including the line to Brixham; the Par-Newquay branch; the Cornwall mineral lines; and the Par-Fowey-Lostwithiel loop line. The scheme proposed using overhead electrification at 3000 V DC which would have required government ministerial approval, for though it was a multiple of the 750 V DC and 1500 V DC recommended by the Weir Report, it was a non-standard departure. British electrification schemes using 3000 V DC overhead systems were in use abroad by the 1930s, an example being the South African Railways system for which Metropolitan Vickers supplied traction equipment. Semmens reports that mixed-traction working would have continued with steam operated services to Exeter, Newton Abbot and Plymouth. He writes that the brief to the consultants

> ... clearly stipulated that they were to work to existing timetables, with no alterations in the average speeds of trains of all classes. . . . The proposed economies resulted from the ability to do away with assistant locomotives on passenger trains. . . . Single manning of all locomotives was assumed.

He states that the report made no allowance for any increase in revenue from a larger number of passengers attracted to potentially faster services. Four classes of locomotive were suggested including a 2550 hp 1-Co + Co-1 of 140 tons; and a lighter, lower powered locomotive of 122 tons and 2100 hp with the 1-Bo + Bo-1 wheel arrangement coming in two versions. A fourth class of four-axled engine was to work light local trains, shunt, and to bank slow goods trains when necessary. Steam-hauled goods trains would work through the electrified section and be banked by this type as a normal routine. The electrification project was planned to meet the peak loads of Summer and Christmas, which considerably increased the estimated capital cost. Even if single manning was permitted by the unions, the return on capital outlay after deducting 1.5 per cent per annum for depreciation came to 0.75 per cent when the bank rate was 2 per cent, which contrasts with the estimated 7 per cent return on the LNER scheme reviewed in the Weir Report. Semmens sees the £2000 spent in commissioning the report as part of an exercise to bring pressure on the coal suppliers to lower their prices. Prices did fall. Export steam coal in South Wales dropped in price by 1s 6d (7.5p) between January and April 1938, and by July it had stabilised at 20p below – a fall which was worth £250,000 per annum to the Great Western Railway in reduced fuel bills. If the electrification feasibility study had helped to achieve this, it was certainly worth the relatively small sum spent on it.

The proposal, like the Weir Report, accepted the necessity for mixed-traction working over the electrified sections and it recommended the electrification of lightly loaded routes not normally considered for this form of traction. Mixed-traction was not the best way to optimise the performance of an electric railway, and in the days of loose-coupled goods trains without continuous brakes, the full benefits of electric traction in freight working could not be realised due to slow speeds and the consequently low line capacity. It was considered more economic to haul such trains with steam traction rather than use electric locomotives, especially when trains worked through electrified sections and engine change would otherwise be necessary. Even where goods trains were worked from one concentration yard to another, over the Shildon to Newport on Tees line of the former North Eastern railway, a fall in traffic made electric working less economic than steam traction when renewal of the system became due in the mid-1930s and the line was given over to steam locomotives (Hoole, 1987, 1988; *RCTS*, 10b, 1990). Electrification of a large system reduced the need for mixed-traction working but was itself uneconomic if it involved electrifying rural lines where traffic loading was light. The oil-engined locomotive, the internal-combustion engined railcar, and the diesel multiple-unit train were better able to work such lines than steam traction, and they complemented electric traction on heavily loaded routes much better than the steam engine did. It is significant that main line electrification in Britain was the result of nationalisation, the introduction of diesel-engined traction, the elimination of steam traction, and the simplification of the railway system following the Beeching report (British Railways Board, 1963). The most important step in upgrading the railway engineering system was the phasing out of steam traction which by 1950 was holding up the introduction of the modern post-war railway.

15.3 The acceptance of the HVAC standard in Britain

The inter-war recommendations of reports on electrification were still in force in 1951: DC systems were to be standard. The long-established DC traction motor had excellent performance characteristics, and worked from a 750 V or 1500 V supply. It was reliable and well understood in Britain, which had little experience of AC traction equipment. The Hungarian K. Kando (1869–1931) had pioneered industrial frequency railway electrification between the wars but his system never enjoyed the widespread success of the HVDC system on the railways in France, Belgium, the Netherlands, the Soviet Union, and the USA.

Electric railways in Britain were DC in 1950. The 'London Standard' system was represented by the Southern Region of British Railways, the Midland Region electric lines out of Euston, and the Metropolitan Railway and other LVDC lines absorbed into London Transport. These gave Britain its largest electrified railway network. The longer Southern Region lines (e.g. London to Brighton) provided the nation with its only electrified main routes. In the period 1945–50 the few electric main line locomotives in Britain were DC units. There was the single prototype no. 13 built for the North Eastern railway in 1922 which was never used and was in store. There were the Metropolitan Vickers locomotives of 1920s design which worked express trains on the former Metropolitan Railway lines, and there were several locomotives built for the NER (later LNER) Newport to Shildon scheme, which were in store following the reversion of that route to steam traction. There were only two classes of British electric main line locomotives running in the 1945–50 period which could be described as modern. One was LVDC, and the other was HVDC. The 1470 hp Raworth-Bulleid CC Class of 1943 ran on a 660 V DC third-rail supply and the Manchester to Sheffield 1500 V DC Bo-Bo dated from 1941. This latter type was joined in 1954 by the 2298 hp (continuous) Co-Co express passenger class to operate the only main line railway in Britain which was electrified according to the recommendations of the Kennedy, Pringle and Weir reports (*RCTS*, 10b, 1990). By 1958, engineers like Warder had achieved a policy change, and all future electrification of main lines was to use the 25 kV, 50 Hz, single-phase standard.

Between 1945 and 1950, electrification was supported by senior railway executives like Sir Eustace Missenden and Sir Cyril Hurcomb. The outstanding executives and engineers responsible for the progressive outlook which triumphed on British Railways after 1956 included Missenden, Hurcomb, Warder and Cock. It is significant that they came from the former Southern Railway where senior management and engineers were experienced in electrification to a much greater extent than colleagues on the LNER, LMS, and GWR. Sir Eustace Missenden, the former General Manager of the Southern Railway, regarded the oil-electric locomotive as a well-suited subsidiary to electric traction on electric systems

Figure 15.1 Metropolitan Vickers 3600 hp locomotive built in 1931 for the 1500 V DC lines of Spanish National Railways (former Northern Railway). Similar locomotives had been delivered in 1930 by Brown-Boveri. British expertise lay in DC traction systems.

Source: General Electric (GB) now Alstom.

because much of their servicing and maintenance could be done in electrical depots, whereas steam traction had to be segregated. By the late 1940s, steam traction was an obstacle to the progress of British Railways towards improved technical and economic status, and Sir Cyril Hurcomb was critical of Riddles' failure to organise the exploration of alternatives to steam traction on lines which could not be electrified. Missenden claimed that:

> ... the economic and operating advantages, and the attractiveness, of electric traction are so great, and the traffic density on a considerable proportion of the system is sufficient, to warrant – or rather render essential – the early and progressive extension of electrification for main-line passengers and freight traffic, as well as for inter-urban and suburban services ... Given the necessary economic stability and reasonable financial resources, a bold electrification policy, coupled with judicious use of diesel traction, and backed by faith in our industry, can as nothing else bring about the rehabilitation of British railways.

At the Convention on Electric Traction in 1950, Sir Cyril Hurcomb remarked on the lack of progress during the inter-war years:

> ... it was possible ... that if some of the railway companies had been so courageous and imaginative as to grasp the advantages of electrification, in spite of its high capital costs, they would have been better off.

The *Times* City Column implied that Hurcomb's appeals for an electrification programme were pleas for statutory control of transport in favour of railways and requests for guaranteed cheap electricity, cheap electrical equipment and financial aid, which had been suggested in the Weir Report. These were perceived as incompatible with the proper relationship between the state, private transport, and the recently nationalised railways. It was quite wrong, the *Times* argued, even to imply legislating to shift private traffic from the road to the railways to make electrification justified and economic, whether by direct or indirect means. The complex issues of rates, electricity charges, financial aid to railways, and the role of the railways within the national transport system, continued to perplex the technical planner as they do to this day, but considerable advances were made in the 1950s. The modernisation plan of 1956 commenced the phasing out of steam traction, a great step forward. It further committed the railways to a state-funded, main line electrification plan which included electrifying the West Coast lines from Euston to Crewe, Liverpool and Manchester, with provision for an extension to Carlisle and Glasgow. Thanks to Warder and like-minded colleagues, this was effected using the single-phase AC system at 25 kV, 50 Hz.

There is no doubt that the economic success of the Southern Railway third-rail LVDC network encouraged interest in general electrification despite arguments about which system should be installed. It was stated in 1951:

> ... in its network round London the former Southern Railway developed the largest electrified suburban system of any single railway in the world, and its prosperity between the wars was in striking contrast to that of other British companies. It would be fallacious to argue directly from this success to a general justification of electrification ...

Nevertheless, the Southern experience does much to justify a faith in further electrification.... A more recent example ... was that part of the old London and North Eastern Line where electrification had resulted in passenger receipts increasing by over 40% and traffic that had gone over to road transport had been attracted back.

In 1954, the International Railway Congress Association held a session in London. This was the third IRCA session in London capital since the Association was founded in 1885. The previous two were in 1895 and 1925 (*Railway Gazette*, 1954). The IRCA reviews of global and national railway practice provide a background against which to assess the later British electrification schemes which marked a change of traction policy.

The most important British electrification scheme reported to the Congress was completed after 1945. This was a project begun by the LNER to electrify the 68 route miles of the Manchester-Sheffield-Wath line using 1500 V DC, to improve heavy coal train movements over a steeply graded line which included a long tunnel where smoke nuisance was notorious. The scheme was delayed by the 1939–45 war, and was essentially a British application of the American HVDC practice which evolved for mineral lines in hilly terrain in the period 1910–20. It was operated by 65 electric locomotives and 8 multiple-unit sets for suburban passenger traffic. This line made an estimated saving of 75,000 tons of locomotive coal per year during a national coal shortage acute enough to lead to some British steam locomotives being converted to oil-firing. Another important scheme reported at the Congress was the electrification of the Liverpool Street Station (London) to Shenfield line, completed in September 1949 on the 1500 V DC overhead system. This electrification from Liverpool Street Station ended steam traction on the suburban services from that terminus which had been given an extended life following the 'Liverpool Street Experiment' of 1920, though steam working of other services continued for years. In 1954, approval was given to extend electrification to Chelmsford, and Southend Victoria station, and to electrify the London, Tilbury and Southend lines from Fenchurch Street station in London. The 'London Standard' network round the capital was upgraded in an exercise which deserves recognition as a great engineering achievement. Although DC systems attracted most attention, the *Railway Gazette Congress Issue* begins by noting the growing interest in single-phase development:

British Railways have closely watched the important development work carried out by the SNCF, and have taken advantage of the need to renew a 45-years old isolated local electrification in Lancashire (the Morecambe-Heysham lines of the London Midland Region) to re-equip it at 6.6kV 50-cycles. Leading equipment manufacturers, and the British Electricity Authority, collaborated closely with the Railway Executive in this experiment, with which trial running began in November 1952, the full electric service being inaugurated ten months later.

The report goes on to compare this project with the much larger conversion scheme in the London area of the Southern Region of British Railways. Here, the LVDC railway was interlinked with a reconstructed modern distribution and supply system which uprated the whole (*RG*, 1954):

272 *Electric railways 1880–1990*

Figure 15.2 Metropolitan Vickers 2490 hp (one-hour) 1500 V DC locomotive built in 1954 for the HVDC Manchester-Sheffield route of the former London & North Eastern Railway, electrified by British Railways after the nationalisation of 1948.
Source: National Railway Museum (York).

Here current has hitherto been supplied at 11,000 volts, three-phase, 25 cycles, to 49 manually-operated rotary convertor sub-stations, which converted to 660 volts (originally 600 volts) direct current at the conductor rail. The whole of the 25 cycle power distribution is being replaced (at a cost of over £11 millions) by a new 50-cycle system, using 71 sub-stations, 70 track paralleling hut equipments, seven switching stations for 33kV circuits, and three additional points of connection to existing 33kV 50-cycle-fed sub-stations; the whole is remotely controlled from three new control rooms.

An excellent and authoritative outline history of post-war electrification is included in Johnson and Long (1981) and is set against a general account of British Railways engineering, and the changes in engineering strategy between nationalisation in 1948 and the decision to electrify using 25 kV, 50 Hz. In 1948 a British Transport Committee group reviewed electrification policy, with members drawn from the Railway Executive, London Transport and Merz & McLellan. The Chairman was C. M. Cock, the Chief Officer, Electrical Engineering, who had pioneered electric and oil-electric traction on the Southern Railway with O. V. S. Bulleid. According to Johnson and Long (1981):

This committee reported in 1950 and endorsed the recommendations of its (prewar) predecessors that the 1500 V. DC overhead system should be the standard for future electrifications, except for certain reservations in regard to the Southern Region.

As with previous reports, multiples of the basic 750 V DC standard were accepted including 3000 V DC, but the possibility of also using high-voltage single-phase alternating current at 50 Hz was provided for:

In considering the AC system, the Committee recognized the advantages it possessed in minimizing the cost of fixed installations, but on information then available considered that this would be counter-balanced by the greater cost of locomotives and motor coaches and by other drawbacks of a technical nature. The Committee estimated that some 6500 route miles (43.4 per cent of the rail network) had a traffic density which would justify electrification on economic grounds. At that time, less than 1000 route miles were electrified or in the course of conversion, about 720 miles on the Southern Region and 240 miles on other regions.

The potential importance of the HVAC system led to conversion of the Lancaster – Morecambe – Heysham line to 6.6 kV, 50 Hz as an experiment by which experience in AC technology would be gained. In Hungary electrification at industrial frequency dated from 1932, when the 56-km route between Budapest to Komarom commenced work using 50 Hz 16 kV. This was extended a further 61 route miles to Hegyeshalom in 1934. In Germany the 35-mile route between Freiburg and Neustadt, through the Hollenthal Valley, was electrified at 50 Hz, 20 kV in 1936. Experience with the Hollenthal line during their occupation of South-West Germany after 1945 convinced the French that industrial frequency electrification was perhaps superior to the HVDC standard. In 1950 the French electrified 48 route miles between Aix-les-Bains and La Roche-sur-Foron using the 25 kV, 50 Hz system. This installation was shown to a party of engineers which included R. A. Riddles and S. B. Warder, who were attending a conference hosted by the SNCF. Trials with this experimental French scheme were successful and the SNCF chose the 25 kV single-phase AC system, with supply at industrial frequency, for the electrification of railways in North East France. This electrification programme ended A Chapelon's projects for a generation of high-performance compound-expansion steam locomotives, which he considered would be superior to contemporary DC electric traction (Rogers, 1972). The decision to use 25 kV, 50 Hz despite the existence of some 2000 miles of 1500 V DC track miles in France was most influential, and 25 kV, 50 Hz became the new international standard for electric railways.

The pioneer French installation, between Aix-les-Bains and La-Roche-sur-Foron, was followed by the electrification of the 173 route miles between Valenciennes and Thionville which began operation in July 1954. In May 1955, the findings of nine months operations of AC main line work was reviewed at a conference in Lille organised by the SNCF. The French experience and the encouraging results of the converted Lancaster – Morecambe – Heysham system caused the BTC to set up a committee to review electrification strategy for main lines chaired by E. Claxton. In particular, this committee was to consider systems proposed for the Euston-Manchester, Liverpool and Birmingham electrification plan. The report presented by S. B. Warder in September 1955 compared the two main alternative systems (HVDC and AC at 25 kV, 50 Hz). Estimates were made of costs of construction and operation over the London Midland routes selected for electrification. Electrification of other main lines was considered, including the East Coast route. This was not electrified in the 1950s or 1960s because it carried less dense traffic than did the West Coast route, and operations were left to diesel traction. The introduction of the 'Deltic' oil-electric locomotive at a time of plentiful, cheap oil from the newly exploited Middle East oil-fields influenced this decision (British Transport Commission, 1956; Webb, 1982).

Warder's report covered four main areas. AC was best for power supply because a DC railway would need a railway owned grid to feed the large number of substations. The AC system used less material in fixed works and required less maintenance of the overhead line and its supports, though it created more problems with clearance than did DC. Signalling and telecommunications, and rolling stock were reviewed, and AC was recommended and accepted, though not without opposition (Johnson and Long, 1981):

One who was opposed to the change was Mr den Hollander, General Manager of the Netherlands Railways, who was a member of the BTC's Technical Development and Research Committee. Electrifications which had been carried out in the Netherlands were on the 1500 V DC system, and den Hollander was a dedicated DC man.

British industry welcomed the opportunity to build up AC expertise, but it feared to lose a major DC demonstration system which might prove necessary for keeping overseas DC markets and winning others. Industry worried how to get the skills necessary for electrifying AC railways with which they were unfamiliar. The decision was not reversed and 25 kV, 50 Hz became the standard for future electrification of main lines and new routes, apart from extensions to the London Standard. Recently constructed 1500 V DC lines, such as those out of Liverpool Street, were converted to the new standard which gradually replaced existing HVDC systems. The Manchester-Sheffield-Wath line was never converted, but was closed to traffic in 1981.

15.4 The impact of the new standard

When the 25 kV, 50 Hz system was accepted as the new standard in 1955, there was a lack of AC locomotive building expertise in Britain. What little there was dated from the early years of electric traction or was limited to overseas schemes (Andrews, 1951; English Electric, 1960). It was not non-existent. British electrical engineering had built one of the first commercially successful AC locomotives in the world. This was a Bo-Bo machine of 160 hp built in 1908 at Trafford Park by British Westinghouse for the metre-gauge Thamshavn-Lokken railway in Norway (Webb and Duncan, 1979). This line was supplied with 15 kV, 50 Hz three-phase power which was converted to 6.6 kV, 25 Hz, single-phase for supply to the contact wire. Two of these engines worked until 1963 and one is now preserved. The London Brighton and South Coast Railway was electrified using single-phase AC at 6.6 kV, and the rolling stock included a 1000 hp van built by GEC and the Metropolitan Carriage and Wagon Co. The motor horsepower was rated at 800 and top speed was 35 mph but in 1923 the Southern Railway adopted the LSWR third-rail LVDC system as standard and there was thereafter little AC development work in Britain. The 50 Hz frequency was itself standardised for the national electricity generation industry in 1929.

The 1955 decision to standardise on 25 kV, 50 Hz was justified by the technical and economic success of the electrified West Coast Main Line which used it. The system was later installed on the suburban lines out of Kings Cross and over the former Great Northern main line, which forms part of the East Coast Main Line and was a route which Dawson wished to be electrified. The electrification of the WCML out of Euston marked the belated introduction into Britain of modern electric railway engineering. The first main line routes, electrified after the new standard, cultivated new engineering practices and attitudes of mind. Together with the oil-engined traction system introduced at the same time, the AC electric railway marked the start of a new phase, during which British Railways was to be reshaped in every department. The transformation of engineering, management and organisation have been described in several works (Bonavia, 1971; British Railways Board, 1969; Johnson and Long, 1981; Joy, 1973; Tufnell, 1979).

New engineering systems foster new attitudes of mind and bring about organisational change. Electric traction and the diesel engine greatly facilitated the reshaping of the British railway system. Electrification and dieselisation required reorganisation if their promises were to be fully realised. The first generation of internal-combustion engined locomotives were introduced into a system dominated by steam practice and performance. In the late 1950s and

276 *Electric railways 1880–1990*

Figure 15.3 English Electric Type 4 2000 hp diesel-electric locomotive first introduced in 1958 which played a major role in displacing steam traction from main line duties.
Source: General Electric (GB) now Alstom.

early 1960s, much of British Railways was a 'steam railway dieselised', with the steam locomotive replaced by a diesel but with duties largely unchanged. The electrified WCML out of Euston was different. On this system electric traction created a high performance railway with operational standards superior to those of the best British steam traction. The Stanier-Ivatt four-cylinder Class 8P 4–6–2 of 1948 and the Riddles three-cylinder Class 8P 4–6–2 of 1954 could not equal the express train performance of the first generation electric locomotives on the West Coast Main Line. The high performance diesel operated railway arrived in 1961 with the production type 'Deltic' which surpassed the best performance of the three-cylinder 4–6–2 steam engines which worked express trains on the East Coast Main Line out of Kings Cross. From these beginnings, a new kind of railway evolved as the steam locomotive was eliminated along with the associated methods of design, manufacture, organisation, and management.

15.5 Traction policy 1945–60

Electrification policy demands that engineering matters be set against a broad background of fuel policy and economics. This was as true in the 1950s as in the 1920s. In the 1940s and 1950s, Britain was in more severe economic difficulties than in the years between the world wars. Coal shortages dominated fuel policy down to the mid 1950s after which exploitation of the Middle Eastern oil fields made cheap oil available and encouraged dieselisation. In the late 1940s coal shortages brought major British industries to a standstill (1947, 1948), and strike action by miners in Europe and the USA caused policy makers to regard oil with greater favour. In the immediate aftermath of the war the shortage of coal in Britain revived the relatively rare practice, for the UK, of oil-firing steam locomotives:

... the scarcity of coal after the Second World War made oil-firing once more practicable in this country, and in 1946 it was decided to convert over 1,200 British locomotives for this purpose, mainly engines with heavy fuel consumption. This was expected to save a million tons of coal a year.

The project was implemented in part only. It was abandoned when operating costs with oil increased and doubts arose concerning reliability of supply from abroad, in the years before the Middle Eastern fields were exploited. During the post-war recovery period of the economy between 1945 and 1955, several factors favoured the retention of steam traction. Steam locomotives could burn both oil and coal. They were established and well understood. Their development did not require a strategic skill which was more urgently required elsewhere, for example by the military – then developing new aircraft and submarines – or by newer industries like nuclear power or electronics. They were cheap to build. A steam locomotive was much cheaper than a gas-turbine-electric locomotive; an oil-electric locomotive; or an electric locomotive and the associated fixed works (Barwell, 1962; Bond, 1964). In his President's Address to the Institution of Locomotive Engineers on 16 November 1950, R. A. Riddles quoted the cost of an express passenger steam locomotive as £18,000 at the end of 1949. A mixed traffic steam engine was £14,000 and a shunting engine £5500. The pioneer gas-turbine Engine 18000, admittedly a single prototype, cost £138,000 in 1950. Riddles gave the cost per drawbar horsepower as follows, presumably using the purchase cost of engines and not including any shared costs of associated fixed works. British Railways two-cylinder Class 5 4-6-0 steam-engine: £13 6s 0d. 1,600 hp Co-Co oil-electric locomotive: £65 0s 0d. 2500 hp A1A A1A gas-turbine-electric locomotive £69 7s 0d. Manchester-Sheffield 1500 V DC Bo-Bo electric locomotive £17 12s 0d. The cheapness of the DC locomotive compared to the oil-electric and gas-turbine-electric locomotive should be noted.

At this time, when there was no AC locomotive to include in the comparison, the British Transport Commission carried out thermal efficiency and performance tests of types representing the major modes of traction using dynamometer cars and the newly opened Rugby Test Station (Carling, 1972). Comparative thermal efficiency trials, the results of which were published in the BTC test bulletins, were conducted with the Southern Region Co-Co third-rail LVDC Raworth-Bulleid electric locomotive no. 20003; the Southern Region 1-Co-Co-1 oil-electric no. 10202; the Great Western Region gas-turbine-electric locomotive no. 18000, and a GWR 4-cylinder 4-6-0 steam engine of the 60XX class. The following costs in drawbar horsepower-hour were recorded. LVDC electric Co-Co no. 20003: 1.28d; oil-electric no. 10202: 1.05d; gas-turbine-electric no. 18000: 2.27d; and the GWR steam engine: 1.18d. One must not draw the wrong conclusions from these figures. The sole representative of electric traction was operating on a LVDC system built by the LSWR and the Southern Railway. The oil-electric engine, as Tufnell describes, worked under 'steam-engine' conditions, and was shedded with steam locomotives. It worked timetables constrained by steam locomotive performance and suffered the disadvantages of an unusual prototype working within an incompatible receiving system. Spares took time to get and in early days 'diesel traction' skills were lacking in repair staff used to steam locomotives. Under these adverse circumstances, the oil-electric locomotive put up a remarkable showing. This encouraged many engineers to regard oil-electric traction as a better future option than electric traction.

The electric locomotive did well but it suffered from being tested on a conductor rail system at a time when power station thermal efficiency was low, and before the Southern Region network was upgraded by the installation of a new supply system. Granted the absence of an AC locomotive to test, the far-sighted decision of Warder to consider single-phase AC traction at industrial frequency, is all the more noteworthy. The high fuel consumption of the gas-turbine engine was never cured, and this mode never became established.

The steam engine, built to the designs of C. B. Collet in 1927 did very well indeed – especially as its design was conservative and little influenced by the work of Woodard or Chapelon. If a 1927 conservative design could show up so well, some argued, what might be done with a modern steam design incorporating all the findings of Chapelon, Woodard and Fry, and working in a rationalised system which could be built for a fraction of the cost of electrification or dieselisation? There were shortages of money, material and manufacturing capacity in the British economy between 1945 and 1955, and there was a case for retaining steam traction in some capacity. Riddles' policy was to electrify heavily used lines and use steam traction elsewhere (Johnson and Long, 1981; Rogers, 1970), but strategic errors were made. Acquisition of expertise in oil-engined traction was delayed, and there was failure to recognise that steam traction had reached its limit and could not meet the requirements of the railway system needed by Britain as it recovered from the war. The case for retaining steam traction was of short-term validity. The disappearance of cheap labour; the decline of slow goods traffic; and the need to uprate the entire railway system in the face of motorway development and ownership of private road vehicles made the steam railway obsolete. This obsolescence was recognised in the 1956 modernisation plan (British Railways Board, 1963, 1965, 1969; Johnson and Long, 1981).

The extent to which main line electrification was carried through was determined by predicted savings on lines to be electrified, by the funds available at a difficult time for the national economy, and by the viability of the proposed traction systems offered in place of electrification. The superior technical performance of electric traction on heavily used main lines and arduous routes such as Manchester to Sheffield over Woodhead was recognised, but there were situations where other forms of traction might be better suited. The alternatives to electric traction explored in the 1945–55 period included a new form of steam engine – the Bulleid 'Leader' CC tank engine (Robertson, 1988); the high-performance oil-electric locomotive; oil-mechanical locomotives of high power; and gas turbine locomotives with electric or mechanical transmission (Cooper, 1985). All these were tried and proved inferior to electric traction on lines with high traffic densities. The diesel locomotive was better suited to lightly used main lines, such as the East Coast Main Line; secondary routes; branches and lightly loaded suburban services. Changes in traffic loading since the 1950s have justified electrification of the ECML between London, Leeds and Edinburgh.

In the 1960s, 1970s and 1980s, the high-performance diesel-electric locomotive dominated many services. The high-performance 'Deltic' locomotives which worked the fastest trains out of Kings Cross proved what this mode of traction could accomplish though at the price of high maintenance costs (Webb, 1982), and the success of the HST, introduced in 1976, continues to this day.

The fate of the more unusual alternatives to orthodox steam and electric traction can be briefly summarised. The atomic powered locomotive, based on the American steam-turbine-electric engines of the late 1940s but with the firebox replaced by the reactor, was never taken seriously by any professional railway engineer. The weight, great size, accident hazard, and total lack of any technical and economic justification ruled it out. The Bulleid 'Leader' of 1949 was a double-ended CC tank engine and was the last attempt in Britain to introduce a new form of coal-fired steam engine but it failed after thorough tests under the guidance of its designer, O. V. S. Bulleid (Robertson, 1988). The Fell 4–8–4 locomotive of 1950 was a high-powered oil-mechanical locomotive for main line work, resulting from a joint project between Lt. Col. Fell of Fell Developments, Shell Petroleum, and the London Midland Region of British Railways. The engine, no. 10100, was completed at Derby in 1950. It was powered by four Davey Paxman engines, each nominally rated at 500 hp, which drove through a mechanical gearbox and transmission to eight driving wheels linked by coupling rods. Test results were not sufficiently encouraging for the design to be repeated (Cooper, 1985; Tufnell, 1985). The Great Western Region trials with hydraulic transmission for moderate and high powered diesel engines did not lead to any change in the status of electric transmission, but the GWR engines enjoyed a useful working life. The Class 41 (introduced 1958, withdrawn by 1967); the Class 22 (1959–72); Class 42/43 (1958–72); Class 35 (1961–75), and the best known, the Class 52 (1961–77), were not successful enough to become an established form but they were not technical failures.

The two gas-turbine-electric units tried by the Great Western Region proved to be uneconomic in fuel and were expensive to maintain (Dymond, 1953; Robertson, 1989). Engine 18000, built in Switzerland in 1949, was withdrawn in 1960 and is now preserved in Britain after serving as a test vehicle in Vienna Arsenal railway engineering research laboratory. Engine 18100, built in 1951, was rebuilt as the pioneer British AC main line engine in 1958. The 4–6–0 gas-turbine-mechanical locomotive, GT3, built in 1961 after a very lengthy development programme, was withdrawn in 1962 after road tests showed that it had no potential for outperforming oil-electric and electric traction at a time of cheap oil fuels. As a method for burning low-grade fuels cheaply with low-maintenance costs, the gas-turbine system has proved a disappointment on rail.

The high-performance oil-electric locomotive capable of high-speed express work was a more serious challenge to main line electric traction. In the 1960s and the early 1970s maximum speeds attained by electric traction in Britain were not markedly in excess of those reached by the best diesel units like the HST, and oil was cheap and plentiful before the 1973 'Energy Crisis'. The 'Deltic' locomotive persuaded many engineers that electrification was not necessary for high performance railway traction, and that it could be got – via the diesel – without expensive fixed works. The demonstration unit 'Deltic' was introduced by English Electric in 1955. It weighed only 106 tons for 3300 engine brake horsepower and was a revolutionary design at a time when many diesels were inferior in speed and acceleration compared to the best steam locomotives. The 'Deltic' changed the status of diesel traction. It led to the Class 55 of 1961, which until 1982 worked the fastest East Coast Main Line passenger trains, keeping to schedules which steam locomotives could not match (Webb, 1982). An attempt to develop the high-performance oil-engined locomotive to run at 125 mph resulted in the Brush-Sulzer 'Kestrel' of 1968, which was a Co-Co unit weighing 106 tons and providing 4000 engine hp. After tests on the Kings Cross to Newcastle service, it was sold to the USSR in 1971, following the British Railways decision to develop the High Speed Train concept for fast passenger work.

Railway electrification in Britain 1920–60   281

Figure 15.4   English Electric 'Deltic' prototype 3300 hp diesel-electric locomotive built in 1955 introduced high-performance diesel traction to Great Britain and led to the Class 55 of 1961.

Source: General Electric (GB) now Alstom.

Since the 1970s the technical superiority of electric railways over alternatives has become marked. The growth of a high-speed railway network in Western Europe linked by the Channel Tunnel to Britain, or by bridges and tunnels across the Great Belt and the Sound to Denmark and Sweden, marks a new era in electrification. The German high-speed ICE (Rahn, 1986) and the French TGV (Revue Generale, 1981) have set performance standards much higher than those set by any non-electric railway, and they have redefined what is meant by the term 'modern electric traction'. It is interesting that the new 'Eurostar' passenger trains developed for operations through the Channel Tunnel between EEC cities operate on the three major electric systems developed as standards for railways in the past. They operate on 25 kV, 50 Hz in Britain, France, Belgium and the Tunnel; on 3 kV DC in Belgium; and on 750 V DC in southeast England on the conductor rail 'London-Standard' lines. The new British Railways goods locomotives designed for the Channel Tunnel service work under the 25kV 50Hz contact wire, and over the 750 V DC conductor rail network. The situation may become more complex if catenaries taking very high voltage DC (15 kV, 20 kV or higher) become established with conversion on the locomotives to supply three-phase motors. Granted the rate of technological progress which sets new standards from age to age, no heavy-duty industrial system is likely to serve as the exemplar for more than 20 years and during this time it is not likely to go unchallenged. The history of engineering in general, and the history of electric traction in particular furnish ample evidence of that.

Chapter 16
Signalling, communications and control 1940–70

16.1 Introduction

The history of the many post-war developments in Europe falls into distinct phases. There was the period of reconstruction of war damaged plant, during which essential repairs were made and worn-out equipment was replaced. Between 1945 and 1955, there were severe financial restrictions on European economies, and priorities in industry were given to strategically important products. Britain gave priority to military matters in view of the country's global responsibilities at a time of confrontation with the Soviet Union. In Britain, certain railway projects implemented during the period 1945–55 were schemes which had been started before the war and then suspended until the late 1940s or early 1950s. The economic depression of the 1930s, the war, and the post-war recovery period delayed some projects by 15 to 20 years. Electrification schemes (LVDC) of the Southern Railway; the HVDC electrification of the LNER Manchester-Sheffield line over Woodhead; and several projects for upgrading power supply systems, substation equipment and signalling come under this category.

The war gave rise to some unusual systems, such as the floodgates installed in the underground railway tunnels to limit flooding if bombs falling in the Thames pierced the bores. These gates were at first left open, to be closed in air-raids, but later they were kept closed and opened to allow the passage of trains, to safeguard against unexploded bombs going off in the river and flooding the system. These gates were interlocked with the signals.

In 1948 the four big British railway companies and a large number of small ones were nationalised to form British Railways, which enabled railway engineering to be planned on a national scale for the first time. Nationalisation of the electric power industry effected plans for signalling and traction by establishing a standard supply network fed from high-efficiency stations, and accessible to most industries wherever sited. The increasing thermal efficiency of coal-fired stations in this period made electrification economic by raising the overall systems efficiency of the electric railway above that of the best steam traction system. The wartime innovations in military signalling, communications and control led to improvements on railways once they were available for general use, but applying them required a period of assessment, modification, experiment, assimilation, and funding by the state. State funding could only come with a marked improvement in the national economy, which was strained by British participation in the Korean War, emergencies in Malaya and elsewhere, and commitment to NATO. By the mid-1950s money was made available for a general re-equipping of British Railways, as part of a widespread modernisation programme and this was followed by the reshaping of the whole system according to the recommendations of Beeching (British Transport Commission, 1963, 1965).

The modernisation plan eliminated steam traction, and introduced diesel and electric traction. It provided for a gradual replacement of manual-mechanical signalling and block-working, on main lines, by the latest, proven electrical and electronic devices. Proven is the operative word. Railway technology had to be reliable and fail-safe above all else. It needed to be as modern as was affordable, yet as conservative as necessary for safety and reliability. The great expense of all signalling, communications and control systems was a further encouragement to conservatism. Consequently, endurable equipment survived long, sometimes to surprise historians of engineering who discovered apparatus which they assumed had disappeared long ago. One example of how older components survived alongside the new in modernised systems may be quoted. In 1910 the Pennsylvania Railroad opened its new station in New York. LVDC services linked the station to transfer points where steam traction took over. The station had its own electric power station, with two 1000 kW Parsons turbines for supplying electrically-driven services: including air compressors and refrigerators. Rotary converters changed the 11 kV, 25 Hz AC supply to LVDC, and three motor generators changed the supply frequency to 92.33 Hz for the signalling system and for a control centre which distributed traction power throughout a 100-mile zone (Milster, ISSES vol.17 no. 3, pp. 23). The rotary converters for traction current were scrapped, but the motor generators feeding the signalling system were in use in summer 1995, though the whole site was then due for immediate clearance. Milster reports that many of the 1910 switchboards were still being used in mid-1995. Another example may be quoted of old practices surviving into another era. The author, travelling by diesel multiple unit in 1991 from Boston (Lincolnshire) to Doncaster along a route where semaphores survived, but along which telegraph poles and wires were being taken down, saw a signalman descend from his mechanical box and close level crossing gates by hand.

There are examples of innovative railway signalling techniques finding application in other fields during this period. A report on interlocking and safety circuits for radiotherapy apparatus (1971) makes clear references to railway signalling practice from which many devices for preventing misoperation of the medical apparatus were derived.

16.2 Centralising signal boxes 1945–70

Powered operations spread in this period, continuing a trend set in the 1920s and 1930s, and gradually manual-mechanical working was phased out, though it was a slow process. The London to Brighton line of the Southern Railway was electrified and resignalled in 1932. It was one of the first long main lines to be fitted with power signalling, though some older mechanical boxes fitted with power signalling were retained, and mechanically worked semaphores were left working in the London area. These were converted to powered operations in the 1950s. Powered working of points also spread between 1945 and 1970. The 1950s modernisation programme electrified and resignalled the former LMS West Coast Main Line and established new systems in and around London, Glasgow, Liverpool, Crewe and Manchester, which were required by improved suburban services in those cities. Extended electrification on the Southern Region (former SR), and the introduction of faster working by diesel-electric locomotives and the high-speed diesel trains, required new signalling throughout the Western Region (former GWR), and the Eastern Region (former LNER). As Kichenside and Williams (1980) relate:

> ... by 1973 modern signalling, with multiple-aspect signals, continuous track circuiting, the automatic track to train warning system (aws), and control from (mostly) large centralised power signal boxes, known by the generic term track circuit block (tcb) extended unbrokenly on lines bounded by Paddington, Bristol, Taunton, to the outskirts of Swansea, from Bristol to Birmingham, Derby, Nottingham, Sheffield and Leeds and from Euston to Glasgow.

At that date the longest continuous length of TCB extended about 430 miles from Taunton via Bristol and Birmingham to Glasgow. At the start of 1977 many lines had been closed following the Beeching recommendations and out of a total (single) track mileage of 22,401 there were 10,094 miles equipped with colour-light signalling. This included mileage signalled from new power boxes at Warrington, Preston, Carlisle, and Motherwell. In the next five years, multiple-aspect colour-light signalling, and track circuit block was extended to cover most of the main lines on the rationalised British network, including the East Coast Main Line. Many of these techniques had been foreshadowed in the projects pioneered by the Southern Railway in the 1920s and 1930s. Techniques proposed before the war and applied extensively afterwards include the 'One Control Switch' system for relay interlocking and the 'Entrance-Exit' system for relay interlocking. The 'One Control Switch' (OCS) system has already been described. It was taken up by the LNER in the 1930s, and used a single rotating thumb switch to select and set routes in signal boxes at Thirsk, Northallerton, Leeds and Hull. The largest example of this OCS system was completed in 1949 in York after a wartime delay. It proved successful, and in the 1960s was widely used by British Railways at Glasgow Central, in the Manchester district, at St Pancras, Newcastle, Chislehurst and elsewhere. Even as it was being installed it was overtaken by another system which had operating advantages, and which

Figure 16.1 Aldersgate signal frame in May 1955 with push-button operation of signals and points. The illuminated board shows the routes set up and signal aspects.
Source: London Transport Museum.

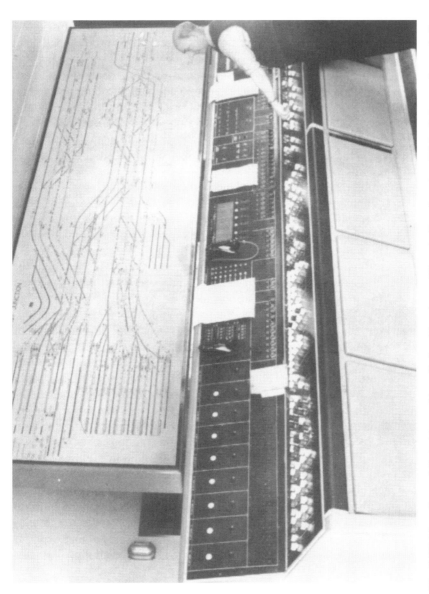

Figure 16.2 The OCS (One-Control-Switch) panel installed at Manchester Victoria East Junction. White lights on the track diagram show the routes selected. The OCS system features in several major resignalling projects carried out under the BR Modernisation Plan of 1955.

Source: British Railways/National Railway Museum (York).

promised further miniaturisation of components. This was the 'Entrance-Exit' system, usually termed the NX system by its developers, Metropolitan Vickers-General Railway Signal. This system used only one switch or button to control an entry signal irrespective of how many routes it applied to. There was an exit switch or button for each of the routes leading from the signal. Operating an exit switch and an entry switch set the route. The operating switches or buttons were mounted geographically on a schematic track diagram and it proved possible to reduce size of panels and indicators below the scale of the OCS equipment. The first NX panel was installed on the Cheshire Lines Railway (LMS-LNER Joint) at Brunswick Goods signal box in 1939. This was a small installation. The first large example was fitted in the new box at Stratford (East London) in 1949, and NX panels were standard for most resignalling projects in the British Railways Modernisation Plan of 1955. The NX became a general standard, made by all the suppliers to British Railways. Though design detail differed, considerable standardisation was achieved. The NX control panels were made to order for a particular location, just as were the old mechanical interlocking frames, but they could be built up from standard components representing sections of plain track, points, signals, etc. This enabled track alterations to be incorporated by changing individual panel sections. Changing an old mechanical interlocking frame was less convenient.

As these techniques spread, it became more common to find long lengths of main line, automatically signalled, without signal boxes, and with control of routes carried out in centralised locations, using powered operations of signals and points. No longer could the crew of a halted train walk to the nearest box. No longer were many signalmen available to observe passing trains for signs of hot axleboxes as they did in prewar days. Access by train crew to signal post telephones became necessary, and radio communication between train crew and control was introduced. Installation of lineside hot-box detectors was adopted.

Interlocking by electrical and solid-state electronics relays enabled signalling equipment to be distributed with much greater freedom than in the days of mechanical interlocking when the frame was large and located under the lever platform in a secure room from which wires and rods ran to signals and points. In the electrical signalling of the 1960–80 period the interlocking relays were near the points and signals they controlled, generally at major junctions,. The control panels often were miles away, with commands sent along a single pair of wires using time division multiplexing (TDM) or frequency division multiplexing (FDM). The larger power signal boxes controlled up to 12 remote relay interlocking systems using these techniques which were developed for telephony between the wars, though some boxes (Trent and Saltley) had all the relay interlocking equipment inside them, with signals and points worked through FDM links. The introduction of computers and better train describers into control and signalling effected a revolution when cheap, reliable solid-state computer systems became available.

The first electronic computer installed by the railway industry is believed to be that purchased in November 1955 by the Association of American Railroads,

Signalling, communications and control 1940–70 289

and the USA companies pioneered use of the computer for simulating locomotive and train behaviour in the 1960s. In 1967, the TeleRail Automated Information Network was set up and in 1970 was used to keep track of US freight cars. This system was expanded in 1973 and by 1975 was employed to estimate future rolling stock requirements. By the mid-1980s, 60 of the US major railways used 725 computers for a wide variety of functions ranging from book-keeping, classification of cars in marshalling yards, car location, and general administration. In signalling and control, the electronic computer was presaged by the electric train describer, and the electro-mechanical controller of automated operations.

One important trend in signal box design was towards complete fireproofing of boxes by eliminating all combustibles as far as possible. This probably began on the London Transport Northern Line following the destruction of the Colindale cabin when an aeroplane crashed on the railway, linked the traction and signal circuits and caused a fire which burned down the signal box. Morden is reputed to be one of the first fireproofed signal boxes and most modern cabins built for electric railways were constructed along fireproofed lines. Brick and wooden boxes, with hand-worked mechanical frames and non-powered operations, were installed and extensively repaired on secondary routes until the 1970s. The last examples were set up in the 1980s. Examples are still to be found on the national railway network, and on the preserved steam-railways (Signalling Study Group, 1986).

16.3 Automatic train describers and Program Machines

From the earliest days of railway signalling based on the telegraph and the block system, signalmen needed to know when trains were approaching their section and which trains they were. Information could be sent by telegraph or telephone. The telephone was most useful if delays caused trains to run out of timetabled order. This information would be passed on to other signal boxes in turn. If a route were lightly loaded, with minutes between trains, simple aids to memory sufficed to remind the signalman of the identity and order of trains approaching his box. In some situations a coded bell signal might serve, but on busy lines more was needed. Trains approached, passed and left sections too rapidly for memory and simple indicators, and needs were met by the 19th-century and early 20th-century describer-recorders.

The most advanced types were found on the London underground lines where services were intense, but the main line companies had need of them on heavily used routes. The Southern Railway (later Southern Region of British Railways) used a simple 'magazine' describer with a display panel listing types of train, and a triple column of lights to identify the first, second and third train approaching a section. The train movements were also shown on the illuminated track diagram. The signalman pressed a button alongside some particular description (where the light showed it to be the first train approaching) to transmit it to the next box when the train passed a particular point. Some systems automatically removed a train from the describer when it had passed the section: others required intervention of the signalman in passing on information and resetting the instrument. Automatic describer-recorders were used in isolated cases between the wars, and were installed as standard in large power boxes after the 1939-45 war during new or delayed electrification schemes.

One of the first post-war automatic describers was used in the signal boxes along the newly electrified Liverpool Street (London) to Shenfield railway. In this 1949 model, the signalman described a particular train in code letters, and this description was passed from box to box by the train activating track circuits. The system could pass a description from the slow line describer to the fast line describer if a train switched tracks. As happened with signalling interlocking frames and levers, there was a steady drive towards miniaturisation of describers, aided by the advent of small valve electronics, followed by solid-state electronics and printed circuits. In the 1960s, cathode-ray tube displays became standard on all British Railways regions, apart from the Western region which used a rotating counter display. In the early 1970s, the electronic computer was used in the new boxes opened at Carlisle, Dartford, Glasgow Central, Motherwell, Preston and Warrington. The solid-state computer stored and recovered much more information about trains than did the best electro-mechanical and electrical types. Computers provided data on train position, track occupancy throughout the system they monitored, and controlled platform announcements and displays.

On London underground railways in the 1960s, signalling and control techniques were in advance of those of the main line railways. By the mid-1960s, the underground system was entirely given over to powered operations and signal boxes were used only at junctions, crossings and sidings. Electro-pneumatic operations were normal. Signalmen worked miniature frame levers which controlled the air-motors via electromagnetic valves governing the air supply. Signalling in these locations was classified as 'controlled' or semi-automatic, because it was normally set to red and was changed to green by intervention of the signalman, with resetting to red done automatically via the track circuits. Sidings could be worked from remote signal boxes. For example, the Arnos Grove box controlled the Wood Green sidings.

Early equipment enjoyed lengthy usage. In the mid-1960s, the pioneer Brown magazine train describer was still being used. This was one of the classic innovations which aided electric railway operations, which originated in the electro-mechanical technology of the period 1900-10. It required the intervention of the signalman who selected the destination of a train on a dial and energised a circuit passing signals corresponding to that destination. These signals activated mechanical strikers which altered the configuration of studs fitted round a rotating drum. Data was stored in the form of stud position. The drum rotated after each activating signal, and so it stored the order and destination of the trains. This information could be passed on to platform displays and signal boxes. Cancellation of data was automatically done by track circuits when a train quit a section. The Brown describer-recorder worked well, but gave way in the 1950s and 1960s to the ribbon storage describer introduced by Robert Dell, one of the greatest signalling engineers in railway history. The Dell describer-recorder used conventional telegraph techniques and Dell effected radical improvements throughout underground signalling and control systems by applying engineering practice long-proven elsewhere.

The Dell instrument used a moving paper tape with holes punched and spaces left blank to encode the data, as in many telegraphic instruments. Feelers made electrical contact through the holes and the completed circuits passed on the information to platform displays and signal boxes. The paper tape provided a permanent record of the nature, order and times of train operations. The ability of train describer-recorders to play a more positive role in control was recognised in the 1920s and tried in the 1930s but was not implemented as normal practice on a large scale until the period 1955–65. At simple facing junctions the automatic train describer controlled the points, and at trailing junctions controlled signals to pass trains according to which one arrived first. Robert Dell introduced the famous Program Machine in the mid-1950s to increase automated control over points and signals. The Program Machines used a plastic roll, 8 feet long, 8 inches wide, into which were punched holes which represented train movements of the day. The movement of this roll past a signal pick-up supplied that detector with encoded information about train working for that day. This controlled points and signals. Actual operation of the points and signals was done by interlocking machines similar to those used in London Transport's orthodox powered signalling systems. The Program Machines acted as robot signalmen working miniature frames: they gave instructions to the interlocking devices. The technique was old when it was installed, though new in automatic working of points and signals. It used punched card data storage and detection by contact-completing feelers commonplace in the business machines, data sorters, and printing telegraphs pioneered by Hollerith, Hughes and others in the 19th century.

The Dell Program Machine was introduced when electronic computers were too expensive, too unreliable, and too unproven. It was relatively cheap, simple and very reliable. It used well understood engineering devices to accomplish what the computers of the day could not be relied on to do with safety. The Dell machine set up routes with points and signals positioned accordingly. It checked the train destination described in the describer-recorder with the one encoded in the Program Machine. If there was a discrepancy, a signal alerted supervisors in the Regulating Room at Leicester Square, who could determine whether or not it was accurately signalled by describer or Program Machine. The Program Machine moved the roll onward as each train was cleared and so it worked through the day's timetable. The Northern Line was the first to be equipped with it and from there it spread to other routes as described below. The working of the Program Machines was supervised by controllers in the Regulating Room, located underground at Leicester Square, where there was an illuminated display of the whole line, showing train positions. If necessary, the supervisors could take over from the Program Machines and resort to conventional route-setting by push-button. The Program Machines stored data about late running trains

Figure 16.3 The electro-mechanical Program Machine used in Dell's system for the automatic operation of London Transport trains. This particular machine was photographed at Watford in 1958.
Source: H. Zinram/London Transport Museum.

and supervisors could change pathing, train order, and accommodate cancellations. In the post-1950s era, it was possible for local supervisors to dial information into Program Machines to effect local changes. Machines at Morden and Kennington worked the platform indicators, and departure signals in addition to controlling the timetable. They were installed on the District line at Turnham Green (junction working); Parsons Green sidings (shunting and routing of empty trains); Putney Bridge (train working; starting of trains); and Hammersmith (train reversal, and movement from fast to local lines). These machines were overseen from a regulating room in Earls Court. The Program Machines spread and the Victoria and Northern lines were controlled by them, aided by the automatic describer, with supervision from a centre in Coburg Street, near Euston. In 1973, small on-line computers were first use to replace local Program Machines and to extend the range of supervision. London Transport replaced Program Machines at Watford on the Metropolitan line with supervision from Rickmansworth. In 1974, the Watford computer was linked to the visual display unit in the Rickmansworth signal box. The visual display unit further effected a reduction in equipment size compared to a conventional display board. Computer supervision was extended to the Victoria and Northern line Program Machines in the Coburg Street centre. A few Program Machines are still in use today.

16.4 The automatic railway

By the 1960s it was possible to work automatically an entire rapid-transit railway. London Transport proved this with the Victoria Line which became the first automated passenger-carrying railway in Britain in 1968–69. It was not the first automated passenger railway in the world because in the mid-1960s there were short sections of automatic metro in New York and Barcelona. Electric traction was essential for automatic operations. There was no point in setting up an integrated, automatic system, capable of fine-scale control of signals and points, if the traction units were incapable of being fine-controlled to the same degree. Steam locomotives were totally unsuitable. All attempts to introduce one-man control, with automatic regulation of fuel supply, steam generation, power and speed, failed decisively. Diesel-electric traction was possible, because units could be controlled precisely from outside the locomotive, and automatic diesel-operated railways were to be found, usually in industry. Signalling on the Underground lines had been automatic from pre-Great War days apart from at junctions and reversing points where signal boxes were necessary until the Dell Program Machine was introduced.

The Program Machine increased the degree of automatic operation when it was remotely controlled from centres like Leicester Square, but the trains were still driven by men. There would be a truly automatic railway when the trains were driverless with all services automated, including issue and checking of tickets, and provision of information in response to passenger enquiry. The

Figure 16.4 Euston Control Centre (Northern Line) showing the interlocking machine (left); the drum-type train describer-receiver (lower centre); and the Dell Program Machine (centre).
Source: C Tait/London Transport Museum.

Victoria line showed that automatic driving, issue and checking of tickets were possible, and contemporary developments introduced automated information provision for passengers. However, the need to aid passengers in emergencies and to provide personal service will probably prevent total automation, and ensure that staff are retained in stations. In technical operations, away from the public, automation will advance to the limit set by engineering possibility. Horne remarks that automatically driven trains were sought for safety reasons, because they were less likely to overrun danger signals than were manually driven ones, and they allowed reduction in number of crew carried, from two to one per train. He states (Horne, 1988, p. 37):

London Transport's particular desire for automatic operation was boosted by the peculiar circumstances of deep-level tube operation in single-track tunnels: in the unusual event of the train driver becoming incapacitated it was expected that the guard (who was trained to drive) would get the train to the next station. If, however, there was no guard then some other means of extricating the train quickly had to be devised – automation appeared to meet this need.

Before being accepted, the automatic driving system was subjected to several years investigation in the early 1960s. Similar experiments were carried out in New York, Paris, Moscow, Stockholm and Barcelona. In London, tests were conducted between South Ealing and Acton Town, using a converted District Line car as test vehicle. The preliminary trials were successful, and resulted in tests with trackside equipment and train-mounted apparatus on the Stamford Brook to Ravenscourt Park section of the District line. These tests involved passenger-carrying service trains. In April 1963 a converted District Line driving car was coupled to the east-facing end of an in-service train which was driven manually for most of its journey. However, when the train reached Stamford Brook, the driver switched to automatic operation between there and Ravenscourt Park. Over this section, operation was fully automatic, including response to signals and speed restrictions. This was followed by a test on a larger scale, on the Woodford to Hainault section of the Central Line which was converted for automatic operations. These began in April 1964, using five four-car trains reconstructed from three eight-car trains of 1960 stock. The success of these trials led to the decision to equip the Victoria Line for fully automatic operation.

On the Victoria Line the single crewman rode in the driver's cab, as a train operator, rather than a guard. He opened and closed doors, and pressed the train start buttons in stations but his main function was to direct passengers in emergencies. The Victoria Line used two types of coded current signals, sent through the running rails, which were detected by induction coils mounted on the front of each train. One type of signal, the continuous safety signal code, used AC coded track circuits with a frequency of 125 Hz. There were four frequencies at which pulses were transmitted in the rails, namely zero and 180, 270, and 420 pulses per minute. These pulses were created by pendulum driven electronic switches interrupting the supply to the safety signalling circuits. The system is still in use, on main line as well as rapid-transit railways, albeit with modifications. The principles were as follows. If the signal frequency was zero (no coded signal), a stationary train could not be started. If the frequency became zero when a train was in motion, it was brought to an emergency stop. A code of 420 pulses per minute allowed unrestricted running up to 50 mph under the driving command system, but it could not start a train. A code of 270 pulses per minute permitted powered operation at up to 25 mph. and it allowed the traction motor control system to switch on a motor which had been switched off by a driving command signal. A code of 180 pulses per minute permitted running at up to 25 mph, but did not allow remotoring after driving command signal shutdown (Bruce, 1968). The other type of coded signal was of much higher frequency, and it transmitted driving commands from 10-ft lengths of running rail known as command spots which were superposed on the coded track circuits. One hundred Hertz 'frequency blocks' were assigned to each 1 mph of speed. A speed of 10 mph required a transmitted control signal of 1000 Hz, and a signal of 5000 Hz set the speed at 50 mph. Command spots were established in all places where speed had to be precisely regulated, at station approaches, junctions and speed restrictions. To start a train, the driver – called an operator on a 'driverless' railway – pressed two buttons to initiate the accelerating sequence. Interlocking between the command signal circuits and safety circuits on board the train prevented the 420 code taking effect until all doors and the cab windows were closed. If a signal had halted a train between stations, a safety signal command of 270 pulses per minute was required to switch on the traction motors for restart. Train speed was controlled using two governors. An electronic governor received signals from a frequency generator mounted on one of the traction motors and held speed within 2 mph of the limits determined by the 180 or 270 codes. As this device was not fail-safe, a mechanical governor worked from the trailer axle. Because the Victoria Line could be worked without drivers, there was no need for lineside visual signals such as are necessary on systems where trains are operated by drivers. However, provision was made for partial or total failure of the pulsed-coded signalling system. Visual (colour-light) signals were installed at stations and junctions to serve in emergencies when the trains are driven manually by the operator. If no signal codes were being received, speed was governed to 10 mph. If codes were being received, but some train-mounted apparatus had failed, speed was kept below 25 mph maximum. These visual signals also served maintenance trains which were not fitted with automatic driving equipment. The Victoria Line was fitted with the electro-pneumatic trip apparatus, common on other lines, which worked to halt trains if speed rose too high, or if there was a failure in the safety signalling system based on pulses in the rails. There were also three emergency plungers fitted on each platform, which stopped automatically driven trains by acting through the safety signalling circuits, and which could be operated if need be – for example if somebody fell or jumped onto the track. The automatic driving system on the Victoria Line worked well, and continues to do so even though it became obsolete in the 1980s. Interestingly, Horne remarks that the automatic fare-collection system, and the passenger information points were not the success they were expected to be.

The automatic driving system worked in conjunction with the orthodox train-mounted control system which in the 1960s was based on an American system developed in the 1920s. Before the 1930s, and for long after on many DC railways retaining obsolete equipment, control was by the series-parallel principle and the multiple-unit system. Sprague did much to pioneer these techniques which became universal practice. The original simple control of DC motors by hand-worked rheostat gave way to multiple-unit control and series-parallel working. At first, switching from series to parallel was by 'open circuit transition', which produced irregular acceleration and wear of the switchgear. This gave way to 'short circuit transition', and finally 'bridge transition' in which resistance stepping and series-parallel connections were made without the need to break large currents. Bridge transition superseded the earlier techniques and was normal in the 1920s. The switching was usually electromagnetic with the wires from the control wheel or handle energising magnetic coils which worked the switches. This system, not unlike electromagnetic controls in signalling networks, enjoyed a long life and is still found in old stock. In the USA, in the 1920s, a new system was successfully demonstrated on the New York subway network. This was an electro-pneumatic system, in which the switching of contacts was by a camshaft rotated by a compressed air engine which was started and stopped by electrical control. The first designs were modified by the inclusion of oil 'buffers' to eliminate shocks in working, and the system went by the name of Pneumatic Camshaft Modified (PCM) which was introduced to London Transport in 1936 where it was known as Pneumatic Camshaft Motor. Bruce (1968) remarks:

A feature of the control which makes for reliability of operation is that the camshaft turns in one direction to bring the switching to full series and then returns to the 'off' position. This also corresponds to full parallel in combination with another switch which performs the transition function. Under normal operation, therefore, after full acceleration has been achieved there is no running back of the camshaft when the equipment is switched off.

This type of control system lasted long and was widely employed throughout London Transport when the Victoria Line was opened in 1969. It could be integrated easily with automatic operations circuits.

16.5 Automatic warning, control and driving systems on main line railways

In the 1950s and 1960s, the steam locomotive dominated the world's railways outside the USA, Switzerland, the Netherlands, and a few other nations. This limited the use of many safety devices developed in the 1920s and 1930s and prevented automatic operations. A steam locomotive was operated by two men who manually worked independent mechanical levers and wheels controlling steam flow; cylinder cut-off governing; water supply; fuel supply, and other functions of the machine. Simple automatic warning systems and elementary cab signalling were tried with steam locomotion before the Great War, but the full potential of the many developments made between 1920 and 1960 required electric and diesel-electric traction, and the abolition of steam traction. By the late 1950s, steam traction was the greatest barrier to the modernisation of railways. Some devices were fitted to all modes of traction, and after the nationalising of British railways in 1948, comparative trials were carried out to identify methods of giving advanced warning of signal setting by visual and audible devices mounted in the cab. Automatic stop devices could be combined with cab signals. The two main systems compared originated in warning devices which were first applied on electric rapid-transit lines in the period 1890–1910. They were used on main lines in the USA and Europe in the period 1920–40 and were accepted as standard in the recovery period after the 1939–45 war.

The Great Western Railway developed its own automatic warning system for indicating signal aspects in the engine cab. This had been proven in service, though expense of installation limited application. This Great Western Audible Signal used a mechanical contact between a shoe on the locomotive and an energised track ramp to indicate the setting of a distant signal. The London Midland & Scottish Railway tried a system dating from the first period of electric rapid-transit railways, 1890–1910, which used magnetic induction to transmit a signal from track apparatus to locomotive. This Hudd-Strowger apparatus gave audible and visual indication of a distant signal aspect and in 1948 was found on the Fenchurch Street-Shoeburyness line of the LMS. The British Railways experiments resulted in a standard system for use throughout BR. This used magnetic induction and gave an audio-visual signal in the cab. It spread throughout the entire BR network, replacing the Great Western system. It used two inductors, one a permanent magnet and one which was energised when a distant signal was set to clear. The permanent magnet was encountered first, and this activated a receiver mounted on the vehicle. The state of the electro-magnet (energised or non-energised) determined whether or not a horn would sound a warning, or a bell would sound clear. The driver had to acknowledge a warning using a cancelling handle. If a warning was not cancelled, the brakes were automatically applied. The system worked with both semaphore and colour-light signals. As Kichenside and Williams remark (1980, p. 93):

> ... in semaphore areas, the circuit is proved through both the signal arm and its lever in the signal box. In colour-light areas all the running signals normally have aws magnets; the electro-magnet is energised only when a green, clear, indication is displayed and caution is given in the cab for red, yellow, or double yellow.

The cab signals consisted of a bell, a horn and a visual indicator in the form of a disk which showed either a black aspect, or a black and white pattern. The use of suppressor inductors, close by the track magnets, enabled the system to be used on single tracks worked in both directions. The standard AWS system was simple and effective for a low speed steam railway, or the first generation of electric or diesel-worked main line railways, but the advent of high-speed traction in the 1960s and 1970s necessitated more advanced methods for cab signalling and automatic driving. British Railways investigated several systems and identified two forms suitable for cab signalling. One used coded signals sent through the running rails, a technique dating from the 1920s. Another used signal circuits independent of the rails, in cables laid a set distance apart or arranged in a specified pattern along the track. Train mounted detector coils pick up the coded signals which were then passed on to the cab signal-indicator, or to the control system if automatic driving was incorporated. Track conductor loops were used to measure speed, and acceleration. Cab signalling was tested under service conditions on electrified main lines on the London Midland Region AC West Coast Main Line at Wilmslow, and in several locations along the Southern Region LVDC conductor rail system between Waterloo and Bournemouth. The investigations carried out in these separate locations were part of a single experiment. The tests on the Southern region were with intermittent cab signalling which displayed the aspect of a signal some 600 metres in advance of it. The aspect was displayed on passing the signal until the train reached the advance indication point of the next signal. The equipment consisted of the standard advanced warning system (AWS) permanent magnet plus a track conductor loop carrying coded currents between the magnet and the signal. This track conductor transmitted a high frequency signal to the cab display unit to show the signal aspect. If the signal was set clear, the track conductor circuit cancelled the automatic braking sequence always initiated by the permanent magnet. Other aspects required cancelling by the driver using push buttons to acknowledge receipt of signal. There was an intrinsic emergency stop mechanism which halted the train if the driver acknowledged a danger signal and then attempted to proceed through it. The cab apparatus transferred the displayed aspect of a signal ahead to a 'signal passed' status when the train passed that signal and approached the next one. It performed the functions of an electric train describer-recorder, as was found in signal boxes, but it described and recorded signals rather than trains. Much of the technology was common and was developed within the general context of signalling and control. This equipment, which was tested on the Southern Region, became known as signal repeating automatic warning system (SRAWS).

The system investigated on the London Midland Region developed the Southern Region continuous cab signalling methods to a more advanced level. A train-mounted computer contained information about load, locomotive power, route topology, line curvature, speed restrictions, and signal location. This information was updated by signals from track located circuits, so that braking distances and advice on speed was to be revised in the light of emergencies or bad weather. Two-way speech communication was provided. The system continuously compared actual speed and permitted speed, and could provide automatic driving by integrating the control system of the traction units with the continuous cab signalling circuits. This had already been done on the more advanced electric rapid-transit railways which were the source of the systems transferred to the main line railways in the 1960s and 1970s. Unfortunately SRAWS and the more advanced developments based on it proved too expensive for general and immediate application and economics determined the extent to which successful and technically-desirable signalling equipment was installed. This has been the case since the 1840s. To facilitate the introduction of the 'London Midland Region' system, it was designed to be installed in stages, each stage providing a more advanced level of performance up to fully automatic driving.

The prohibitive costs of SRAWS prompted a search for cheaper systems to meet the signalling requirements of the high-speed operations which in the 1960s followed the electrification of main lines and the introduction of high performance diesel-electric locomotives. The simple safety trip for stopping trains overrunning stop signals was widely used on London Underground, and on a few British Railways LVDC electric lines such as the Waterloo & City Railway, the Euston to Watford line, and the Liverpool Central to Birkenhead route. It was not installed elsewhere. For main line service something more advanced was required which dispensed with mechanical contact. The transponder was an important innovation which enabled signals to be transmitted from trackside equipment to cab-mounted displays without contact. The transponder was fastened to the sleepers and was a passive (unpowered) electronic circuit energised by induction when a train-mounted inductor passed over it. Induction generated 24 V DC which powered the coded signals which transmitted information about location and speed limit. A train-mounted computer compared actual route and speed with the planned route and speed, and the comparison was displayed in the cab. The transponders first tried on the WCML in anticipation of 150 mph operations by the Advanced Passenger Train could not transmit variable information about lineside signal aspect and were limited to passing fixed information about location and speed set during manufacture, though in some ways the system was a form of cab-mounted speed signalling. Normal high-speed operations on the WCML established that it was possible to run regular services reaching 125 mph without cab signalling, using multiple-aspect colour light signals supplemented by the BR standard AWS. The cancellation of the Advanced Passenger Train (Potter, 1987) reduced the signalling requirements and avoided the need for extensive investment in some form of SRAWS which it was thought might be required in addition to transponders.

16.6 General trends 1940–70

Throughout this period there was steady improvement, limited by the size of state funding. The British railway system like railways abroad, including those of the USA, lost traffic to the private motorcar, the long-distance lorry, and the aeroplane. The building of motorways, the containerising of road, river, and sea freight, and cheap, mass travel by air put many railways into the category of loss-makers. Even with state aid, drastic economies were essential. Branch lines, secondary lines, and once-important main lines were closed in Europe and the USA. Combination of private companies into larger systems and nationalisation of others transformed railways as passenger and goods carriers. The emphasis was on container and unit freight operations; on rapid-transit passenger services, and on profitable main line passenger operations. In the USA, air and motor traffic destroyed the long-distance passenger train service, but this survived in Europe. The need to reduce costs was overwhelming, and it impelled traction towards electrification and dieselisation; and signalling, control and communications towards labour-saving devices. A massive reduction in numbers employed justified the spending in North America and Europe of very large sums of money on railways which had passed their period of greatest profitability. Some railways could never be profitable but were deemed to be necessary by the state.

The latest technologies cut losses by reducing manpower and maintenance costs, and by increasing capacity of existing equipment. There was a general drive towards powered operations and automation in traction, signalling and control. The full potential of the equipment was not always realised at once because of political and union opposition to innovations leading to redundancies. Despite this, the increased productivity of the newer engineering systems was essential in saving the railways in the post-1950s world. Gradually, manual methods gave way to powered operations. Many signal boxes and control points were replaced by fewer centralised cabins or control rooms. Human intervention surrendered to automatic electro-mechanical action. Improved communications by telephone, television and radio increased centralisation and supervision from a distance. The computer became everyday apparatus. Mechanical methods of detection and control were replaced by electro-mechanical or entirely electrical systems, which were themselves transformed by the solid-state electronics devices of the 1950s and 1960s. A few examples are provided to denote the kind of change taking place between the mid-1950s and 1970. British examples are in the main provided, but they stand for developments taking place elsewhere.

Generally, the USA led the way. It could afford to because its national economy did not suffer the set back experienced by the European economies in the 1940s and 1950s. By the 1960s, the main European systems, both in the Warsaw Pact and NATO blocs, were re-equipping along much the same lines though at markedly different rates. In Europe, the railways of Britain, France, Germany, Benelux, Italy and Scandinavia represented best practice during this period. France was pre-eminent following its commitment to the high-performance electric railway developed and built with state backing. The spread of automatically worked multiple-aspect colour-light signalling, with the four-aspect signal and continuous track circuiting, speeded the abolition of the traditional block worked from its own box and increased line capacity speed. Electrical detection of the setting of points blades was normal with powered operation, and could be applied to mechanically worked points in place of the purely mechanical detector and interlock dating from the 19th century. Continuous welded rail became standard though it required insulated fishplated joints to separate the individual track circuits corresponding to separate sections. Experiments were carried out with frequency-receptive track circuits, using high frequency AC coded currents, with a view to eliminating rail joints. There was a general application of automatic signalling to rapid-transit and main line railways. For example the signalling of the Central Line of London Transport was greatly improved between 1946 and 1952 when work halted by the war was finally carried through and new construction undertaken. By 1940, the old three-rail system of the Central Line had been converted to four-rail (Horne, 1987, p. 38), and automatic signalling was installed between Shepherds Bush and east of Bank, though some boxes were retained and a new one installed with an electro-pneumatic frame to control the British

Figure 16.5 Camden Town Control Room, London Transport, in 1955 after construction of new signalling centre.
Source: C Tait/London Transport Museum.

Museum siding. Shoreditch box (LT) was the first air-worked interlocking frame in Britain in 1943.

Typical of the steady improvements which transformed the railways was the 1946 replacement on the Central Line of the automatic semaphore signalling west of White City by colour light signalling and the use of remote control of signal boxes using 'slave' cabins worked from another box. For example, North Acton box was controlled from a new cabin at White City, and Ruislip Gardens box was controlled from West Ruislip. In such techniques, London Transport was about ten years in advance of the best practice on British Railways' main-line system. In 1948, line capacity at Liverpool Street station on the Central Line was improved by signalling controlled by the speed of trains, which enabled closer spacing of trains approaching this stop where congestion was severe. The economic situation grew better in the 1950s and the need to increase the carrying capacity of London Transport demanded technical improvements. The few remaining mechanical frames at Queensway, Marble Arch and Bank, dating from the opening of the Central London Railway, were replaced by miniature electro-pneumatic frames between 1951 and 1958. Powered operations were introduced at South Woodford in 1961. These and later developments on other London Transport lines are described by Horne. In the 1970s, push button control of slave cabins, rather than control from a lever-frame, was installed in White City box which took over push-button control of Ealing Broadway signalling in 1974, with closure of the latter cabin. Similar improvements were repeated throughout rapid-transit railways and general purpose railways in developed countries at this time.

Centralising control with remote operation by electricity brought about a shift from electropneumatic operation of points and some semaphores, to electric operation of points and replacement of semaphores by colour-light signals. It was simpler and cheaper to provide electricity for working remotely located points than to install a local compressor and air motor, itself controlled by electric signals. Most manufacturers of signalling and switch apparatus introduced electric points motors, and some railways – including British Railways – undertook development work of their own. Alternatives included electro-hydraulic systems, such as the clamp-lock points machine developed by BR and first used on a large scale at London Bridge, Cannon Street, and in other resignalling work of the late 1970s and early 1980s. Hydraulic motors, fed from reservoirs pressurised by electric powered pumps, moved the points. Electrical detection of relative position between switch and stock rails was employed. Signalling by route-setting became the norm as funds became available for installing modern versions of systems which were proven in the prewar period. The advent of industrial solid-state electronics transformed signalling, control and communications engineering even where the basic principles of operation dated from the 1930s, by miniaturising apparatus, increasing reliability, and facilitating remote operations from centralised points.

Chapter 17

Main line direct current traction in Britain

17.1 The electrical industry and the railways after 1945

Large British general manufacturers like English Electric, or General Electric (GB) were well established in the business of electrifying railways and supplying electrical equipment to industry. These manufacturers were well placed to build diesel, gas-turbine and electric locomotives should British Railways need them. These general manufacturers urged the home railways to order diesel and electric locomotives to create a home market within which British industry could develop the skills required for winning foreign orders.

Those British companies which were purely locomotive builders, and which were not big combines, were less well placed. There was a relatively large number of small companies, some with a history going back to the Stephensons, which did not have the resources of the larger British combines to sustain their own research and development programme. Some of these companies had been building petrol, oil, battery and electric locomotives for many years, but these were small industrial units for mines, quarries and contractors' railways. They had a long history of supplying mechanical components to the general electrical industry for use in locomotives built by the latter, and some erected complete locomotives from components provided by the bigger firms. They lacked the resources and the expertise to lead electrification in Britain and abroad. They went out of business in the 1950s and 1960s, or were taken over by one of the big combines which in turn passed into the present-day dominant concerns, GEC-Alsthom (now Alstom) or Brown-Boveri-Daimler Benz (now Bombardier). Some of these companies made strenuous efforts to enter the diesel and electric locomotive business, but they enjoyed no more success than the American companies which were in the same situation following the dieselising of US railways. There was a 'boom' following the 1939–45 war when colonial railways restocked with steam traction at a time when British railways themselves, before and after nationalisation, were also committed to it. The 1956 announcement by British Railways that steam traction was to be abandoned over a period of about 15

306 *Electric railways 1880–1990*

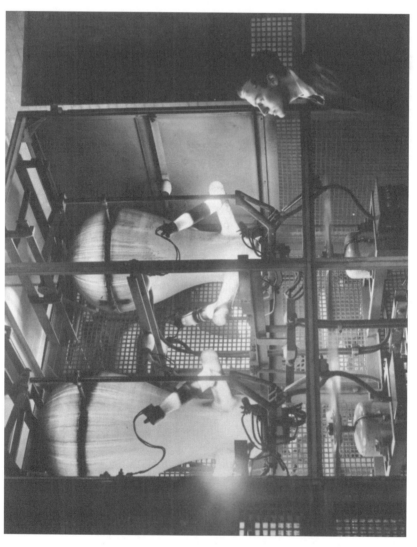

Figure 17.1 Air-cooled glass-bulb mercury arc rectifiers, Stepney substation, London Transport, in 1952. British electrical industries supplied them to LVDC and HVDC railways. They found use in substations and on rectifier locomotives.
Source: London Transport Museum.

years transformed traction policy throughout the Empire and Commonwealth. This forced a too-rapid change on the British locomotive building industry, which could not make the transition and survive, save by becoming sub-units in the larger electrical combines.

A third group of locomotive manufacturers was British Railways' own workshops. This was a very large industry indeed, as the former private companies had built a high percentage of their own locomotives and other rolling stock. The Southern Region (former Southern Railway), with its network of LVDC lines centred on London, had a considerable ability to design and build its own stock, and it produced noteworthy electric locomotives of its own which used components supplied by the established electrical industry. Towards the end of their independence in 1948, the private companies established links with the British locomotive building industry, and with the larger manufacturers, to build prototype electric and diesel electric locomotives. Excellent units were produced, with the railway workshops building some of them for trial on company lines. Despite these achievements, the railway workshops and the design teams of British railways could not carry electrification through on their own. Electric traction and dieselisation ended that considerable degree of independence from outside industry and state regulation, which the railways enjoyed during the era of steam traction.

The electrification of British main lines was retarded by goods operations of many diverse kinds. These were 'inimical to economic electrification', because of the legal requirements laid on the railways to be common carriers. The associated 'pick up goods' work, of which there was a large amount in the 1950s, was labour intensive and inefficient. It involved shunting of individual wagon loads, sorting, marshalling, and the 'picking-up and setting-down' of loads made up of any kind of merchandise in any quantity. This service could not be worked with electric traction, and mixed traction was needed with electric operations supplemented by steam or diesel traction. The diesel alone could not make pick-up goods working economic in the post-war period, and the solution was to rid the railways of it when the law allowed, once the 'Beeching reports' of 1963 and 1965 indicated the full extent to which these services were losing money. The Beeching reports and the subsequent reshaping of British Railways were essential for producing the kind of railway in which economic electrification could be conducted to a maximum effect. Though the pioneer AC electrification of the West Coast Main Line was an engineering success, the costs of working goods and electrifying marshalling yards and sidings reduced profitability and further electrification was opposed in favour of diesel working.

The situation between the end of the war in 1945 and the Modernisation Plan of 1955 was dominated by economics and finance. The government controlled spending by nationalised concerns because the economy was in a parlous state and the country faced bankruptcy in the late 1940s. Priorities were given to rearming the armed services during years of international political tension. Money assigned to the railways was required by a backlog of repair and renewal dating from wartime. Major capital expenditure was impossible and large-scale electrification was a long-term future project dependent on state aid. Riddles' decision to retain steam traction by introducing the 'Standard' classes and building selected pre-nationalisation steam locomotives needs to be judged against this background. New locomotives were required and steam locomotives were cheap. Money shortage prevented buying diesels from the USA (politics apart), and home industry could not supply large numbers of them. Riddles therefore adopted a policy of renewing a modest proportion of the steam stock (4 per cent of the total) and looked to electrification of main lines and heavily used rapid-transit routes when economic recovery permitted it. By 1955, the British economy had recovered sufficiently for the government to finance a modernisation plan of £1660 million which was spent between 1955 and 1965. This money enabled the steam railway to be eliminated; the system to be reshaped; diesel traction to be introduced and electrification to be extended. Electrification included the West Coast Main Line, which was the most important project carried out. The planned East Coast Main Line electrification was not implemented until the 1990s.

The Modernisation Plan was necessary for creating a railway able to meet the demands of the economy which in the 1950s was becoming radically different from the prewar economy served by the steam railway. Unfortunately, the launch of the plan coincided with a rapid fall in coal and heavy goods traffic, which cut profits from freight movements. The growth of motorways; the rise of prosperity for all classes; and the availability of second-hand and new cars obtained through hire-purchase began a steady decline in passenger use of trains. Although electrification and high-speed long-distance services increased passenger-miles on some routes and checked an overall decline, there was a decisive loss of traffic on many secondary, branch, and urban lines. These closed. This favoured electrification of the surviving system but during the 1950s and 1960s there were objections to extending electric traction. No real investment appraisal of the Modernisation Plan was carried out, and in 1960 the House of Commons Select Committee on Nationalised Industries was adversely critical of the plan's expectations which assumed increased revenue and falling costs.

The plan to electrify services between Euston, Birmingham, Manchester, Liverpool and Glasgow was not the only one. Earlier reviews, such as that of Lord Hurcomb (1948–1951) suggested the electrification of the East Coast route. The East Coast scheme was favoured in the 1957 report of the Electrification Committee which proposed electrifying the ECML by 1970, and the WCML between Weaver Junction and Carlisle by 1974. The Eastern Region produced a plan of its own in 1959 for electrifying 415 miles, estimating return on investment of 11 per cent rising to 20 per cent. Unfortunately the sharp decline in traffic between 1957 and 1959 continued into the 1960s and the railways turned to high-performance diesel power represented by the English Electric 'Deltic' locomotive, and the later high-speed trains, which matched the anticipated performance of electric services without the cost of fixed works. Only later, following the introduction of very fast trains in Japan and France, did the performance of electric railways gain a level which diesel traction could not match. This was not the case in Britain in the late 1950s and 1960s.

It has been said that London Midland Region management were set on electrification at all costs and that the fragmented structure of British Railways, which assigned considerable autonomy to the separate regions, enabled the 'LM electrification lobby' to get its way. Dr Beeching argued that electrification would not be economic north of Weaver Junction. The LM scheme was severely criticised because no opportunity-cost comparison with diesel traction was carried out, and the LM justified its scheme on the grounds of ton-miles rather than investment cost appraisal. The argument that further electrification should be carried out simply to keep together a team which had developed strategic expertise in electrification vital to the nation's railways and manufacturing industries was scorned by many economists. There never was irrefutable evidence that the electrified WCML was more economic than a diesel worked railway using the high performance diesel-electric locomotives demonstrated during the late 1950s and 1960s. These included the Brush 'Falcon' D0280 of 1961; the BRCW Sulzer-AEI 'Lion' D0260 of 1962; and the remarkable Hawker Siddeley-Sulzer-Brush 'Kestrel' HS4000 of 1968 which developed 4000 hp. However, the WCML electrification was carried out before the 'great inflation' of 1970–83, and 'negative real financial returns in this period gave a premium to construction works in the 1960s'. The costs of the scheme looked cheap in the 1980s. Long distance passenger traffic on main lines had grown throughout the 1970s and 1980s and the WCML was undeniably an engineering triumph.

17.2 LVDC electric locomotive development after 1945

In July 1937, the Southern Railway completed the electrification of its main line to Portsmouth. This was the end of the third phase of a plan intended to electrify all main lines using the LVDC conductor rail system. It began with the conversion of the old LBSC company's overhead AC system to third rail to comply with the LSWR standard. It was continued between 1932 and 1935 with extension of this network to South Coast towns, including Brighton, aided by government money. By 1937, the SR had 610 route miles and 1550 track miles electrified, worked with multiple-unit stock. The standard of the best multiple-unit trains, like the 'Brighton Belle' used on the longer services was high and the equal of steam worked express trains elsewhere, but the heavier boat trains, goods trains, and trains working through electrified sections, remained steam hauled through lack of an electric locomotive. Tufnell mentions the Kentish coalfields trains, sometimes loaded to 1000 tons, and the boat trains to Dover, Folkestone, Newhaven and Southampton which were made up of 'steam' coaching stock, including Pullman cars. An extended electric traction policy required locomotives for these trains. At this time, the only electric passenger locomotives at work in Britain were the Metro-Vick machines on the Metropolitan lines of the London Transport network dating from 1922.

The first electric locomotives for the Southern Railway originated with a design planned by the 'special development department' at London Bridge (Tayler). They were jointly designed by Alfred Raworth who became Chief Electrical Engineer in 1938, and R. E. L. Maunsell, the Chief Mechanical Engineer. The design dated from 1936, and was originally for a Bo-Bo machine, with a weight of 81 tons, and four 375 hp motors. Design revision increased the weight to 84 tons, when the wheel arrangement was changed to Co-Co to reduce axle-loading. The original scheme was for a goods engine but in 1938, when Maunsell retired, O. V. S. Bulleid became Chief Mechanical Engineer and he revised the design to one for a Co-Co mixed traffic locomotive. General design was done at Waterloo Station with bogie design carried out in Ashford Works (Tufnell) The number of motors was changed to six, one per axle, which were English Electric type EE519A rated at 245 hp each. The engine was designed to work with current taken at 660–750V DC from conductor rail (its usual mode), and from an overhead contact wire in sidings where this arrangement was employed to protect shuntsmen from being shocked through contact with third rail. In the days when loose coupled wagons were coupled by men with poles, running alongside the moving trucks, this was a necessary precaution. It made electric traction unsuitable for marshalling and sorting and steam shunting engines were retained until diesel shunters became available. In the late 1930s, the first diesel-electric shunters were tried but shunting by steam survived into the 1960s.

Tests under the overhead LVDC catenary drawing power through the pantograph were conducted on sidings on the Brighton line, near Balcombe tunnel, and proved successful. The traction motors were of the axle-mounted type. The locomotive rating, with voltage drawn at 660 V, was maximum tractive effort: 40,000 lbsf; continuous tractive effort: 6000 lbsf at 67.5 mph with current of 1630 A and 1080 hp at rail; maximum rail hp, 2200 hp at 35.5 mph. Top speed was 75 mph. Two problems stood in the way of success. One concerned the bogies, and the other related to possible interruptions to the current supply caused by gaps in the conductor rail at junctions. The designers of the bogies had to fit three axle-mounted motors between the frames and to provide a system of supporting the locomotive body, and pivoting the bogie, but the central motor occupied the place where a conventional pivot would go. There were two possible solutions: either to place bolsters between each motor and construct a double bogie bolster, or to use a large, segmented bearing, with the support surfaces taking the form of pads fitted to the frames on the arc of a large circle. The segmented bearing pattern was chosen, either by P Bolland in charge of bogie design at Ashford, or by Bulleid: perhaps both share the credit. The segmented bearing worked well, and was later used on the Southern region diesel-electric locomotives nos. 10201–10203 and on the British Railways type 40, 44, 45 and 46 locomotives, built by Brown-Boveri and English Electric.

The first locomotive, CC1, was completed in 1941 at Ashford Works. The buffing and drawgear was mounted on the main frame of the locomotive body and not on the bogies as originally intended. The bogies were coupled together in articulation but this was abandoned as it caused poor riding.

Main line direct current traction in Britain 311

Figure 17.2 LVDC 1470 hp locomotive built in 1941 to designs by Raworth and Bulleid for Southern Railway 630-750 V conductor rail and overhead system.
Source: National Railway Museum (York).

The most interesting feature of the locomotive was the method of avoiding 'gapping' that is stopping with all four collector shoes out of contact with a live rail in a lengthy gap. Two English Electric motor generator sets were carried in the body of the locomotive, taking their driving power from the conductor rail, resembling a system proposed by Ward Leonard in the 1890s, and tried by J J Heilmann on a DC test locomotive in 1898. These two sets were used to control the locomotive. They were known as 'boosters'. Each motor generator set was coupled to the three traction motors in the bogie beneath it, and the output voltage, which was of the same order of magnitude as the live rail voltage, could be applied across the three motors per bogie. Each set could be isolated if it failed, and the locomotive could run on half power using the other set. When the locomotive started, the output voltage of each motor-generator opposed (i.e. was deducted from) the voltage applied across the traction motors from the third rail. If the voltage output from the motor-generators equalled the voltage in the third rail, there would be no voltage across the motors. By reducing the voltage of the motor-generator sets opposing the third rail voltage, the voltage across the traction motors could be increased to full live rail voltage when output volts from the sets were at nought. In the next phase of control operations, the output voltage from the motor generators was added to the live rail voltage across the traction motors until a maximum of 1320 V DC was applied across the three motors per set. Tufnell (1985) remarks:

The controller had 26 notches, and for the first 16 of these, the booster output opposed the line voltage, gradually increasing this across the three motors from zero to 660V. After this, the next seven notches boosted the output to twice the line voltage thus increasing the motor voltage to 440V across each; the last three notches brought in weak field operation.

All the controller notches could be used continuously, and the 26 positions compared with the ten continuous running notches provided by the controllers of the LNER Bo-Bo HVDC locomotive built in 1940 for the proposed Manchester-Sheffield-Wath project. The boosters were used to overcome the problem of gapping which had troubled the Metropolitan Railway's Metro-Vick Bo-Bo locomotives. A one-ton flywheel was fitted to the shaft of each traction motor-generator, which rotated at 1800 rpm. The energy stored in the flywheel kept the sets turning and maintained power supply to the traction motors so that gaps in the conductor rail could be traversed. The system worked well, and was used in the next generation of SR conductor rail locomotives, though there were problems with it in the first days of trial. Tufnell gives some interesting details of trouble experienced with the pioneer engine, CC1 (Tufnell, 1985):

The first problem was with flashovers on the booster motors due to short circuits on the live rail caused by hanging brake pins or other portions of freight stock. The rate of current rise was found to be of the order of ten million amps per second, requiring a circuit-breaker opening speed of 0.0035sec. Tests on the London Transport at this time had shown voltage surges up to 15,000V due to this problem of short circuits on the live rail.

The original overload circuit breaker had a response time of 0.2 sec which was too slow. A current limiting relay of the Branchu type was fitted in each booster motor circuit which operated on the rate of current increase, and a response time for the circuit breaker system of 0.0023 sec was obtained. This solved the problem. Sometimes, the locomotive ran onto a section of conductor rail in which the current had been switched off for purposes of repair. It was discovered that the booster motor voltage could energise this rail, to the hazard of repair staff. A differential relay was fitted which ensured that the line breaker was open unless the line voltage was 15 per cent greater than the booster motor back EMF.

The first locomotive did well on trials in 1941 and proved capable of hauling 14 coach passenger trains and 1000 ton goods trains. A second locomotive was completed in 1945 and a third in 1948. These worked the first locomotive hauled long-distance main line electric express passenger services in Britain when they took over the London Victoria to Newhaven boat trains in May 1949. They were successful and worked reliably until the last one was withdrawn in January 1969. Tufnell tells of one incident which typified the close co-operation between the Southern Region and English Electric, the supplier of equipment for these locomotives. A burned-out control frame was removed on Monday evening; transported by road to English Electric in Preston; and was rebuilt and returned to Brighton Works by Thursday night. The locomotive was returned to traffic that weekend (Tufnell, 1985):

This sort of co-operation paid off in the subsequent supply of electric and diesel-electric traction equipment to the Southern Region, all of which, except for the Class 33 locomotives, is of English Electric (now GEC) manufacture.

The locomotives were conservative in design and weights of 99 tons 14 cwt for the first two, and 104 tons 14 cwt for the third, were heavy for a power output of 1470 hp. Their successors, introduced in 1959, showed the advances made in LVDC traction technology in the 20 years since the first locomotives were conceived. By 1959, the extensions to SR third rail along the North Kent routes to Ramsgate, and from Gillingham to Dover via Canterbury, had increased the need for electric locomotives to work goods trains and passenger trains which included the famous Golden Arrow and Night Ferry expresses from London to Paris and Brussels. Fifteen Bo-Bo machines of Class 71 were introduced, which incorporated the English Electric booster system of the pioneer locomotives, but which differed considerably in other ways. Only one booster set was fitted, serving four traction motors which were fully mounted on springs in bogies of Swiss design, using SLM flexible drive built under licence at the Doncaster Works of British Railways. A pantograph for use under the catenary over sidings was retained. They were more powerful and lighter than the pioneers. They were rated at 2552 hp, weighed 77 tons, and exerted a starting tractive effort of 43,800 lbsf. A new type of 33 notch controller was fitted, which enabled the driver to notch step by step, and allowed notching-up be carried out automatically – a feature included in the London Midland Region AC locomotives built for the

WCML. The type was successful, but traction requirements on the Bournemouth electrification, completed in the 1960s, resulted in ten of the Class 71 locomotives being extensively rebuilt as dual-mode 'electro-diesel' units in 1967. The five engines of Class 71, and the ten rebuilt as electro-diesels (Class 74) were all withdrawn by December 1977.

The electro-diesel locomotive arose from the need to work trains over both electrified and non-electrified routes under circumstances where electric traction on the electrified lines was desired. If the latter requirement was absent, diesel traction would do, as in the case of the SR diesel-electric multiple-unit sets used on the Hastings lines. An early requirement arose when third rail reached Dover in 1959, and Continental boat trains needed working to the quayside where conductor rails could not be installed, and where overhead catenary would interfere with cranes unloading ships. Motorised luggage vans were constructed, carrying batteries for powering the vans to the quay, but they could not haul a train there. This required an electric locomotive fitted with a diesel engine of sufficient power to move the boat train from the end of the electrified section to a quay. Such a machine could work trains through dead sections where current was off for maintenance, or take a train onto a non-electrified route where maximum power was not needed. Because performance away from the electrified sections did not need to be equal to that under electric traction, the diesel engine output could be a third or a quarter of the electric power rating. The first electro-diesels were six units built in 1962 which were followed by 42 others resulting from a build begun in 1965. English Electric supplied the diesel engine, a four-cylinder type, rated at 600 hp at 850 rpm. There were four traction motors, and the rating under electric traction was 1600 hp. They were of Bo-Bo type, with a starting tractive effort (diesel) of 36,000 lbsf; and (electric) of 40,000 lbsf. Weight was 75 tons, 12 cwt. The later build differed in detail.

Changeover to diesel mode from electric mode could be done in motion, and the electro-diesels could operate in multiple with each other, with Class 33 diesels, and with electrical multiple unit stock. Some electrical multiple unit stock cabs had controls for changing traction mode between electric and diesel in an electro-diesel locomotive working the train. This class worked until recently, the Gatwick Express shuttle being one of its duties. The Class 74, rebuilt from the Class 71 in 1967 only lasted 10 years. The booster control system was retained, with the flywheels removed, and a six cylinder Paxman diesel engine with its own generator was fitted. The plan to drive the booster set with the diesel engine in the diesel-electric mode was not implemented. This Paxman engine was rated at 650 hp at 1500 rpm. The original electric traction motors were retained, and the electric mode rating was unchanged at 2552 hp. The rebuilding work was carried out in British Railways own shops, in Crewe, and resulted in a much more powerful electro-diesel than the Class 73. Starting tractive efforts were 40,000 lbsf (diesel), and 47,500 lbsf (electric). The economics of a complex and extensive reconstruction were questioned but SR reckoned that an electro-diesel of 2550 hp was needed on the Bournemouth lines. The Class 73 could not meet this need unless worked in multiple, so rebuilding the

Class 71 was ordered. During the rebuilding, solid state electronic control was installed, and difficulty was experienced with the printed circuit boards suffering damage. The electronic control gave stepless increase of motor voltage which maintained a steady current so that the motors operated on a particular determined tractive effort versus speed curve. Like the Class 73, they could work in multiple with Class 33 diesel-electrics and electric multiple units. The decline of Southampton as a passenger liner port reduced the need for running boat trains, which had been one of their main duties, and they became superfluous to requirements – another example of unforeseen changes in traffic patterns upsetting traction policy. The Paxman engine was non-standard in a region where EE equipment was the norm and it proved unreliable. The locomotives were scrapped in 1977 along with the unrebuilt units of Class 71.

17.3 The high-voltage direct current system

The only example of the High Voltage DC main line railway in Britain was the Manchester-Sheffield-Wath system, projected by the London & North Eastern Railway in 1936 to overcome the problems of working heavy coal trains by steam traction over a steeply graded line with sharp curves, which included a tunnel at Woodhead in which smoke nuisance was a hazard to engine crews. These were the circumstances in which American railways like the Norfolk & Western; the Great Northern; the 'Milwaukee Road' and the Boston & Maine, had introduced electric traction. The objective was not to electrify a section from which a network would spread, but to remove an operating bottleneck. Steam traction was adequate on services leading into and out of the electrified section. The heavy coal trains from the Wath concentration yards needed two eight-coupled locomotives at the head of the train, and on the steepest slopes an additional two in the rear were often required. The largest steam engine in Britain, the LNER 2–8–8–2 Garrett was built to replace two banking engines in pushing coal trains, very slowly, up the hills to Woodhead. In winter, the weather could be bad, which added to operating difficulties.

Following American experience with lines containing similar bottleneck sections, and drawing on its recent practice with HVDC on the Shildon-Newport line (dismantled in 1935), the LNER decided to electrify the Woodhead tunnel section in 1936, using 1500 V DC. This latter decision was in part influenced by the storage of locomotives from the Shildon-Newport electrification, which could be used on the planned route which was 73 miles long. The projected locomotives were of a simple and reliable design, resulting from co-operation between Metropolitan Vickers, and the LNER engineering staff, with mechanical components designed under the supervision of Sir Nigel Gresley. The first designs were for a Bo + Bo locomotive, with bogies coupled, and buffing and drawgear on the bogies. There were two overhead pantographs, taking DC current at 1500 V from the catenary to supply four nose-suspended, axlehung traction motors with spur gear drive. An electropneumatic control system was fitted.

Figure 17.3 Main frames and running gear of prototype HVDC locomotive no. 6701 during construction in Doncaster Works during 1940 showing plate frames, and four axle-mounted nose-suspended DC traction motors. The bogies were coupled together.
Source: National Railway Museum (York).

The locomotives as built were fitted with compressed air brakes, regenerative brakes for use on long inclines, and (after 1959) rheostatic brakes. The fixed works included special substations for dissipating regenerated power which was surplus to requirements and could not be used. Maximum starting tractive effort was 45,000 lbsf, and the engines were rated at a continuous power of 1300 hp at rail at 56 mph. Maximum power was 3000 hp at 27.5 mph, though the usual rating for the class was 1868 hp. They weighed 87 tons 15 cwt. Maximum speed was 65 mph. They were simple machines, with outside plate frames to the bogie and springing which reflected steam railway practice.

The pioneer locomotive, LNER 6701, was built at Doncaster in 1940, and exhibited at York in 1941. It ran tests on the HVDC Manchester-Altrincham line, and was then stored following wartime cessation of work on the electrification of the former Great Central line over Woodhead. This work was resumed when the post-war economy recovered. After 1950, 57 locomotives were built in the former LNER (Great Central) works at Gorton, Manchester. They were known as Class EM1, later Class 76. The prototype ran on the Netherlands Railways after the end of the war, providing experience with HVDC main line locomotion and helping the Dutch rebuild their devastated system. The Dutch trials revealed bad riding at express passenger train speeds, and radically different bogies and springing were designed and incorporated into the post-war locomotives. The electrification of the Manchester-Sheffield-Wath line was completed in 1954, delayed by the construction of a new double-tracked tunnel at Woodhead to replace the old twin single-bores, which were too small to accommodate overhead distribution. Collapse of this new tunnel during construction delayed the opening until 1954, by which time the project was 18 years old, though it still merited the title 'the most modern railway in Britain'. The stored locomotives from the Shildon-Newport line were not used, apart from one employed in shunting. The locomotives were conservative in design but they worked well until the line closed through shifts in traffic movements in 1981. The pioneer engine, LNER No 6701 had been withdrawn in 1970. Control of the Class EM1 was manual and there were ten normal running speeds. In the later 1960s they were fitted to work in pairs with multiple-control, and saw service on heavy coal trains serving Lancashire power stations.

The Co-Co express passenger locomotives were a more advanced design, being planned well after the war. Like the EM1 locomotives, they were erected at Gorton Works, Manchester, with mechanical components made by British Railways and electrical components from Metropolitan Vickers. There were seven of them, known as Class EM2 (later Class 77), and when built in 1953–54, they were the most modern and powerful motive power units in Britain. The original scheme was for 27 Co-Co locomotives, but only 7 were constructed, and some passenger duties were worked by EM1 locomotives, specially fitted with train heating boilers. Passenger services failed to expand over the route and in the 1960s declined sharply. Manchester to Marylebone express trains over the Great Central route were withdrawn and Manchester to Sheffield traffic was diverted to the Midland Railway route through Hope Valley and worked by

Figure 17.4 Prototype HVDC 1868 hp locomotive no. 6701 built at Doncaster Works in 1941 for proposed 1500 V DC electrification of LNER main line between Manchester and Sheffield via Woodhead Tunnel.
Source: National Railway Museum (York).

diesel traction. Manchester-Harwich boat trains were re-routed. The Class EM2 (Class 77) was withdrawn between 1968 and 1970 and sold to the Netherlands Railways in 1971 where they worked until the mid 1980s. One was returned to the Manchester Museum of Science and Industry for public display. The EM2 type were designed for higher powers and better riding at speed than the EM1. The six wheeled bogies were similar to those which had given good service on the pioneer diesel-electric locomotives built for the LMS in 1947 and 1948. The bogies were not articulated, and the buffing and drawgear were mounted on the locomotive main frame, not the bogies as in the case of the EM1. This improved the riding, and a top speed of 90mph was permitted. Six Metropolitan Vickers type 146 traction motors were fitted. These were axle-hung nose-suspended with direct spur gear drive. Cooper (1985) remarks:

There was considerable resistance to flexible drives at that period although in Continental practice they were general for speeds of that order (90 mph). It was recognised, however, that lateral movements of the heavy motors could increase the stresses on the track at high speeds and in these locomotives they were restrained by struts with rubber-bushed attachments to the motor and bogie frames.

They were very powerful for the time of their construction (1954) though conservative in concept. Starting tractive effort was 45,000 lbsf, and 15,600 lbsf was provided at 57 mph, giving 2300 hp at rail. Maximum rail horsepower was 4600 hp at 38 mph. They were given a general rating of 2490 hp. These figures were far in excess of any performance of a contemporary steam or oil-engined locomotive in Britain. Air brakes and regenerative brakes were fitted for the locomotive and vacuum brakes for the passenger stock. A train heating boiler with a capacity of 1000 1b/hr was carried – a necessity in an era when most train heating was done with steam drawn from the locomotive. The EM2 type was a large locomotive, measuring 59 ft over buffers, and weighing 102 tons 10 cwt. Control was electro-pneumatic. Following the decision by British Railways in 1955 to adopt 25 kV, 50 Hz single-phase AC for all future new electrification other than extensions to the SR conductor rail system, the Manchester-Sheffield-Wath line became an isolated anachronism, representing technology and practice defined in the period 1910–20 when electric traction was used on difficult routes where steam traction couldn't cope. When both passenger and mineral traffic declined, and the remnant was switched to other trans-Pennine routes, the system was scrapped and the line closed. Shorter HVDC passenger lines in other parts of the country were converted to single-phase AC.

Chapter 18

Main line alternating current traction in Britain

18.1 Advantages and disadvantages of the standard systems

In the early 1950s, when the comparison between the HVDC and the AC system was made there was anxiety about the clearances which the HVAC system required for safety. It was thought that the 25 kV AC railway would require 11 inches static clearance, and 8 inches passing clearance. The recommendations for 1500V DC, dating from 1932 recommended 10, 6 and 4 inches depending on circumstances. The greater clearances needed for AC might involve raising bridges, lowering tracks or including sections at 6.25 kV. It was eventually shown that electrical clearances could be reduced and three standards were identified. For static clearance these were 10.6 in/270 mm (normal); 7.9 in/200 mm (reduced) and 5.9 in/150 mm (special reduced). For passing clearance these were 7.9 in/200 mm (normal); 5.9 in/150 mm (reduced) and 4.9/125 mm (special reduced). The clearance between a pantograph and a brick or masonry wall or tunnel could be further reduced to 3.2 in/80 mm with permission from the Railway Inspectorate and it proved possible to electrify the West Coast Main Line using 25 kV throughout. The extensive 6.25 kV sections installed in the first AC schemes in areas within cities were converted to 25 kV, and the need to install 6.25 kV as a routine requirement in cities, to meet the clearance requirement, was obviated. These expensive restrictions, which would have been a severe though not fatal drawback to AC, were avoided. It was further argued, based on SNCF experience, that AC rectifier locomotives enjoyed advantages over HVDC locomotives controlled by resistances. These included better adhesion and less weight for equal power, though they were more expensive in first cost.

As a result of Warder's report, and other recommendations, the British Transport Commission accepted 25 kV, 50 Hz AC as the future standard (apart from extensions to the Southern Region LVDC network) and the government approved this in March 1956. A policy of electrification had been decided on earlier. In a report to the Railway Executive in 1952, R. A. Riddles had written that:

Electrification would gradually replace steam as justification could be demonstrated, route by route, and capital became available. This assumed the co-existence of both forms of motive power for many years ahead.

Riddles and his staff did not then think that diesel motive power was sufficiently developed for widespread introduction, so that coal-fired steam traction would be used on non-electrified lines. These is strange when one of the arguments used in favour of extensive electrification was that it would release up to 14 million tons of coal for export. Oil-burning steam locomotives had not been economic, and if diesels were unsuitable (a questionable claim) there was no alternative to coal-burning steam motive power. However the Modernisation Plan of 1955 decided to eliminate steam traction within 15 years and replace it on non-electrified lines by diesel traction.

18.2 Gaining AC expertise

In 1952, the former Midland Railway branch line between Lancaster and Heysham, which had been electrified in 1908 6.6 kV, 25 Hz was converted to 6.6 kV, 50 Hz to investigate electrification at industrial frequency. Trials were carried out using former London & North Western Railway multiple-unit conductor rail stock from the Earls Court-Willesden LVDC line, which was taken out of store, converted and fitted with mercury-arc rectifiers. Experiments on this branch were carried out between 1955–56 with germanium rectifiers. This was a very early use of the solid-state rectifier in traction. The Heysham branch trials confirmed Riddles and Warder in their choice of 25 kV, 50 Hz single-phase AC for future electrification. Experience on the first AC systems to be installed, such as the 1959 Colchester-Clacton-Walton scheme, justified this decision. The 1955 modernisation plan advocated that AC traction be applied to selected main lines and to suburban networks which included the Glasgow district and lines in the London area running out of Kings Cross, Liverpool Street, and Fenchurch Street. Near cities, where clearances were limited, voltage was to be 6.25 kV. The electrification of suburban networks using AC was carried through by 1962 even though it meant converting lines which had been recently electrified on 1500 V DC. These included the Liverpool Street-Shenfield; Shenfield-Southend Victoria; and Shenfield-Chelmsford lines which totalled 51 route miles, compared with the 73 miles of HVDC route between Manchester, Sheffield and Wath.

There were plans for electrifying three longer routes over which passenger and goods trains would be worked and for which locomotives would be needed. The plan included lines which were not electrified until the 1990s, probably because of national economic difficulties, inflation, and a decline in traffic following the completion of the WCML electrification to Glasgow in 1974. The development of the high performance diesel train deferred electric traction on the East Coast route until performances were required which only electrification could provide. Two of these projects, recently completed, were for the Kings Cross-Doncaster-

Leeds-York project (extended to Edinburgh); and for the Liverpool Street-Harwich route. The major scheme was for the WCML from Euston to Birmingham, Crewe, Liverpool, and Manchester. The decision to adopt the AC system was remarkable because British Railways had little AC expertise. As Haresnape (1983) writes:

Sir Brian Robertson, the then Chairman of the British Transport Commission (BTC) said that the AC system must be regarded as new and as yet untried on a large scale under British conditions, which were in many ways more exacting than those existing on any other rail system worldwide ... the decision implied ... acceptance of the risks ... attached to the adoption of something new.

At this time, there was not a single AC locomotive under design suitable for working heavy passenger trains and goods trains over the busiest main line in Britain. In the first phase of electrification many of these trains would be handed over to steam or diesel traction at Crewe and multiple unit passenger stock could not deal with all the passenger traffic. Locomotives able to match if not surpass the performance of the most modern steam locomotives would be needed to work the line. In fact, the first AC locomotives on the WCML completely surpassed the achievements of the most powerful steam locomotives and performance was raised to still higher levels within a few years. The 1959 Colchester-Clacton-Walton line provided experience in working an AC line with multiple-unit stock, and further information became available when the Crewe to Manchester route was electrified as a pilot project for the 25 kV, 50 Hz WCML scheme.

The first AC main line locomotive was not completed until autumn of 1958, and proved to be an historic machine in many ways. The first electric locomotives for the WCML were to number five classes. More had been considered. It would take time for the British electrical and locomotive building industries to overcome the lack of AC expertise and produce units able to meet the requirements of a staff training programme. A pilot machine was needed at the earliest possible date which could be used to train staff and reveal potential problems in main line AC locomotive working. The pioneer locomotive, no. E2001 was built using components from one of the two Great Western Region gas-turbine electric units, number 18100. This machine, originally built by Metropolitan Vickers at Trafford Park in 1952 was withdrawn in January 1958 and converted into an electric locomotive, number E1000, renumbered E2001 in 1959. The gas-turbine-electric locomotive had lain unused in Dukinfield, Manchester, and was taken to Stockton on Tees, to the works of Metropolitan Vickers- Beyer Peacock Ltd., a new company formed by Metro-Vick and Beyer Peacock at a time when the latter was trying to gain entry into the diesel and electric traction business. Reconstruction was completed by October 1958. A brief description of the rebuilding is provided by Webb and Duncan. It was an extensive job, which lowered the weight from 130 tons to 105 tons, and the wheel arrangement was changed from Co-Co to AlA-AlA. The gas turbine, the DC generator, air filters and fuel tanks were removed, and the cabs changed from the Western region

324 *Electric railways 1880–1990*

Figure 18.1 The first main line electric locomotive constructed in Great Britain for single-phase 25 kV 50 Hz operations. Rebuilt in 1958 by Metropolitan Vickers from gas-turbine-electric locomotive. Fitted with glass-bulb rectifiers and rated at 2500 hp. Extensively used for instructional purposes during electrification of the West Coast Main Line.

Source: Metropolitan Vickers/GEC (now Alsthom).

right hand drive to the left-hand drive adopted as standard for British railways. Much of the mechanical equipment was retained, as were four of the six Metro-Vick DC axle-hung, nose-suspended traction motors, which were left to drive the outer axles of each bogie. The electrical equipment needed for the conversion included two Stone-Faiveley pantographs; a Brown-Boveri air blast circuit breaker; and Hackbridge & Hewittic mercury-arc rectifiers for converting the AC line supply to DC for the traction motors. Provision was made for moving the locomotive when away from the catenary using the batteries, though for short distances only. There was a cab at each end, and the body contained a small messroom to cater for crew awaiting training.

By 1960, when the AL1 and AL2 classes were entering service, Engine E2001 had helped to train 1300 locomotive crewmen. It started its duties in 1958 on the first electrified section of the Manchester to Crewe route, between Mauldeth Road and Wilmslow and was used to train staff with the help of multiple unit stock intended for the Tilbury line of the Eastern Region, but sent to the London Midland Region until the latter got stock of its own. It was later used to instruct crew in the Liverpool area, in Crewe, and in Glasgow. After being used on the Glasgow suburban AC system, it was stored for periods at Crewe, Goostrey and Rugby with brief spells of training duty. It was withdrawn in 1968, stored until 1972, and then stored for scrapping Granted its experimental nature, E2001 worked well. It demonstrated that the AC/DC rectifier locomotive using mercury-arc rectifiers would work from a supply at industrial frequency, and it did much to build up AC expertise before the production classes of AC main line locomotive were introduced.

18.3 The first production classes: AL1–AL5

The first production classes were introduced between 1959 and 1960. They represented a joint design and construction exercise between the British electrical combines; the private locomotive building companies, some of which began in the days of Stephenson; and British Railways workshops. They were designed when electrification of both the ECML and the WCML was planned, for which an A series of mixed traffic locomotives and a B series of goods locomotives were envisaged. The original plan called for 80 type A and 20 type B. The type B was geared for lower speeds than the type A, and was for heavy goods trains fitted with continuous brakes to permit speeds higher than usual for such traffic. The introduction of fitted goods stock did not proceed with the rate expected, and in the 1960s there began a serious decline in goods traffic which was never reversed and which caused serious economic setbacks to the railways. Slow 'pick-up' goods trains were eliminated following the Beeching reports and freight traffic was concentrated as much as possible into faster container trains or unit trains carrying one type of load like coal, ore, clay or minerals. This reduced the need for goods engines and the initial plan for 20 of type B was reduced to 5. Only two were constructed as B variants of

Class AL3 and the other three were completed as A units of the AL1 and AL3 classes.

All the designs were for high powered Bo-Bo locomotives weighing 80 tons. Flexible drives and fully spring-mounted traction motors enabled four axles rather than six to be used. Five types, classified AL1 to AL5 were built between 1959 and 1964. Later these were re-classified as types 81 to 85. They set new standards on British main lines for power, power-to-weight ratio, availability and speed. They effected a 'traction revolution' by surpassing by far the best that steam traction had done. They outperformed contemporary diesel traction though there were engineers who claimed that diesel-electric locomotives could be developed to match the electric locomotives. At the time of their introduction, the DC traction motor was the favoured norm and in accordance with practice pioneered in the United States, Germany, and France, and in Britain on E2001, static rectifiers were used to provide motor current. Classes AL1 to AL4 were fitted with ignitron mercury-arc rectifiers of the steel tank type, specially developed for locomotive service. The ignitron, and other locomotive rectifiers, caused much trouble in the USA., France and Britain. The AL3 and AL4 classes suffered severely from failure and rectifier locomotives were generally unreliable until solid-state semiconductor silicon-diode rectifiers were introduced in the 1960s.

These first five classes were designed and built to the British Railways specification as a joint enterprise between the electrical industry, then undergoing combination, and the former builders of steam locomotives which were amalgamating with the electrical combines or approaching close-down. In some cases British Railways workshops erected the machines. Economists adversly criticised the order for five classes within a general specification along with much of the WCML electrification programme and preferred the American practice of relying on two very large traction combines for supply: General Motors and General Electric. Cooper remarks that manufacturers were at liberty to develop their own ideas but there were components which had to be used in common. These came from overseas suppliers because there were no proven British equivalents. He lists air-blast circuit breakers; pantographs and flexible drives. The first air-blast circuit breakers came from Brown-Boveri, though AEI (Associated Electrical Industries) developed one for locomotive use which was used in Britain. Pantographs were of the Faiveley type employed by the SNCF and built in Britain by Stone & Co. Both the Alsthom, and the Brown-Boveri/SLM flexible drives were installed (Cooper, 1985):

Two types of flexible drive were specified. One was the Alsthom drive with rubber-bushed links connecting drive arms on a quill shaft to opposite corners of a 'floating ring', the other corners of which were similarly connected to the driving wheel. The second was the Brown-Boveri/SLM drive in which torque was transmitted by spring-loaded pads within the main gearwheel to the arms of a 'star' wheel mounted on the axle.

In these systems the entire weight of the traction motors was fully sprung-mounted on the bogie frames. In the classic axle-hung arrangement, half the weight was carried by the axle which caused severe damage to track. The flexible drive systems permitted movements of the axle with the primary suspension, whilst keeping constant the distance between motor pinion centre and final drive gear centre.

Two types of control were fitted. Tap changing on the low-tension secondary of the transformer was used on classes AL1, AL3 and AL5.

In Classes AL2 and AL4 the tappings were on the winding of a high-tension auto-transformer, and a variable voltage was applied to a fixed-ratio transformer with its secondary connected to the rectifiers.

The advantage of high-tension tap changing is that small currents are passed through the tap-changer and switches compared to those in the low-tension type. A well-illustrated account of the equipment of these engines is given by Webb and Duncan. The first class was the AL1, later British Railways Class 81. It was constructed by the British Thomson-Houston Company, then part of Associated Electrical Industries (AEI), and the Birmingham Railway Carriage & Wagon Company. The first engine, no. E3001, was delivered to British Railways in November 1959 and was used for crew training over the Styal line on the Manchester-Crewe route. The class was of the mixed traffic type and numbered 25 units including two which were designed as type B, but which were completed as type A. There were four AEI type 189 motors, fully spring-mounted in the bogie, driving through Alsthom flexible gears. Control was by tap changing on the low-tension side of the transformer. With supply rated at 22.5 kV, performance was as follows: maximum tractive effort, 50,000 lbsf; continuous tractive effort, 17,000 lbsf at 71 mph; 3200 hp at rail at 71 mph; max. rail power, 4800 hp at 44 mph. Maximum speed was 100 mph. The class gained the reputation for being rough riding caused by the original Alsthom bogie. In 1977, comparative trials were conducted between Glasgow and Beattock with a newly repaired machine and two engines still in service. These revealed defects in the bogie which were then corrected. The engines were built with two Stone Faiveley pantographs and three air-cooled, six-anode pumpless steel tank mercury arc rectifiers, diametrically connected in parallel. These were replaced by silicon rectifiers in a programme of refurbishment which began in 1970. One pantograph was removed and dual braking fitted for working trains with air or vacuum brakes at a time when British Railways was moving from the vacuum brake to the air brake standard.

The Class AL2, later Class 82, was introduced in 1960. It was constructed by Metropolitan Vickers, then part of Associated Electrical Industries, and Beyer-Peacock, the Manchester based builder of rolling stock which once supplied steam locomotives to the Empire. There were ten locomotives in the class. There were four AEI type 189 traction motors, fully spring-suspended in the bogie, with the Alsthom flexible drive. Control was by high-tension tap-changing. The Class AL2 had multi-anode mercury-arc rectifiers like the AL1. Performance at 22.5 kV was as follows: maximum tractive effort, 50,000 lbsf; continuous tractive effort, 17,000 lbsf at 73 mph with 3300 hp at rail, and a maximum rail

Figure 18.2 Prototype AL2 3300 hp rectifier locomotive built by AEI-Beyer Peacock in 1960 for the West Coast Main Line electrified at 25 kV 50 Hz.
Source: GEC (now Alsthom).

horsepower of 5500 hp at 47 mph. Maximum speed was 100 mph. The prototype locomotive proved overweight during construction at Gorton Works, so fibreglass and aluminium were substituted for steel in non-loadbearing parts of the structure and some electrical components had to be replaced by lighter units. This delayed delivery of the class, which took two years for the ten machines. Not all delays were due to the weight-reducing changes. Some were caused by problems with the igniters of the rectifiers and with the tap-changers. Beyer-Peacock built the class with a heavy-load bearing underframe, Commonwealth bogies and a light bodywork. The first two features were probably responsible for the excessive weight of the design. The Commonwealth bogies were a great success and gave a good ride. They had a suspension based on that used in the Great Western Railway gas-turbine-electric locomotive no. 18100 which was converted into the first AC locomotive E1000 (later E2001).

The major electrical components were supplied by Metropolitan Vickers. MV was then part of AEI, and supplied the mercury-arc rectifiers which were fitted in the AEI Class AL1. Two engines were destroyed by fire, one in 1966, and another in 1971. The surviving 8 locomotives were refurbished after 1971. Engine E3049 was the first of the prototype AC locomotives to have its mercury-arc rectifiers replaced by silicon rectifiers. The class gained the reputation for being more reliable than the other four classes, because their electrical circuits were 'by far the simplest', and they enjoyed good riding (Webb and Duncan, 1979).

The Class AL3, later Class 83, was introduced in 1960. It was built by English Electric at the Vulcan Foundry, Newton-le-Willows, which had been an independent builder of steam locomotives which was taken over by the EE combine. The class numbered 15 mixed-traffic locomotives constructed between 1960 and 1962. Three of them were originally planned as type B locomotives. Two of these were built as type B but converted into type A by regearing in 1962. The other was not completed as a type B machine but was finished as a special type A engine, number E3100, and used in important experiments with transductor control, wheelslip protection, automatic control of tractive effort, rheostatic braking and a silicon rectifier. All other engines, as built, had ignitron rectifiers. The Class AL3 locomotives were fitted with four English Electric EE535A motors, with fully sprung suspension in the bogie and Swiss SLM flexible drive. Control on the first 14 engines was by low tension tap changing. With the supply rated at 22.5 kV, the maximum tractive effort was 38,000 lbsf; continuous tractive effort was 15,260 lbsf at 73 mph with 2950 hp at rail; maximum power was 4400 hp at rail at 50 mph. Maximum speed was 100 mph. The class had several striking design features such as the extensive use of weight saving construction techniques and materials. Fibreglass and light alloys were employed and steel castings avoided wherever possible. The body and underframe were of integral stress bearing steel construction. In 1960, a complete body was proved by tests in the English Electric mechanical laboratories at Whetstone, Leicestershire and then furnished to experimental locomotive E3100. The bogies were equipped with a new form of secondary springing, and the locomotives rode well provided the weight distribution between body and bogies was corrected by ballast.

The chief electrical engineering feature were the single-anode rectifiers in place of the multi-anode type found in the classes AL1 and AL2. Groups of single-anode rectifiers, either ignitrons or excitrons, enabled a failed rectifier to be replaced individually, whereas a failed anode in the multi-anode form often necessitated replacing the whole unit. There were two main types of single-anode rectifier on the market, known by trade names which were loosely applied. There was the General Electric (USA) ignitron, in which the arc was ignited at the start of each rectifying half-cycle, and the Allis-Chalmers excitron which maintained a continuous arc when the ignition coil was energised, so that the cathode was ready to conduct current when the anode fired. The Class AL3 used ignitrons. Each locomotive had eight liquid-cooled ignitron rectifiers, arranged in two bridge-connected groups, with each bridge supplying two traction motors in series, with the centre point connected to the transformer to keep the voltage stable during wheelslip. Each ignitron was about 8 inches in diameter and 32 inches high, and was water cooled. The main control was by low tension tap-changing with camshafts driven by air motors. These ignitron rectifiers gave trouble to such an extent that the existence of the class was threatened. Webb and Duncan (1979) report:

The troubles with the rectifiers were intermittent, and one could start an 83 (AL3) from cold, commence a run with a train and the rectifiers would heat up and work without trouble. However at the end of the run the locomotive would lay over with pantographs down, and on the next run as the rectifiers heated up back-firing would occur. Again, various causes such as mercury condensation, the inability of four rectifiers out of balance to share the load correctly, were quoted as possible causes.

Persistent rectifier failure resulted in proposals by British Railways and English Electric to remove the ignitron rectifiers and fit solid-state silicon rectifiers similar to those tried on E3100. The rectifiers were to be roof-mounted for cooling in the natural air stream. British Railways approved of this scheme but English Electric demurred. The entire AL3 class of ignitron locomotives was withdrawn during 1967 and 1968 and stored at Bury shed until 1971, following the introduction in 1966 of the Class AL6 which created a motive power surplus on the WCML. In 1971, it was decided to rebuild the Class AL3 – including the experimental engine E3100 – with silicon rectifiers, and dual-system braking equipment. This was done at Doncaster Works between 1971 and 1973. Earlier classes AL1 and AL2 were similarly reconstructed with solid-state silicon rectifiers. Unfortunately, removal of ballast and the mercury-arc rectifier frames during reconstruction upset the trim of the locomotives and this needed to be corrected by adjusting the secondary suspension. By 1982, the class were in store again, as newer types made them surplus to operational needs.

Engine E3100 was the most successful of the AL3 Class and its role as a test bed for new techniques was important. This engine was built with the lightweight body developed by Vulcan Foundry and was selected to test the transductor method of control, which gave smooth variation in motor voltage rather than the step increases of the normal tap-changing system. Engine 3100 differed from the rest of the Class 83 which had ignitron rectifiers, in having silicon rectifiers which proved more reliable. The transductors were in the secondary AC circuit between the conventional tap changer on the secondary transformer coil and the silicon rectifiers. The transductors were magnetic windings which produced low resistance when unsaturated and high resistance when saturated. The effect was to smooth the sudden changes in voltage which took place when the tap changer was notched up through each step. With this stepless tap changing, the driving motor torque varied smoothly and there was less chance of wheel slip when tractive effort was increased. The increase of tractive effort of Engine 3100 was automatically regulated and it pulled heavier loads with greater acceleration than the rest of the class 83. The engine was completed at Vulcan Foundry, Newton le Willows, and delivered to Crewe Works in April 1962 Riding was improved by adjusting ballast weights and by lubricating bogie sidebearers which made it about one-and-a-half tons heavier than the other class 83 locomotives. During 1962 it ran with bogies geared for express speeds, and with bogies from the only remaining Class 83 locomotive geared for 80 mph maximum.

During the tests with the transductor system, it was planned to send the locomotive to Europe to compare it with German, Swiss and French locomotives in trials scheduled for May 1963. The British electric traction industry were anxious for this to happen because failure of the first AC multiple unit sets on British Railways was receiving widespread attention and damaging the industry's reputation. An excellent performance by E3100 in competition with foreign manufacturers would win welcome publicity and restore confidence. It was returned to Vulcan Foundry in November 1962 to be prepared for the overseas trials. It was fitted with a system of notchless control, which employed constant current control of the traction motors during acceleration and braking, with the reference current set by a master controller. Automatic regulation of motor current and tractive effort was provided to halt wheel slip and excellent adhesion with high tractive effort was obtained. English Electric further modified the locomotive by fitting weight transfer equipment, operated by air cylinders on the bogies, to compensate for the shift in load between axles during braking and acceleration. The proposed May 1963 tests were postponed until October, and the modified E3100 was successfully tested between Stafford and Crewe in May. S B Warder remarked:

... this design performs as well as, or even better than, designs incorporating single-motor bogies, without the mechanical complications of the latter, while preserving the simplicity and easy maintenance associated with individual axle drive.

Unfortunately the proposed European comparative trials of electric locomotives from several countries, rescheduled for October 1963, were cancelled and E3100 remained in Britain. The trials with E3100 suggested ways of removing the chief defects of the class which were the unreliable nature of the ignitron rectifiers and the trouble with weight distribution. In 1963, English Electric and

British Railways considered fitting the 14 standard machines of Class 83 with silicon rectifiers and transductors, or modifying them to resemble E3100 but without notchless control and the weight transfer equipment. Other proposals were considered. In 1967, it was suggested that E3100 be converted to thyristor control, and fitted with dual brake and rheostatic brake, for development tests. This was not done and in 1968 the entire Class 83, including E3100, was stored at Bury engine shed until 1970 when modification of the class began at Doncaster Works. Between 1970 and 1973 the locomotives were fitted with silicon rectifiers and dual braking systems. By then, notchless control was available using thyristors without a tap changer and was preferable to the experimental system used in E3100. By 1982, the class was in store again.

The Class AL4, later Class 84, was introduced in 1960 and withdrawn by 1980. It consisted of 10 mixed traffic locomotives. The type was a joint product of General Electric (GB) and the North British Locomotive Co., Glasgow. The locomotives had four General Electric DC traction motors fully spring-mounted in the bogie, driving through Brown-Boveri flexible, single-reduction gear. The maximum tractive effort was 50,000 lbsf; continuous tractive effort was 17,600 lbsf at 66 mph, when the power at rail was 3100 hp. Maximum rail power was 4900 hp. Maximum speed was 100 mph. The class was distinguished by weight-reducing body construction, and an improved form of mercury-arc rectifier.

To reduce total weight, the North British Locomotive Company developed an integral body structure at its Hyde Park Works, Glasgow, which took the form of a largely mild-steel, welded Vierendeel truss. This did not make the AL4 the lightest type, though it was lighter than it otherwise would have been. The working-order weights of the classes compare as follows: AL1, 80 tons; AL2, 77 tons 9 cwt; AL3, 74 tons 10 cwt; AL4, 77 tons; AL5 79 tons 13 cwt.

The rectifier was of the mercury-arc type and was designed to eliminate the problems encountered in locomotives in the USA and Europe. It went under the trade name 'Com-Pak' and General Electric (GB) claimed that it combined the best features of the ignitron of GE(USA) and of the excitron of Allis-Chalmers. It was a water-cooled, single-anode rectifier. Compared to the types in regular use like the ignitron, which was the most common form in locomotives, the arc length was reduced, the voltage drop lowered, and efficiency raised. The 'Com-Pak' was supposed to be resistant to shock and surge of the mercury pool, but most manufacturers made such claims for their locomotive rectifiers. Unfortunately the claims made for the 'Com-Pak' rectifier were not justified and the Class AL4 was plagued by breakdowns of this component. In 1962 a locomotive was sent to GEC (GB) for investigation and by 1963 the whole class was in the GEC works at Dukinfield, Manchester for modifications. The defects were not remedied and following repeated in-service failures the class was stored after 1967 in Bury shed, along with the Class AL3. In 1971 and 1972 they were rebuilt at Doncaster Works with solid-state silicon rectifiers and dual braking systems but return to work was followed by repeated traction motor failure. Withdrawal of the class started in 1977 and was complete in 1980.

The failure of the rectifiers on Class AL4 is strange considering the very thorough research done by General Electric (GB) to produce a shock-proof unit. GEC fitted two former parcel vans with a diesel generator in one car to provide electrical power and transformer and rectifiers in the other. This was shunted into buffers; run in loose-coupled goods trains; and rapidly started and stopped. Careful measurements were taken of rectifier behaviour during these operations, which were more severe than a main line locomotive would experience. The rectifier promised well, yet within 18 months repeated failures began. The problem was ended rather than solved by the switch to solid-state silicon rectification in the early 1970s. The British experience paralleled the American, and the locomotive mercury-arc rectifier did not perform reliably anywhere. Locomotive mounted rectification required the solid-state system which was installed from the start in the class AL5.

The class AL5 of 1960 (later Class 85) was designed and built by British Railways at Doncaster Works using AEI electrical equipment which included solid-state rectifiers. This class of 40 locomotives was the first in Britain to be built with solid-state rectifiers and it proved to be reliable and successful. Apart from the rectifiers, the AL5 design resembled the AL1. Maximum tractive effort was 50,000 lbsf. Continuous tractive effort was 17,000 lbsf at 71 mph when the power at rail was 3200 hp. Maximum power at rail was 5100 hp at 45 mph. The four AEI motors were fully spring-suspended in the bogie and drove through Alsthom single-reduction, flexible drive. The AL5 had Alsthom bogies, like the AL1, and riding was poor. This was never entirely eliminated and the engines were mainly used on goods work or slow overnight sleeping car trains. The rectifiers worked well and solid-state devices became the norm. Britain pioneered locomotive-mounted germanium rectifiers which were fitted to AL5 engines E3056 to E3085 but the silicon rectifier, pioneered in the USA, and in France since 1958, and fitted to E3086 to E3095 became standard.

During overhaul of the class between 1968 and 1971, the germanium rectifiers were replaced by the silicon type, which also supplanted the mercury-arc valve in locomotive work. The germanium rectifiers used in the Class AL5 required a total of 1280 diodes. The silicon type needed only 336 diodes which improved reliability. The germanium rectifiers were made up of 20 racks or trays, arranged vertically, with each having 64 diode cells. The silicon rectifiers had 12 trays of 28 cells. Cooling was by fan.

The success of the first five classes of electric locomotive was varied. Classes AL3 and AL5 were partial failures, but these first British AC electric locomotives effected a revolution in traction. On the electric railway linking London Euston with Birmingham, Crewe, Manchester and Liverpool, the single-phase AC locomotives set new standards of performance on British Railways. The first phase of the electrification of the WCML in some respects was a 'steam railway electrified'. Many operations dated from the steam traction era; there was shunting, the working of semi-fast trains and mixed goods trains. However, a marked improvement was evident on those services where little changed apart from using an electric locomotive in place of a steam or a diesel engine. The work of the best steam and diesel locomotives was completely surpassed. Locomotives of the 4–6–2 type, like the LNER A1, the LMS Class 7P, the BR Class 8, and the first generation diesel-electric types, could not approach the performance of electric traction, not even in the case of the famous English Electric 'Deltic' – then the most powerful single-unit diesel locomotive in the world.

When the reshaping of British Railways took place, following the work of Dr Beeching, much inefficient goods working was abandoned. The accent was placed on long distance, through working of passenger and container trains, which better suited electric traction. Developments after 1970 favoured electrification, and despite remarkable developments in diesel traction, no other form can match the performance of the electric railway.

In the early 1960s, it was suggested that future electric traction might use the linear induction motor. This device had been understood since the origins of electrical engineering, but practical application to traction had been held back by the common assumption that the windings would be stationary and fixed to the sleepers. This would result in an expensive and obstructive structure. In the early 1960s, a form was developed in which the windings were on the vehicle. The poles and windings on the vehicle were energised by current picked up from conductor rail, overhead contact wire, or supplied by a diesel-electric generator. Currents were induced in a metal strip laid between the rails, and reaction drove the vehicle forward. Anticipated advantages were high tractive effort, high power, good acceleration and braking, all of which would be independent of adhesion factor and locomotive weight. The linear induction motor was identified as a potential motor for very high speed traction, and a small experimental prototype, in the form of a light four wheeled carriage with a single seat, was set up on standard gauge track in Gorton Locomotive Works in 1962 with cooperation between Manchester University and the British Transport Commission. The test model gave a tractive effort of 1000 lbsf, and reached 30 mph in 3 seconds. Schemes were proposed for its use on railways at 120 mph or 200 mph. It was also suggested that small units with linear induction motors could replace conventional shunting engines in marshalling yards and banking engines on steep hills. The complexity and obtrusive nature of the reaction plate told against the scheme, but the most serious matter discouraging its use was the contemporary developments of electric trains to run at regular speeds of 120 and 200 mph. The French and Japanese projects for very high speed electric and gas turbine trains, using proven technology, showed the obvious way forward and this was the path followed by high performance electric traction since the 1960s. The linear induction motor train; the tracked hovercraft; the magnetic levitation vehicle; the jet-powered hovertrain remained test models or solitary examples of expensive prototype systems which offered no advantage over developments of the conventional railway.

In Britain, the next phase of electrification extended the system north of Weaver Junction to Carlisle and Glasgow. New classes of locomotive were needed to work services over difficult routes which included Shap and Beattock inclines. The first of the new classes was the AL6 type, later termed Class 86. The first locomotive was built in July 1965 to a joint British Railways-English Electric design. The class was made up of two main sub-classes. One was rated at 3600 hp at rail at 67 mph and the other rated at 4400 hp at rail at 62 mph. Maximum rail powers were respectively 5900 hp at 38 mph and 6100 hp at 39.5 mph. The class was not equipped to work from a 6.25 kV supply which had been considered necessary for the AL1–AL5 classes. These were built when reduced voltage sections were planned where clearances were restricted on the WCML In practice, no reduced-voltage sections were installed on the line. Only one pantograph was fitted instead of the two provided on earlier locomotives. Electrical equipment was supplied by English Electric and AEI. Mechanical design was by British Railways at Doncaster. Forty units were built at Doncaster Works and sixty units by English Electric at Vulcan Foundry. Silicon rectifiers were used instead of mercury-arc valves. Each traction motor had its own rectifier, smoothing choke and blower, connected to an individual secondary winding on the fixed-ratio transformer. The tap-changer followed the design used on the Class AL2, later Class 82. The traction motors were of the axle-hung type widely used in diesel-electric locomotives on British Railways and in the multiple-unit Southern Region electric stock. Unfortunately very poor riding was experienced in the high-speed services which were normal for the AL6. It was worse than the riding of the AL1 and AL5 classes which had caused trouble. Engine E3173 of the AL6 class was subjected to experiments with the 'Flexicoil' suspension system which had given good service in Germany in electric locomotives of the E103 class, introduced in 1965, which ran at 125mph in regular traffic. This suspension was successful and was applied to a major proportion of the Class AL6. Before this was done, bogie frame fracture and broken rails placed the WCML operating authorities in difficulties granted that reliable high-speed operations were being publicised as features of the electric railway. Fifty-eight locomotives of the former Class AL6 (which was reclassified as 86), were rebuilt as Class 86/2 with the 'Flexicoil' suspension and SAB resilient wheels. These provided the high-speed, good-riding locomotives which were required. Class 86/1 were fitted with new bogies and fully spring-mounted motors in preparation for introducing a new class – the Class 87. A third sub-group, the Class 86/3 had SAB resilient wheels but not Flexicoil suspension and were limited to speeds below 100 mph. The unmodified locomotives, in Class 86/0, were limited to speeds below 80 mph.

The Flexicoil suspension, the resilient wheels, and the spring-mounted motors, showed that speeds of over 100 mph could be regularly worked without track damage. This was essential if British Railways were to run express trains at the very high speeds being accepted as normal in France and Japan. By the early 1970s, the high-speed electric railway which had emerged in the 1950s was outclassed by the new kind of electric railway pioneered by the French and Japanese in the form of newly built, segregated systems. To investigate high speed working over conventional track, tests were carried out with E3173 fitted with a streamlined fibreglass nose. Trials between Berkhamstead and Leighton Buzzard in 1973 showed that 125 mph was easily attainable with existing technology. E3173 reached 128 mph on one occasion. In the same year, two Class 86/2 locomotives worked a four coach special train with one engine in front and the other in the rear and reached 129 mph near Nethercleugh in Scotland.

The Class AL6 (Class 86) proved successful once the rough riding was cured. Later, engines of sub-classes 86/0 and 86/3 were fitted with Flexicoil suspension (the 86/0 types were given resilient wheels) and transferred to the single-phase 25 kV, 50 Hz electric system installed on the former LNER Great Eastern lines out of Liverpool Street Station, London to operate services to Norwich and Harwich. These locomotives then became Class 86/4.

In their account of the AC locomotives of British Railways, Webb and Duncan refer to a 1966 proposal by AEI for a CoCo electric locomotive of 8000 hp for service through the proposed Channel Tunnel. This locomotive was designed to work over both the existing LVDC third rail system of the Southern Region and the 25 kV, 50 Hz system through the Channel Tunnel. The outline proposal shows a machine owing much to WCML locomotive practice in the 1960s, which in turn was influenced by contemporary developments in Switzerland and Germany, especially in the matter of suspension, bogie design, and transmission.

18.4 The Class 87 and its significance

The Class 87 was sometimes known as the 'Royal Scot' class. It was introduced in 1973 to provide high powered traction for express passenger services over the West Coast Main Line when the electrified section between Weaver Junction and Glasgow was finished in 1974. Thirty-five units were built at Crewe Locomotive Works. The electrical equipment came from GEC (GB). An additional unit of Class 87/1 was built to investigate thyristor control of traction motors. They were of Bo-Bo design with silicon rectifiers. They had four series-wound direct current motors fully spring-suspended in the bogies with flexible drive. Secondary suspension was by 'Flexicoil' springs. Control was by high tension tap changing. The practice of dividing the electrical equipment into four sets, one per motor, which had been used on the Class 86 was continued. Each set or 'pack' had a transformer secondary winding, bridge-connected silicon-rectifier, and smoothing inductor per traction motor. Rectification was simplified by reducing the number of cells per cubicle from 16 to 96 compared to the Class 86. This reduction in the number of electronics sub-components needed became even more marked in the 1980s and did much to improve the reliability of locomotive mounted electronics systems. Coupled with a marked reduction in bulk, this was decisive in determining their use. The axle hung motors of the earlier type were not used following the trouble they had given at high speed, though they were used in high-speed locomotives in Germany and elsewhere. Some trouble was experienced with the driving gear. The motor was mounted in the bogie on a three-point suspension with rubber mounts. The single-reduction spur gear drive were enclosed in a gearbox supported on the axle by tapered roller bearings. The gearbox was supported in the bogie by a rubber-bushed, torque reaction link. The driving gear wheel was mounted directly on the axle. To accommodate relative motion between the motor and the driving axle and gearbox, the driving torque was transmitted by a connection through the hollow motor shaft. To reduce unsprung mass, the locomotive axles were hollow. In 1976, a gearbox pinion failed in service and drew attention to housing bolt breakage, and new bolts were fitted. Axle bearing failure resulted in the hollow axles being replaced with solid ones. Some problems with riding necessitated enlargement of clearances and modifications to the damping. In 1978 rewelding of the bogie-yaw damper bracket welds proved necessary. Wheel slip on gradients was caused by weight transfer during acceleration, and high velocity airsanding gear was fitted to some units.

Despite these difficulties the class was a success. Starting tractive effort was 58,000 lbsf, with a continuous output of 21,300 lbsf at 87 mph. For the time, power output was very high. The performance showed how far the rest of British traction was outclassed by the latest HVAC traction. A continuous power output of 3750 kW (5000 hp) at rail at 87 mph was rated, with a maximum of 7860 hp. Even these excellent figures were exceeded by the outputs of contemporary German and Swiss units, such as the German E03. Weight was 81 tons 19 cwt, and maximum speed was 110 mph. The bogie design (BP9) had performance characteristics similar to the bogies of the power car of the diesel-electric high-speed train, and incorporated improvements made to the Class 86/1. It proved satisfactory at high speed after some problems in the first months of service, and the Class 87 set new standards of performance. Before it appeared, 3000/3300 hp at rail represented high powered traction and would have been remarkable in a diesel locomotive anywhere. After the Class 87 appeared, 5000 hp at rail defined high powered traction.

In the 1970s, there was great interest in solid state control systems. In the 1980s, thyristors were used to control DC traction motors, and tests were carried out with 87.101, built in 1975. Cooper describes the operation of the thyristors as follows:

Each of the four motor armatures is fed by two thyristor/diode bridges in series, and the field circuits are separately excited by similar circuits but with one bridge to each field. From zero to half armature voltage, the firing of the thyristors in one bridge is advanced while those in the bridge in series with it are cut off, the diodes in that bridge simply providing a path for the rectified current. When the thyristors in the first bridge are fully advanced, those in the second bridge come into action and supply a gradually increasing voltage in series with the voltage from the first bridge. When both bridges are fully advanced the motors receive their full operating voltage: the firing of the thyristors in the field circuits is then retarded to weaken the fields and allow acceleration to continue to full speed. Rheostatic braking effort is also controlled by the field circuit thyristors.

The advantages of thyristor control included better tractive effort versus speed characteristics; control of wheelslip; and weight reduction. Compared to the other members of the Class 87, Engine 87 101 weighed some 5 tons less: it did not have the high tension tap changer, and its transformer weighed 6 tons rather than the 11 tons of the type used in others. As Webb and Duncan indicate (p. 88), thyristors offer reduced maintenance, simpler transformers, and greater hauling ability because tractive effort troughs arising from tap changing were eliminated. Continuous power was 4850 hp, with a maximum short term rating of 7250 hp. The engine worked well and fears about interference between the solid state electronics system and power supply, signalling networks and telecommunications proved unfounded though protection equipment had to be installed. The locomotive hauled loads 19 per cent greater than the rest of the class, due to the thyristor control system; the separately excited DC motors, and the wheelslip control. Engine 87101 was said to be the best AC electric locomotive on BR.

The locomotive was so successful that it was the starting point for 50 later locomotives introduced between 1987–88 to work WCML services and to replace older units. These machines were originally classified 87/2, but this was changed to Class 90. These engines incorporated the latest thyristor control systems and other electronics components, but many features remained the same despite considerable differences in external appearance. After initial difficulties which delayed final delivery until 1990, the class worked well and the engines saw service over all the British 25 kV main lines though most are employed on the WCML.

The class 87 marked the ultimate development of the 'first phase' AC main-line railway in Britain. This railway grew out of the French (SNCF) work with industrial frequency single-phase AC supply to the overhead contact wire used with locomotives fitted with mercury-arc rectifiers and DC motors. This French work owed much to experiments with General Electric ignitron locomotives on the Pennsylvania Railroad in the late 1940s; and to prewar investigations into rectifier locomotives in Germany and Hungary. The WCML became much more than 'a steam railway electrified' or a railway with steam traction replaced by electric locomotives. Right from the start in the 1950s performance was raised far above that possible with the best steam locomotives however well the latter were organised and managed. It is true that the economic justification of electrifying north of Weaver Junction to Carlisle and Glasgow can be questioned. Traffic did not expand as anticipated. Goods traffic declined, and improvements to motorways and airline shuttle services threatened passenger traffic so that the WCML did not build up the regular, well used express traffic between Euston and Glasgow, as anticipated. The WCML faced rivalry from the East Coast route with its high-speed diesel-worked trains using the 'Deltic' locomotives and the later diesel-electric High Speed Trains introduced in 1976. Its role in demonstrating the technical advantages of the HVAC railway is unquestioned.

Chapter 19
Solid-state electronics, motor control and the locomotive

19.1 The advent of solid-state control systems for locomotives

The single phase 25 kV, 50 Hz railway became a global standard because of rectifier locomotives with DC motors. Without them, HVDC railway would have remained the standard and might have evolved into a 6000 V DC railway as the Russians proposed. The position of the AC system was helped by solid-state power electronics systems, which of course can also improve DC alternatives. Solid state rectifiers and thyristors greatly improved the performance of the rectifier locomotive and dispensed with the troublesome mercury-arc rectifiers.

Solid-state rectifiers and control systems were made possible by the thyristor, which introduced a new phase in locomotive design. Several countries contributed to this development. The USSR played a major part in applying electronics to traction In 1966 the Tbilisi works constructed an electronically controlled 6000 V DC prototype based on the VL8 Bo + Bo + Bo + Bo type introduced in 1958. This machine, VL8 V-001, was tested under 3000 V and 6000 V DC supply, and was the first in the world to work under 6 kV DC. The Russians experimented with current inverters and asynchronous motors but without the commercial success of the German companies.

British firms pioneered the solid state (germanium) rectifier which was tested on the Lancaster-Morecambe-Heysham line in 1955–56 during investigations by Riddles and Warder into industrial frequency electrification. Solid-state switching systems and solid-state rectifiers led to solid-state inverters and frequency control systems. It took longer to devise switching systems to control three-phase motors in a convenient and flexible manner. In the 1930s, there was simple inversion from DC to fixed frequency, fixed voltage AC. There was frequency changing between one supply and another. However, inverters and frequency changers were insufficiently flexible and responsive to control a three-phase motor under variable loads. There were no compact and reliable switches available, despite many experiments with valves and electromechanical systems. The equipment used in pioneer experiments by Kando and others was large, heavy and inconvenient for locomotive use.

Solid-state rectifiers, valves, and switches revolutionised locomotive-mounted circuits, bridges and networks for rectifying AC to DC; inverting DC to three-phase AC, and controlling all kinds of motor, including the three-phase induction type. Principles which were understood in the era of mercury-arc rectifiers and thermionic valves, and which were established in stationary practice, could now be applied in locomotives. Between 1963 and 1964, Brush of Loughborough experimented with solid-state control of three-phase motors, perhaps the first trials of their kind. Brush used a diesel-electric locomotive to avoid the interference between control systems and power supply which might have occurred had an electric locomotive been used. The locomotive chosen for the experiment was no. 10800, which was built by the North British Locomotive Company, Glasgow between 1949–50 for the London Midland & Scottish Railway following an order placed in 1946 as part of a diesel traction pilot scheme. The locomotive was a Bo-Bo type with an engine brake power of 620 kW. It weighed 70 tons in running order. It was intended for light passenger duties and goods working over short distances at moderate speeds. It had a single cab near one end, and followed the configuration of the contemporary American 'roadswitcher'. It served on the Midland Region of British Railways, with periods on the Southern Region in 1952, and on the Birmingham-Norwich route in 1955. In 1959 it was withdrawn from service when British Railways reduced the number of non-standard diesel locomotives. After storage outside Doncaster Works, it went to Brush in Loughborough to be a test vehicle. The original Davey-Paxman engine was replaced by a Bristol-Siddeley-Maybach engine, rated at 1007 kW (1350 hp), which drove a Brush three-phase alternator. The AC output was rectified by a Westinghouse silicon diode rectifier, which supplied direct current to the inverter system. This supplied the three-phase asynchronous motors, which were controlled by frequency variation using thyristors. The engine was referred to as the 'Hawk' though this does not seem to have been an official name carried on a plate. In 1963 and 1964, a few test trips were made over the Nottingham-Leicester line. The engine was tested at the Rugby locomotive testing station in 1965 and was returned to Brush in Loughborough later that year where it remained until 1972. Tests were then ended and during the miners' strike in 1972, the engine and alternator were removed and used as an emergency standby unit for the plant. Valuable components were removed for storage or reuse and the remainder were scrapped.

Difficulties were encountered because the control systems were complex. A large number of components were required in the first generation of current-inverters and their auxiliary equipment, which made them prone to breakdown and costly to maintain. The 'Hawk' project was abandoned too early. Brush gave up research into the system for controlling three-phase traction motors when it was moving from the experimental to the promising stage, and thereby lost ground in a technique which became standard in the 1980s. The initiative was never recovered and leadership in the field passed from Brush to Brown-Boveri when the system was proved. Several research projects in this field were under way in the late 1960s. In 1968, the USSR carried out trials with inverters and frequency control of three-phase locomotive traction motors, but with no immediate success.

19.2 The Brown-Boveri system

In 1971, the German state railways and private enterprise began systematic experiments to perfect a locomotive-mounted system for controlling asynchronous three-phase motors fed from a single-phase, industrial frequency supply. Reconciliation with other supplies was a project goal. The project was well resourced and supported by the necessary industrial and academic expertise throughout lengthy trials and a long-term development programme. From the start expert management was provided to see it through to success. It realised that a new traction system, even in the form of a single unit, demanded sustained, expensive support to succeed The persistence of the German programme is one of its noteworthy features and accompanied a shift in research and design initiative from the railway company engineer to the staffs of the big combines which alone can supply the expertise and the financial support to take a project from inception to success.

In the late 1960s, advances in solid-state electronics made it possible to combine inverter-control of three-phase traction motors with rectification of contact wire current in a locomotive fitted with asynchronous motors running over a wide speed range and supplied with single-phase AC at industrial frequency. The first regular use in railway traction of solid-state power electronics was to rectify AC current to DC. In electric locomotives, power electronics was used to rectify single-phase supply from the contact wire to provide DC current for the motors, as in the Class 90 of British rail. Likewise, the output of a diesel driven alternator was rectified to supply the DC traction motors of diesel locomotives such as the Class 60. Solid state rectifiers enabled the DC generator of the diesel-electric locomotive to be replaced by the smaller alternator, with a considerable reduction in weight and size of components. DC traction motors were practically universal for oil-electric traction, hence three-phase AC had to be rectified to DC. Solid-state power-electronics components were small enough by the 1960s to achieve an overall saving in weight and volume of machinery despite the need to include a rectifier on a diesel-electric locomotive.

The project to develop a system for asynchronous traction motors began in 1965, in the Brown-Boveri plant in Mannheim, Germany. Between 1965 and 1971, Brown-Boveri devised methods for controlling the speed of asynchronous traction motors which were good enough for extensive road-trials. The experiments with test and service trains were first carried out with a Co-Co oil-electric locomotive, no. 202 002–2. This engine was used to introduce solid-state rectification, flexible control systems, and three-phase motors into modern traction. It was built by Thyssen-Henschel in 1970. Diesel traction was chosen to test the frequency control of the three-phase traction motors in place of electric traction to avoid interference with the contact wire supply by the current-inverter system. During trials, the frequency of the motor supply varied between 0.5 Hz and 165 Hz, which gave a speed range wide enough for operating regular traffic. It ran as a demonstrator in Germany and in other countries and was followed by new prototypes which incorporated the lessons of experiments conducted between 1965 and 1970.

Locomotive trials of the new control and motor system began on 30 June 1970 when the new diesel-electric locomotive, 'Henschel-BBC-DE2500' (Federal Railway number 202 002–2) performed its first trial run. It was a Co-Co unit, weighing 84 tonnes, with a maximum speed of 140 km/h, and a power rating of 1840 kW (2500 hp). The diesel engine drove a three phase alternator, the output from which was rectified to direct current, then inverted to three phase AC to supply the motors. There were three banks of thyristors and the frequency of the motor supply varied between 0.5 Hz and 165 Hz. There was therefore a minimum frequency (0.5 Hz) and a corresponding minimum wheelspeed below which the inverters could not be used. At start, when rotational wheelspeed was zero, the current was pulse wave modulated at constant frequency. This got the locomotive moving slowly until inverter control was brought in at 0.5 Hz or some other low frequency corresponding to a slow rate of advance. The locomotive established the current-inverter system in diesel-electric traction and was presented at the Hannover Exhibition in 1971 as 'the first operational locomotive with modern three-phase drive technology in the world'.

In 1974, it was modified to demonstrate the suitability of the Brown-Boveri (BBC) system for locomotives taking power from the single-phase 15 kV, 16.66 Hz supply which was widely used in Germany. The diesel engine and the alternator were removed and replaced by ballast. The control apparatus was modified to prevent harmful interference with the power supply system and the signalling network. The control and interference-suppression equipment occupied too much space to be accommodated in the locomotive alone, and a control coach was attached, equipped with a pantograph, which contained the transformer and a four-quadrant controller. All axles of the locomotive were motored, so that it retained the Co Co wheel arrangement. Power was 1550 kW, and weight of the locomotive only was 80 tonnes. Test were carried out in 1974 with single-phase AC power drawn from the contact wire through the control coach pantograph. The locomotive did well pulling heavy goods trains. The four-quadrant controller was perfected during these trials. This was to form a basic component in the three-phase locomotive developed from the DE2500. The trials showed that for driving and braking within a wide speed range, power factors of unity were attained, which was a great improvement over the performance of the conventional thyristor technology of the time. Electric braking and energy return to the contact wire during braking were demonstrated. The BBC system was proved reliable and responsive to suppression of interference with the power supply and signalling circuits. At this stage of development, the control apparatus was very bulky and required constant maintenance to function properly.

The asynchronous motors provided good adhesion, and worked well over a wide speed range. The trials were successful, though they indicated that considerable development was needed to reduce complexity, weight, and volume of the whole system.

The locomotive was further modified in 1977, to demonstrate the BBC system in locomotives working under the 1500 V DC contact wire of which there were extensive networks in the Netherlands. A single 1400 kW asynchronous motor was fitted to drive one axle of one bogie through BBC cardan transmission, giving a wheel arrangement of 3 – 1A1. Tests in the Netherlands were satisfactory and showed that interference with signalling systems could be kept below permissible levels. These road trials between 1970 and 1978 showed that a system was in the offing suitable for diesel-electric or electric traction. The inverters were fed with DC, usually at 1000 to 2800 volts. This could come from an HVDC contact wire, a LVDC conductor rail, or from an AC supply via a transformer and rectifier. Demonstrating the system on DC railways was an important achievement because there were HVDC railways in Netherlands, Belgium, France, Italy, Russia, South Africa, India, South America and Japan. The system worked well with LVDC railways, of which the most extensive example was the conductor rail network in Southern England. The advent of compact, solid-state rectifiers enabled the system to be used on any AC system used on railways.

By 1975, the new technology challenged the two world standard systems: the single phase AC railway, with locomotives fitted with AC motors or rectifiers and DC motors; and the HVDC railway, with locomotives powered by DC motors. After 1975, there was the prospect of locomotives with three-phase motors working over both systems and over any other system. The BBC system allowed asynchronous motors to be used in diesel traction. For the first time, the supremacy of the DC traction motor was threatened on heavy duty railways, and on light rail when tramways adopted the asynchronous motor.

The rebuilt pioneer locomotive was numbered 1600P and ran over the Netherlands DC railways. It established the two-quadrant controller in DC traction, working with three-phase drive and with inputs from overhead HVDC, from LVDC conductor rail, and from batteries. The two-quadrant controller, used with direct connection to a DC supply, required a large inverter but permitted electric braking with energy feedback to the catenary. Electric braking could also supply energy for auxiliary services on the locomotive or train. The BBC techniques for DC railways were developed for supply voltages as low as 600 V. The dual-mode electro-diesel locomotives of the former New York, New Haven & Hartford Railway, were rebuilt to provide the Metro North Commuter Railroad with motive power able to run from the LVDC conductor rail supply, and to work in the diesel-electric mode elsewhere. These used gate turn-off inverters, three phase motors, and the BBC system, and other multi-system vehicles made use of it.

In 1974 engine no. 202 002–2 was followed by two diesel-electric inverter locomotives fitted with the BBC system. Each marked a stage in the development of a flexible, electronic traction system. One was a Co-Co locomotive, 202 004–8, and the other was a diesel-electric locomotive of the Bo-Bo type. Both were classified as the DE2500 type along with the pioneer unit, and, apart from differences in wheel arrangement, each carried the same main items of equipment. Each had the same Henschel 1500 rpm engine rated at 1838 kW. Starting tractive effort was 270 kW; top speed, 140 km/h; length over buffers, 18 m; wheel diameter, 1.10 m; gear ratio, 5/16. The total mass was 76 tonnes for the Bo-Bo and 80 tonnes for the Co-Co, but 84 tonnes is sometimes quoted for the latter. The Bo-Bo was the remarkable 'Umkoppelbare Antriebmasse', or as it was generally known, the 'Um-An'. This was built with a conventional body, but was later fitted with a streamlined nose for high-speed trials with an experimental bogie and suspension. It carried the number 202 003 0. It was a high-speed locomotive designed to reach 200 km/h unaided and to attain 250 km/h when coupled to one of the Class 103 electric locomotives. The 'Um-An' showed that the current-inverter system and asynchronous motors were suitable for speeds which were very high for the 1970s and it demonstrated experimental bogies in which the weight of the traction motor could be shifted between suspension on the bogie and suspension on the locomotive body. Though not as successful as expected, the 'Um-An' facilitated the development of high-speed bogies. On the rolling test stand in Munich, the 'Um-An' was driven up to wheel rotational speeds equivalent to 350 km/h to investigate suspensions for the German 'Inter City Express' (ICE) trains which in road trials reached 406 km/h. The class 103, introduced in 1970, was used to establish the limits of maximum power outputs from single units and no. 103 118 provided 14,000 hp or 10,400 kW for short periods. The 103 class was also used to test remote control of locomotive worked trains. Eventually, the several lines of development were drawn together into a 'total traction system' built round a standardised, universal locomotive able to meet a wide variety of operating conditions and engineering environments. In the past, 'standard' often meant inflexible, but today electronics integrates components and subsystems into ensembles which are both 'standard' and 'flexible'.

The three locomotives of the DE2500 class are preserved in museums. Engine 202 002–2 is in the Werkmuseum of Thyssen Henschel; engine 202 004–8 is in the Technisches Landesmuseum, Mannheim; and engine 202 003 0, the 'Um-An' with its streamlined nose, is in the Verkehrsmuseum, Berlin.

19.3 The Deutsches Bundesbahn Class 120

In the early 1970s many engineers judged the Brown-Boveri system for controlling asynchronous traction motors as being too complicated and unreliable for normal regular use. The simplicity of the asynchronous motor was recognised but the associated electronics control equipment took up several times the volume of conventional DC motor control gear and was less reliable. Failures were common in the 1970s but the potential of solid state electronics was recognised.

Solid state systems were better than mercury-arc rectifiers for converting AC to DC on board locomotives, but the Brown-Boveri system was not reliable until the late 1970s, and without it, there could be no modern asynchronous traction motor. Most traction engineers waited for the BBC system to be proven beyond doubt. There was general recognition that the system was proven by 1984 when prototype designs fitted with it were successfully demonstrated in Germany, Sweden, Denmark, and the USSR. These prototypes were followed by production classes. After 1984, the Brown-Boveri system ceased to be experimental and was used by the conservative British and Americans in their electric and diesel-electric locomotives.

Alternatives to the BBC system were advocated. These included the synchronous motor and the separately excited DC motor but the Brown-Boveri system and equivalents became regular practice. They are now used in a wide range of diesel-electric and electric locomotives which includes shunting engines, railcars, rapid-transit trains, heavy haulage goods locomotives, express locomotives, very high speed train sets, and Channel Tunnel stock. The Class 120 locomotive of the Deutsches Bundesbahn, introduced into traffic in 1979 demonstrated the reliability and economy of the system. It convinced engineers that the defects had been eliminated. This class incorporated the lessons of the experiments with the three pioneer 1860 kW diesel-electric locomotives first fitted with the BBC system, which were tested between 1970 and 1974. In 1979, the German Federal Railway took delivery of the first 'universal locomotives' of the new E120 class. It incorporated improved thyristor-based controls developed after the trials with the pioneer DE2500 machines.

The Class 120 was an 84 tonne Bo-Bo electric locomotive with an output of 5600 kW which was intended to work all types of Deutsches Bundesbahn trains. Five prototypes were delivered and were tested for four years during which they covered 4 million km. The Class 120 locomotive could haul 2200 tonne goods trains up gradients of 5 per cent at 80 km/h and 550 tonne express passenger trains up 2.5 per cent gradients at 200 km/h without change of gear ratio. The universal locomotive was meant to replace existing six-axled goods and passenger electric locomotives and after four years of tests an order was placed for 60 units. In the early tests the control system interfered with signalling and communications but this was cured by tuning the BBC system's circuits. On trial, one of the locomotives (no. 120 005) reached 231 km/h (143.45 mph), and broke the record for three-phase motors set in 1903 on the Zossen-Marienfelde military railway by high-speed cars from AEG and Siemens.

In the 1980s the concept of the 'universal locomotive' was re-assessed in the light of experience with the Class 120 and similar machines. The 'modular' concept (qv) was developed by rivals of ABB-Henschel, such as the AEG Daimler Benz company which in 1994 launched the 12X modular locomotive. These companies have now merged into the giant combine ABB-Daimler Benz, known as Adtranz. The ABB-Henschel company responded to the AEG-Henschel challenge by developing the E120 type into the E121. This was based on E120–004 and 005, and had GTO equipment, microprocessor control,

Figure 19.1 Class 120 locomotive of the German State Railways, built by Krauss-Maffei. This class, introduced in 1979, was greatly developed and helped to establish the asychronous motor and solid-state control in modern traction.
Source: Krauss Maffei Verkehrstechnik.

'Flexifloat' bogies, and integrated complete drive. Power outputs were 6400–7000 kW and speeds up to 230 km/h were obtained within the Bo-Bo configuration. This technology was also developed for use in the latest diesel-electric locomotives. The E120 locomotive is still in service. It did much to allay adverse criticism of the BBC system and equivalents produced by other companies. Reliability was improved by a marked reduction in the number of electronics components within the system.

Since 1970, when the pioneer experiments began, the switching power of thyristors has been greatly increased and the number of components required to serve a locomotive has been reduced. In 1970, the number of semiconductor elements for the traction inverter were 96 thyristors, 48 diodes. In 1980, these numbers were 24 thyristors, 12 diodes. In 1987, they were 6 GTO, 6 diodes. The inverter could deal with 3000 hp or more by the latter date. Between 1970 and 1990, the number of semiconductor components required for a diesel-electric locomotive fitted with the BBC system was reduced by 92 per cent to 8 per cent of the 1970 figure. The transition to GTO equipment effected further improvements. A class 120 locomotive number 120 004 was fitted with thyristor inverters and number 120 005 was fitted with GTO technology. Compared to the thyristor fitted engine, the GTO-fitted machine had 17 per cent the number of semiconductors; 66 per cent the electronics system weight; 50 per cent the volume of electronics components. Its control equipment could deal with 55 per cent more power. Engine 120 005 (renumbered 752 005–9) was the first oil-cooled GTO locomotive on the German Federal Railways which were formed after the Union of the former East and West Germany. Compared with other engines in the E120 class, the energy losses in the GTO control system of number 120 005 were 40 per cent less than in the thyristor control system of the other engines, largely because the commutating circuits and the motor input choke were omitted.

19.4 Solid-state electronics and locomotive drive systems

The success of the E120 locomotives established the BBC system in general service. In 1985, it was used in the Thyssen-ABB-Henschel power car, built as part of the German high-speed electric train project. The first power cars of the InterCity Experimental (ICE) trains delivered 4200 kW each, for 5 minutes, and provided a continuous output of 2800 kW. Two cars were provided per train. The first sets used the drive, suspension and bogie design derived from the high-speed tests with the 'Um-An' diesel-electric locomotive. On the Munich rolling test facility the drive and bogies were tested at speeds up to 350 km/h. Tests on the track were extended up to 406 km/h (Week 17 of trials, 1988) and high speed performance was modelled in theory up to 500 km/h, in a project promoted by the former German Federal Ministry for Research and Technology.

Examination of the papers published in the railway engineering press suggest that 1984 was the year in which the new control and drive system became established. After 1984, the articles treat it as proven technology, with an emphasis on its advantages, and how best to introduce it. The number of locomotive classes using it increased rapidly and asynchronous prototypes marked the spread of the system through Europe. In 1984, Denmark saw its first locomotive with asynchronous motors, working under the 25 kV, 50 Hz system. In 1985, the Skoda Works in Czechoslovakia announced that asynchronous motors were under consideration for designs planned to replace its standard 25 kV AC locomotives and its 3000 V DC locomotives. Skoda argued that use of the inverter and three-phase motors greatly reduced the length of HVDC locomotives of equal power which used DC motors and which were built for the HVDC system found throughout the Soviet Bloc. In 1985, Netherlands railways announced that in future its oil-electric 1180 kW light duty locomotives would use asynchronous motors. In 1989, *Railway Gazette International* reported that South African Railways had ordered a new locomotive with power electronics, microprocessor control, and asynchronous motors. In the same year, the system was tried on a diesel-electric locomotive in the United States. In 1989 in Germany, Krupp-MaK built a Co-Co diesel-electric locomotive, number DE1024, which was rated at 2650 kW, and which used the asynchronous drive with GTO thyristor-inverters for traction and auxiliary power supply. In Britain, the first fare-paying passengers were carried on a train powered by asynchronous motors on 4 November 1989. It was announced in 1989 that all future orders for the London suburban electric multiple-units would use the system.

The British were slow to try the new technique, but tests with a converted diesel multiple unit, no. 45 7001, which was fitted with equipment from Brush led to its acceptance for rapid-transit duties in the London area. After 1990, new diesel and electric stock used current inversion; asynchronous motors; and microprocessor control to an increasing degree in Continental Europe. The DC motors fitted to the classes 60 and 90 are probably among the last of their kind.

Some engineers criticised the system. They suggested that recent improvements to solid-state devices and electronic microswitches made the synchronous motor a better machine than the asynchronous motor. The French promoted the synchronous motor, which remains a minority option though because of the stature of the authorities arguing its case it deserves attention. In 1985, the French constructed two dual current prototypes with synchronous motors. In February 1984, an article in *Railway Gazette International* remarked:

The French decision to adopt synchronous motors which are also likely to be installed in the next generation of TGV trains for the Atlantique routes, is a direct result of successful experiments with a modified 25 kV 50 Hz test locomotive, No 10004, which demonstrated during trials in 1982 and 1983 that it could replace the three types of 4100 kW electric locomotives currently used in freight and passenger service ... the self-regulating synchronous motor powered by a current inverter has truly remarkable advantages compared with dc or asynchronous motors.

This claim was contested by Wallace F. Powers, former control engineer at General Electric Locomotive Equipment Department (USA) who criticised the eminent French authority, F. Nouvion who argued for the synchronous motor.

The synchronous motor was simple but in the 1980s its control systems were more complex and less reliable than those for the asynchronous motor. Present practice favours asynchronous motors for high performance rolling stock. Because of the simplification in circuits, Y. M. Machefert-Tassin argued that nothing prevented use of both the induction (asynchronous) and synchronous motor in traction. The synchronous motor was used in the TGV-Atlantique trains for the SNCF. The TGV-E (Sud-Est) 23000 Series, introduced in 1978–1986, and built by Alsthom/Francorail-MTE/de Dietrich used DC traction motors, and worked off both 25 kV 50 Hz AC and 1500V DC supply. The sets for the Atlantique services, of the 2 + 10 formation, are fitted with eight synchronous motors, of the self-commutating type, each giving 1100 kW continuous output. Like the earlier TGV trains, they are fitted to run from bi-current supply, either 25 kV, 50 Hz AC, or 1500 V DC. The Shinkansen trains of Japan, which operated from 25 kV, 60 Hz AC supply were fitted with DC series wound motors for the 100 and 200 series, but the 300 series was fitted with three-phase asynchronous motors. The German Intercity Express sets introduced from 1990 onwards use three-phase asynchronous motors. The Class 91 locomotive which works East Coast Main Line services, uses separately excited DC motors suspended from the underframe and driving through cardan shafts. The Swedish X2000 tilting body train, built from 1989 by ASEA/ABB uses three-phase asynchronous motors. The Italian tilting body ETR450 train, running under 3000 V DC supply has 16 DC traction motors with thyristor chopper control driving through Cardan shafts. The Eurostar trains introduced in 1994 to work from 25 kV, 50 Hz AC, 3000 V DC and 750 V DC supplies uses asynchronous motors. In April 1995, the Southern Pacific railway in the USA took delivery of 25 diesel-electric locomotives which were part of an order for 282 General Electric units. All these locomotives have asynchronous three-phase AC traction motors and are of the 4400 hp AC44CW type, apart from three which are of a new 6000 hp class. They replaced DC motored locomotives on a three to five basis. In March 1995, the Sante Fe system took delivery from General Motors of the first locomotives of an order for 25 Co-Co units of the 4300 hp SD75M class fitted with DC motors. The new Japanese E2 and E3 Shinkansen trains, designed for supply at 25 kV and both 50 Hz and 60 Hz use asynchronous motors. In France, the TGV Duplex double-decker train, has 4400 kW power cars designed for 1500 V DC and 25 kV 50 Hz supplies, and employs asynchronous motors. The Super-TGV power car were designed for 6000 kW and commercial speeds of 360 km/h, with current supplied from four supply systems: 15 kV, 16.66 Hz; 25 kV, 50 Hz; 1500 V DC, and 3000 V DC. Asynchronous motors were chosen. In 1991, China began research and development of a 4000 kW electric locomotive with three-phase traction motors as part of a programme to introduce the asynchronous system. Granted the variety of motor systems and mountings, the present situation is far from being dominated by one philosophy.

Figure 19.2 The Common Bloc control centre designed by GEC-Alsthom in early 1990s to enable Eurostar trains to operate from three different supply systems.
Source: GEC-Marconi/GEC-Alsthom.

19.5 The modular system

The concept of the 'universal' locomotive, able to operate the full range of passenger and goods duties, has been questioned in recent years. The railway engineering press reported that German Railways were 'not . . . entirely happy with its Class 120s. . . . The search for a "universal" loco has therefore been abandoned . . .'. Design strategy shifted towards the modular philosophy by which several derivative forms were associated with one master type or source design. Alternative types were built up using common components and sub-systems. In the 1990s German Railways announced plans for updating the fleet of 3890 electric locomotives. Several locomotive engineering combines prepared modular locomotive families, which make extensive use of components which integrate and provide several different forms. The first locomotive designed to this new specification was the 12X, which was first shown at the Henningsdorf plant of AEG Daimler Benz in Berlin in June 1994 (AEG Daimler Benz, 1994). This machine was a demonstrator, built as a speculative venture costing 50 million DM. It was hoped to derive from it such diverse forms as high-speed power cars, high performance locomotives, lightweight locomotives for local duties and twin-unit types for heavy goods service. Many of the components employed in the 12X were tested in the Class 120, including the main traction converter and the AEG prototype drive. Three-phase asynchronous motors were used as they were practically standard for new, advanced designs in 1994.

The 12X is a Bo-Bo locomotive, weighing 84 tonnes, with a continuous rating of 6400 kW up to 220 km/h, and with a short-term output of 7200 kW. Design maximum speed is 250 km/h. The latest electronics control system from AEG is installed. The main transformer is underneath the locomotive and has six secondary windings supplying six four-quandrant controllers, and the auxiliaries and the train bus for 'hotel' power. The power converter is made up of two GTO thyristors; two diodes; a protective circuit; transducers and two GTO gate control circuits. It can work as an inverter; a four-quadrant controller; and as a chopper for DC traction supply. The transformer and inverters are cooled by a Voith system using water and glycol. Other locomotive manufacturers prepared their own prototypes with which to launch a modular family. Recent types include the SLM (Swiss Locomotive and Machine Works) Lok 2000 family, which uses ABB components; and the Krauss-Maffei/Siemens Bo-Bo Class 127 (DB) called the 'Eurosprinter'. The latter type worked in Portugal (25 kV, 50 Hz, 5600 kW); Spain (25 kV, 50 Hz, 3000 V DC, 5600 kW) and Germany (15 kV, 16.66 Hz, 6400 kW) and was advertised as suitable for different operational programme, current systems, gauges and signalling systems. In April 1995, German Railways (DBAG) orders for locomotives included 145 Eco 2000 units from ABB Henschel; 195 Eurosprinters from Krauss Maffei/Siemens; and 80 12X units from AEG Daimler Benz. Deliveries started in 1996/97. In 1994, Norwegian Railways announced that after comparative trials, they had chosen the SLM-ABB Lok 2000 to update their high performance fleet rather than the Krauss-Maffei/Siemens Eurosprinter.

The gate turn off thyristor and the microprocessor are essential components in the control systems of rolling stock, and have dominated the design of electric and diesel-electric locomotives since the mid-1980s. They provided improved operational performance, reduced costs, greater reliability and an increased ability to meet a broad range of operational demands. The basic converters which underlie these recent advances are the DC drives with GTO four-quadrant choppers and three-phase AC drives; AC drives with AC single-phase two- and four-quadrant converters, pulsed converters, and three-phase drives. These modern systems rely on microprocessors because it enables the performance characteristic to be matched to the operational demand, thus providing flexible response, and leading to projects to introduce (first) universal locomotives, and (more recently) modular locomotive families.

The microprocessors are also used to protect the inverter systems from interference from outside and to suppress the harmful interference arising in the control system which threatens supply, signalling and telecommunications (White, 1994).

Where an existing line is to be upgraded with the new traction equipment, it is financially more cost effective to design the new traction equipment such that it is more compatible with existing telecommunications and signalling systems. This is due to the fact that the cost of replacing the signalling and telecommunications network far outweigh the cost of modifying the traction control system.

The microprocessor ensured the success of new electronics control systems by eliminating harmonics which would interfere destructively with other systems. The techniques became established in diesel-electric and electric traction throughout the world. In 1993, in the conservative USA, 350 SD70M-AC Co-Co locomotives were ordered by the Burlington Northern Railroad (USA) from General Motors and Siemens Transportation Systems. These locomotives enabled five 3000 hp DC motored locomotives to be replaced by three of the new type fitted with AC asynchronous motors, so that 350 new AC units were replacing 580 DC locomotives. This Burlington Northern order was worth $675 million and represented the largest locomotive order in US history. It announced the arrival en masse of the AC-motored locomotive in American goods service. Burlington executives believed that DC traction technology could advance no further. The DC traction motor will become much less common than before unless innovations revolutionise its design.

In the late 1990s, the dominant philosophy in advanced systems design was to develop 'universal technology' which does not require a universal locomotive. Modular systems like the ABB-Henschel integrated complete drive and bogie; multisystem traction components; and multipower systems, combine elements to meet most requirements. The Henschel Bogie is applied to heavy goods and inter-city express work. The two and three axle version have many parts in common. The three-phase technology and the ABB-Henschel 'Flexifloat' bogies influenced all the later advanced developments by that combine. These included the E120 class, the ICE power cars and a range of both electric and

Solid-state electronics, motor control and the locomotive 353

Figure 19.3 Electro-Motive (General Motors) – Siemens diesel-electric locomotive of Class SD70M-AC with asynchronous motors and solid-state control ordered for Burlington Northern Railroad in 1993.
Source: Burlington Northern & Sante Fe.

diesel-electric shunters, road switchers, goods and express locomotives. The ABB-Henschel rubber-cardan drive, introduced from 1974 onwards, is a subcomponent in the system. So are modern nose-suspended three-phase motors. The part played by nose-suspended three-phase motors in advanced traction shows how unwise it is to make any general statement about likely trends in such matters as motor mounting or type. Three-phase drive and integrated complete drive technology spread rapidly and co-operation between the combines became common, leading to combination of the combines. The equivalent electrical systems from major electrical manufacturers are found in combination with the mechanical components of major mechanical manufacturers. For example ABB components were found in locomotives by Henschel, SLM, Krauss-Maffei, Krupp, etc. Lessons still are being learned from the Class E120 as ABB experiments continue to reduce the number and size of electronics components. The work done with E120 004 suggested that the number of semiconductors could be reduced from 400 to 80, using GTO elements, and the power was increased from 5600 kW to 6400 kW with no addition to locomotive weight. The control systems have been developed to take DC voltages of 600, 1400, 2800 and higher in the DC link between the supply side and the inverters, so that the GTO technology can be used in locomotives running over railways with any class of supply, AC, or DC, or with diesel and battery locomotives.

The transformation of post-1965 railway traction occurred when solid-state electronics, microprocessors and improved mechanical components, created an harmonious system, full of long-term development potential by uniting features of older systems which were once conjoined with difficulty. Complex two-wire distribution systems; phase-splitting; electro-mechanical converter locomotives; locomotive-mounted mercury-arc rectifiers; and concatenated three-phase motors given way to compact solid-state control systems and asynchronous motors able to work over a wide speed range. Integration of the components continues to improve. The microprocessor better harmonises the control system with power supply, signalling networks and telecommunications. Flexible drives, self-steering wheelsets, integrated complete drives, wheelslip control, and composite materials integrate the rolling stock and permanent way more closely. There is no reason to suppose that this evolving integration of traction, power supply, signalling, control and permanent way is near its end. Electronics may ensure that it is a ceaseless progression.

Chapter 20
Solid-state electronics in signalling, communications and control

20.1 Centralised traffic control and interlocking 1965–85

Trends continued towards extending centralised control; reducing the size of all components, and improving the accuracy of monitoring operations (Hawkes, 1986). Automation and unification of systems were sought everywhere. These trends originated in the 1920s but solid-state electronics realised their promise and effected a transformation of signalling. In many places older technology sufficed and survived. In Britain, semaphores and mechanical boxes are still found on secondary lines. The speed of introducing new technology depended on state funding, or getting bank loans at favourable rates. Where these were lacking, the old equipment worked on. Complex and extensive systems like railway signalling networks are rarely shaped by just one engineering philosophy, and usually there are several in evidence. Amalgamations and mergers bring different design traditions into single systems, and recent combinations in the USA were in part decided by the compatibility between the signalling equipment of the companies seeking to unite In many cases it is nevertheless possible to pick out a dominating trend, though this is sometimes overtaken by a newer philosophy which modifies policies based on earlier practice (Bailey, 1995; Kerr, 2000).

In Britain, by 1980, there was a widespread acceptance of the route-setting principle in signalling. Routes extended from one signal to the next and the system did not always involve points operation. Equipment and displays were much smaller than in the pioneer boxes of the inter-war years. Routes were usually selected on the entrance-exit (NX) principle by means of small push buttons placed next to a representation of the signal in question. Automatic reset from track circuits replaced the manual cancellation of selected routes, prior to resetting, which in older boxes has been by push-pull buttons, switches or levers. A good example of 1970s technology is the control centre at London Bridge, commissioned in 1976, which controlled the movement of 2000 trains per day using route-setting by push-button action on a miniature panel. An

Figure 20.1 The central control room at Euston, London Transport, in 1970 showing the Victoria Line controller (left-hand desk) and the Northern Line controller (right-hand desk). The train regulator of the Victoria Line is sitting in front of the illuminated track diagram in the background.

Source: London Transport (now London Underground Limited).

illuminated track diagram, 18m long, showed the dispositions of trains over 240 track-kms. Electronics transformed the physical aspects of interlocking. The older electrical interlocking was usually electro-mechanical rather than purely electrical. Electric circuits worked mechanical locks on some component within the control system (such as the signalman's miniature levers) which governed the setting of points and signals. In modern devices, interlocking was applied through signalling and points relays governed by the setting of the control room panel switches. These could always be moved but the electric circuits would not respond to an incorrect setting. Miniaturised electrical equipment allowed plug-in relays to be arranged on pre-wired panels, with each set governing particular points with their signals. Relay rooms in the 1980s used 'geographical circuitry' with the sets of relays distributed according to the physical layout of the track. Coded pegs and holes in the fitment of relay groups prevented sets from being replaced in the wrong position after removal for checking or repair work.

Joint research by British Railways engineers and manufacturers produced failsafe relays in which the armature always returned to the de-energised position if the operating current were interrupted. British Rail specifications required relays to be immunised against 'false operation' caused by induction, leakage, or other interference from traction supply. The British Signalling Safety Line Relay was commonly used in the 1980s in the interlocking circuits for signals and points, and in the control of colour-light signal display. In Britain, these were termed safety circuits, but in the USA, they were called vital circuits. In Britain, circuits such as route selection button circuits and panel displays, were termed non-safety circuits, but in the USA, these were called non-vital circuits, and less-stringently specified relays were used in them. In recent years, the American terminology has become the international norm. The advent of solid-state devices enabled equipment to be much reduced in size and the first British Railways trials of solid-state logic devices to serve in place of electromechanical relays in interlocking networks, were at Henley-on-Thames (Western Region), and at Norton Bridge (London Midland Region). These were removed, but by the mid-1980s it was decided to move towards solid-state interlocking following in part from work done at British Railways Research in Derby.

In the 1970s in the United States, electrical and electromechanical relays provided interlocking, using much the same practice found on first class railways elsewhere. In the USA, vital-circuit relays interlocked points and signals, and protected crossings on the level of lines of rail. In the USA, the law sometimes required that a train halt, or a flagman be employed, if a crossing of two rail lines was not protected by interlocked signals. 'Time and approach' interlocking ensured that trains likely to be signalled to stop by some permitted route selection, could in fact do so. Interlocking can protect level crossings of rail and road if traffic levels warrant it. Route locking over systems where speed signalling was normal ensured that nothing could be changed to reduce the route's speed rating or capability from the time an approaching train was committed to enter that selected route, until it cleared the last set of points. Any technique which enabled the speed control of the trains to be exercised automatically, rather than through the driver, promised further improvements and these have been introduced (Bailey, 1995; Kerr, 2000).

Sectional route locking was introduced to speed up the handling of trains in rapid sequence, to unlock sections of route once a train had cleared it, and to enable other routes to be set up with minimum delay. These interlocking signals also functioned as block signals to cover the rear-end of trains (Armstrong, 1978, p. 107). By the 1970s, automatic interlocking was widely employed at isolated level crossings of lines of rail to control passage of trains across each others path, and to set signals to 'halt' after a train had passed. In busy terminal sections, route interlocking was automated to allow control to be exercised by one signalman or supervisor rather than several. Other line sections protected by electrical interlocking based on relays included sections of 'gauntleted track' found on some bridges or through limited clearance double-track tunnels. In the USA, there was a trend towards 'train operation by signal indication', which switched the priority in the role of controllers from the trains to the dispatcher. Armstrong (p. 107) describes this technique, which better suited the American system as it was dominated by long-distance goods movements, rather than the European railways with their much more frequent timetabled passenger trains. Armstrong writes:

Where all trackage in a territory is controlled by block signals and interlockings, it is common practice to institute operation by signal indication, superseding the superiority of trains and eliminating the necessity for train orders in moving trains on designated tracks in the same direction ... or in both directions ... In effect, all trains become 'extras' and instructions for their movement are conveyed directly by the signal system as supervised by the dispatcher, directly or through the block and interlocking station operators.

Clearly, the efficiency of this concept would be increased by using cab signalling and automatic driving. In Britain and the USA centralising of traffic control over very wide areas was increased, using route-selection with electrical interlocking and power signalling. In the USA, there were long sections left unsignalled and many single track routes carried infrequent traffic. This led to differences with British practice. In the USA in the 1970s, CTC increased the traffic capacity of single track to about 70 per cent the capacity of double track with automatic block working. It was used to handle increased loads over single track; to reduce double track to single and so reduce maintenance costs and taxation, and to increase the capacity of double-track by working trains in both directions over each line. This practice is found today on most first class railways (Hott, 1994). By the late 1970s, about 50,000 miles of line in the USA were single track with CTC and passing loops, much of it converted from double track. Actual CTC control systems in the USA were much like those in Britain and elsewhere, in consequence of the international nature of modern railway engineering, and the dominance of signalling and control by a few global manufacturing companies. The American controller used route selection, display boards to show train location, and long-range automatic working of interlocked switches and signals. As in Britain, these developments were influenced by engineering innovations which promised economies. Much of railway telecommunications is expensive, and widespread applications often follow developments which enable a known principle to be applied with economy. Centralised traffic control (CTC) is a case in point. The pulse-code technology of the 1920s and 1930s enabled a geographically extensive district to be controlled over two wires only. It was also used in track circuit signalling. Without it, the communications lines would have been very expensive, perhaps prohibitively so. These non-vital message-carrying circuits have been transformed by the telecommunications practice developed since the advent of solid-state electronics. The 'vital-circuits' which directly controlled and activated points and signals were subject to stricter safety regulations. At first cautious conservatism delayed the acceptance of solid-state systems in the place of electromechanical relays, but once reliability was established, they were used throughout the vital circuits.

By 1970, United States engineers were considering extending CTC to control an entire railway from one centre and in the late 1970s, single centres were controlling what would have been a large railway before amalgamations of the period created giant systems larger than were known previously in the USA (Schmidt, 1997). The Fort Worth centre on the Burlington Northern & Sante Fe has realised the forecast of 20 years ago. The 1970s saw a transition from control panels with individual switches, positioned on a 'geographical network diagram', to television screen displays of any section of an 'extended territory' controlled through CTC. The TV monitors were programmed to show the track sections which the controller was supervising through the console control buttons. Greater use was made of computers as they became cheaper, more reliable, and more able under the stimulus of the US space programme. In the 1970s, computer control was used in rapid-transit railways where it helped to work an intensive service with short headways between trains. On both sides of the Atlantic the computer took over the functions of speed regulation and length of train stop periods; identification of empty platforms into which trains should be pathed; automatic changing of platform train indicator displays and announcements. This could be done by the conservative electro-mechanical technology of Dell, or by electronic computers proper. However, as Armstrong (1979, p. 110) remarked:

... all successful systems of this type still separate the functions of automatic train operation (ATO) from those of automatic train control (ATC); that is, the computer-controlled ATO tells the train how to proceed, but it will only do so to the extent that the vital-circuit ATC has independently assured the computer that it is safe to do so.

European developments followed the same pattern. The strength of the American economy in the period 1945–70 compared to European nations gave the US a considerable lead and its space programme provided a source from which many innovative electronics systems came to be applied in railway signalling, control and communications. However, as European economies recovered from the wartime setback and passed through the long period of economic revival, the American lead narrowed and was partly closed when some European nations funded railway modernisation. Today, the European, North American and Japanese systems are equally advanced. In the near future, China, India and other nations will develop a general railway engineering capacity of the same degree of competence. In Britain, CTC was developed to include remote control of interlockings by time division multiplexing, a technique originally developed for telegraphy and telephony which enabled all components to be directed over one pair of wires. Solid-state technology began to replace the once-universal electro-mechanical relays and in the 1980s the first experiments with optical fibre cable in CTC were conducted between Birmingham New Street and Coventry. This had the advantage of not requiring screening against electrical interference. The technique of frequency division multiplexing enabled a common two-wire circuit to carry the control signals for many functions. In the late 1970s and early 1980s, a system was developed for British Railways that carried 80 control and indication channels on different frequencies, all able to work at the same time. In extensive signalling systems, both kinds of multiplexing were used together (RGI Review, 1984).

20.2 The computer in signalling and control before 1985

The computer was used as an advanced train describer-recorder. The electro-mechanical describer-recorders of the inter-war period had been replaced by devices which passed a description from signal to signal using logic circuits based on ordinary relays. The signalman who first selected the route and entered the train description on the set-up panel relied on the description being passed automatically from box to box, and from signal to signal within any one box's district of control. Ordinary relay logic circuits could do this, but the computer had an advantage: it could be reprogrammed to meet changed circumstances, including modifications to the signals, points and track layout. The older describers based on relay logic circuits required rewiring when layout was altered. The computer undertook many tasks in addition to stepping descriptions from box to box. It provided information about train location, state of traffic, changed information displays, operated platform displays, and sent regular reports to various managers, staff and departments. It printed out reports; displayed them on TV monitors and recorded events for future perusal. It reduced printed-out material by comparing actual train timings with the working timetable and printing out the discrepancies only.

In the mid-1980s, the West Hampstead signal box was operated entirely by solid-state computers, and represented train description at its best. The identity of a train was indicated by the timetable head-code number which in earlier British Railways days was carried on the front of the train. This number was displayed on the illuminated track diagram, where it was passed from signal to signal. The computerised system was divided into four sections: data collection, data processing, information display, and communication with 'fringe' signal boxes associated with the adjacent control areas. A brief description is given by Hawkes (1986). Information was collected from the signalling system, and from control units in the central signal box and from boxes on the fringe of the district. This data was supplied to two computers via two identical systems of input printed circuit cards, one per computer. The state of signalling relays and operator's push buttons was transferred to data input cards, and information from the fringe boxes was stored by data set receiver cards. The input to the printed circuit cards was transferred to the computer when the cards were addressed through a 16-bit input data highway. The computers scanned the state of signalling inputs, track sections, points and signals with a frequency high enough to monitor all train movements. The response to a change, such as operation of push-buttons on a signal box control console, was practically instantaneous – which was one reason for using electronic computers. Two computers were used for data processing, with both normally on line. Outputs were interleaved and each computer served as standby to the other. Information was displayed in the main box on illuminated diagrams using small cathode ray tubes in the early installations. Later equipment used light-emitting diodes. Large screen cathode ray tube monitors were developed, with electronic control modules, to back up the main display and to show additional information such as platform availability. Using push-button controls, signalmen inserted new descriptions and questions into the system. Setting up a signal number identified the train approaching that signal, and setting up a train number identified the signal which that train was approaching. The system was developed to operate platform indicators, to make pre-recorded station announcements, and to exercise automatic train control. Automatic route selection by the train numbers themselves was possible but as Hawkes remarks (p. 13):

... whilst the programming required is relatively easy to produce and the cost of implementing it is comparatively low, the task of defining the rules relating priority to position, speed, destination and punctuality of a multiplicity of trains is mammoth ... as far as a complex junction or a large station is concerned.

However, for simple cases, such as a single diverging junction it was both possible and economic. In signalling, the computer set routes automatically. Shortly after it became a component in CTC systems in the USA, the computer was used in an automatic route setting installation in Britain to control three junctions from the new signal box commissioned in 1983 at Three Bridges on the Southern Region of British Railways.

The electronic computerised train describer system was first used in automatic driving and automatic control of other operations in rapid-transit service where electro-mechanical describers – serving as early computers – were installed during the first decade of the century. The integration of the train describer computer with the signalling system control network created the core of the automatic driving and control system, which since the 1980s has been developed for all kinds of railway. In the 1980s, the whole integrated system was known as Automatic Train Control (ATC), which was made up of three sub-systems.

There was Automatic Train Supervision (ATS), which centralised and displayed train and track information in a central control room and was used in the automatic or manual control of the railway. Time division multiplexed links provided remote operation of relay rooms which worked local points and signals through interlocking units, under manual control in a central box. This technique was widespread in the 1970s on rapid-transit and general-purpose railways. Many of the principles predated the solid-state electronics.

Automatic Train Protection (ATP) was old in principle. It used coded track circuits to transmit control signals via inductive links to equipment on board the rolling stock. Speed control, distancing of trains, and emergency braking were provided for and were found on many railways, including London Transport.

Automatic Train Operation (ATO) provided automatic driving, anticipation of signal and speed restriction orders, and correct rates of acceleration and deceleration. It provided automatic shunting operations at terminal stations, safety control of door opening, and minimum headway running. In the ATO equipment carried on London Transport stock in the 1970s and 1980s, commands were given via loop aerials by the track. The ATP system acted as a fail-safe back-up system. These systems served rapid-transit railways well, where passenger trains of the same weight and kind were worked to regular timetables over a limited number of routes. The considerable economic pressure to increase line capacity kept rapid-transit railways in the vanguard of progress in electric traction, signalling and control. The development of these techniques for general purpose main line systems was more difficult and economic returns were uncertain but the advent of the very high speed electric railway and the constant need to reduce staffing levels in all grades of job necessitated their general application. Great advances in electronics engineering improved, revolutionised and extended these 'rapid-transit' techniques to suit operations on all kinds of railway.

In the 1980s, recent advances were rendering obsolete the once-revolutionary systems installed on the Victoria line. In 1988, the original type of automatic driver box, fitted under the passenger seats of the rolling stock, was supplanted by the 'Replacement Automatic Driver Box' (RADB) which used solid state electronics and a microcomputer to exercise control functions. The train mounted system continuously monitored the working of the train and stored data about failures to inform maintenance staff in depots. These facilities were incorporated in new stock delivered to London Transport after 1980. The new ATO system improved performance in approaching stations and in braking by selecting the appropriate braking rate to coincide with a predetermined best braking curve identified for each station. Improved techniques of this type were used in many electric railway vehicles built after 1980, with rapid-transit and high-speed practice setting the standard.

20.3 Reversible working

American practice was to increase the capacity of single track routes by combining CTC with absolute permissive block (APB). This influenced British Railways which developed 'tokenless block' working to dispense with the traditional physical token of permission to proceed. This generally took the form of a staff, released from an electrically interlocked frame which prevented more than one being issued for a section at the same time. The system permitted several trains to follow each other down one track, each with its token, once used tokens were stored in interlocking frames at the other end. Tokenless block working relied on preceding trains working track circuits and signals so that all movements were protected – similar to the American system. As in the USA, the British version of CTC and automatic interlocked signalling was used to allow each line of a double tracked route to be worked in both directions. This was known as reversible working in Britain. It was first used on the Western Region between Didcot and Bristol on the routes through Chippenham and Badminton during preparations of trackwork for regular 125 mph services with the high-speed train. Reversible working is now found elsewhere and it greatly reduces delays when repairs to one line of rail are being carried out.

Reversible working is a feature of the recent upgrading of the North-East Corridor between New York and Boston, USA, which is operated by Amtrak (National Railroad Passenger Corporation). Route-speeds were raised to 240 km/h by extending electrification, resignalling, improving control methods and modifying the permanent way with high-speed crossovers with swing-nose frogs and long switch-blades (Stangl, 1996). The upgrading did not incorporate the advanced techniques of the ECTS or the ACTS systems, and relied on coded track circuits, transponders and cab-signalling to increase line capacity over a busy route. Dispatchers pathed 240 km/h express passenger trains and 80 km/h goods trains through the New Haven-Boston corridor aided by the improved control systems. Before the improvements most of the 253 km route between New Haven and Boston was signalled to take trains running in one direction only on each track of a double line and despite work carried out in the 1970s and 1980s, control was from local signal boxes connected by telegraph and telephone lines strung on poles. The modernisation programme resignalled the route for reversible working on all main lines and introduced a nine-aspect cab signal in place of the four-aspect cab signal already in place. Control for the entire route was centralised and the pole lines eliminated. The Centralised Electrification & Traffic Control Centre in Boston controlled all interlockings, power supply, traction, and the five movable bridges on the route. Microprocessor-based track circuits were installed to generate the cab signals to allow reversible working, using standard industrial products from Harmon Industries and General Railway Signal Co. All interlockings were equipped for microprocessor control. The four-aspect cab-signalling equipment was expanded to nine-aspect by adding a 250 Hz additional frequency to the 100 Hz of the original system. Some lineside signals were retained, namely the home and distant signals which protected various interlockings, but the cab signals controlled all the intermediate block occupations. Information about speed restrictions is now passed to trains by passive transponders. Portable transponders are used where temporary restrictions apply. It is hoped to extend transponder-based speed signalling and nine-aspect cab signalling, throughout the Amtrak Corridor.

20.4 The locomotive-mounted computer

Throughout the 1970s and 1980s, the size of components was reduced, reliability was increased, and maintenance made simpler by the availability of relatively cheap solid-state components. It became possible to mount advanced systems on rolling stock. Systems far in advance of those carried on the automatically controlled London Transport trains were available and new methods emerged for cab signalling, driverless trains, and the integration of locomotive-mounted control systems with automated CTC networks. During the 1970s and 1980s, the research programme for the Advanced Passenger Train investigated locomotive mounted systems, as did the high-speed train projects in France, Japan and Germany. The British team developed a system called Control-Advanced Passenger Train (C-APT) which relied on two microprocessors in one computer which independently decoded messages received from the signalling network. One microprocessor displayed the data to the train crew, and both checked the display against the data received. Any discrepancy triggered isolation of the computer and switched in a second computer carried as stand-by. The APT programme was successful, and more advanced on-board control, signalling, information and safety systems, are carried as standard items on the French TGV, the Japanese Shinkansen, the German ICE, and the Channel Tunnel Eurostar trains (Satoru Sone, 1995; Tamarit, 1997).

The development work on the German Inter-City Express illustrates the advances made in locomotive control and regulation. Loessel, Falk and Winden (Rahn, 1986, p70) describe the original control system as having two parts, a train control level, and a vehicle control level. In later vehicles, subdivision has been carried further. Each power unit in first generation stock has a train control computer for combining automatic motoring and braking control, and high level sequence control. Input-output computers provide the interface with the motoring and braking control units. The train control computers and the input-output computers interface with a fibre-optics station through a data transfer computer made up of an input-output processor, and a 'first-in, first-out' buffer memory. The fibre optic data bus links up to 16 data transfer computers. Continuous automatic train control provides a continuous link between trackside safety systems and the train. Transmission is in both directions. Data about set speed, length of clear track ahead, and revised speed orders are passed from lineside units to the train. Data about distances covered, brake power and train length are passed from train to lineside units. Data transmitted to the train is in part displayed in the drivers cab, and in part passed on to the

Solid-state electronics in signalling, communications and control 365

Figure 20.2 Unpowered electronic transponder, activated by inductive signal from passing trains and employed to gather information necessary for automatic or semi-automatic operations. The one shown was intended for signalling the Advanced Passenger Train between Euston and Glasgow.
Source: British Railways (now Railtrack).

automatic motoring and braking units which exercise automatic train control. The automatic train control equipment includes a train data setter into which information relevant to that train is put before the journey begins.

The functions of the locomotive mounted computer are displayed by the products of any first rank international manufacturer such as ABB-Daimler Benz (Adtranz). The modular philosophy of developing an internally consistent and compatible range of equipment, which can be combined in many particular kinds of vehicle to meet all traction requirements, is being pursued by this largest of all rolling stock manufacturers. Computers are used in all stages of design and manufacture (Godward and Gannon, 1994). CAD systems optimise solutions to design problems, and computers perform physical and mathematical modelling of riding, distortion, vibration, pantograph-catenary interaction, and rolling stock performance. Without the computer, finite element analysis could hardly be carried through to present levels of reliability, and designs would be coarser than they are. Monitoring the complex electronics networks which control motors and auxiliary equipment relies on computers and the driver's console makes use of displays derived from the control room technology once limited to signal boxes.

A typical traction control system is the ABB-Daimler Benz Micro-Computer-Automation-System, Variant S (MICAS-S) which is suitable for transport requirements from commuter traffic to long-range high performance work (Teich, 1994). These systems enjoy wide application because electric street tramways and light railways are benefiting from the electronics revolution along with rapid-transit and main line railways. The MICAS-S system controls, regulates, diagnoses, and informs on three levels. There is the Train Control Level installed throughout an entire train or through several locomotives working in multiple; there is the Vehicle Control Level, within the vehicle scale, and there is the Drive Control Level, governing the drive systems of traction units. What makes electronics such a powerful agent for progress on the systems scale is the ability to integrate such 'train scale' networks with even larger systems – including those for signalling. The MICAS-S equipment, and equivalent devices from other manufacturers, can be used with both electric and diesel-electric traction.

Early attempts to govern rolling stock performance through signals sent from lineside stations by inductive transfer or electromechanical contact were hampered by the nature of the components to be controlled. Electric locomotives had large switches and relays, and steam locomotives had valves and levers which were massive. Rather than fit electro-mechanical motors to shift heavy pieces of equipment, it was easier to send a simple signal to the engine crew and instruct them to move some lever or switch by hand. This set a limit to the fineness with which the locomotive and the traction system could be controlled. The ability to control the modern high-performance railway with a fine precision was made possible by solid state electronics and the miniaturising of much traction equipment. Today, the speed and precision of control extends to all components of the traction system and effects all operations. For example, electronic actuation of air brakes in lengthy goods trains is being introduced in the

USA to increase operating speeds (Carlson and Peters, 1996). An American goods train can be some 3 km long and the air brakes are generally pneumatically controlled. The speed with which the brake actuating signal propagates has been increased from the 100 m/sec of the late 1930s, to circa 200 m/sec, but it still takes about 17 seconds for a full service brake application on the leading locomotive to take effect on the tail vehicle of a 3 km train. This limitation is preventing heavy, fast goods trains working at full track speeds because the speed of brake actuation throughout the train effects the time needed for the train to stop, and therefore the stopping distance. If speeds and train lengths are significantly increased, the stopping distance will exceed existing signal spacing. In order to work faster and heavier goods trains within existing signalling limits, some American railways are introducing Electronically Controlled Pneumatic Brakes (Carlson and Peters, *RGI*, Sept. 1996, pp. 583–86). The pressure in each brake cylinder on each wagon is set directly by computer-controlled, electrically-operated valves rather than by a pneumatic signalling system. The idea of controlling air brakes electrically is old, but a universal electrically operated air brake system never was established, despite suggestions and experiments going back to the 19th century. The electronic system, after rigorous testing by the Association of American Railroads, has been recognised as a potential standard. Each wagon has a single battery-powered neuron chip, and is treated as a node in a computerised network, which is in constant communication with the control unit (usually in the leading locomotive) and with each other wagon. The driver is informed by display of brake application mode selected and of other important system parameters. The system tested in 1997 has a capacity for 160 vehicles (locomotives and wagons) and requires 3650 m of cable. Cable was selected because it was simpler and more cost-effective than other methods of communication which were considered, which included short range radio, and infra red and optical fibres. Each wagon carries a battery to supply the system which is designed for 230 V DC. These batteries can be charged from the locomotive or from a local vehicle mounted generator, and the use of DC enables the power line to be fed from locomotives distributed throughout the train without the problems of matching phases in an AC system. An alternative system, favoured by some American railways, uses radio repeater valves at intervals along the train which allow the brakes to be applied simultaneously from several points. The ECP system is said to be the best because the pneumatic signals of the rival system are propagated from each radio repeater to the wagons between and there is slower build-up of brake cylinder pressure. The ECP system works faster and promises a greater reduction in stopping distances, but general introduction will be expensive, especially if provision is made for working trains with mixed braking systems – some wagons with ECP, some without. The expectations are that ECP will be extensively used in the USA, and on railways where American conditions of operation are usual.

20.5 Radio communications, signalling and control

Radio communications have long served railway signalling and control. In the USA there was readiness to use radio as widely as possible in railway communications. After 1945, the portable equipment developed during the war was adapted to railway needs, and VHF FM frequencies were allocated to railway communications. There was then a considerable increase in the use of radio between train crew and the dispatcher, and between train crew on the very long goods trains where the caboose or brake van could be several kilometres from the leading locomotive. In Britain in 1980, radio communication was used to carry the codes for the block telegraphs in signalling and was replacing the land lines which had served the signalman from the origins of railway telegraphy in the 1840s. The first use, as an economy measure, was in 1980 between Inverness and Wick, in the Scottish Region of British Railways. Radio control of trains or radio dispatching, was a way of lowering operating costs in areas where distances were great but trains few and lightly loaded. Any technique which permitted manned signal boxes to be closed was welcomed, an example being the replacement of physical single-line tokens by a radio-controlled cab-signal giving authority to proceed.

In 1980, British Railways were using radio to communicate between train crew and signal boxes. This was an essential part of one-man operation of trains and was used between trains on the St Pancras (London), Moorgate and Bedford electric services and the signal box at West Hampstead. Today, radio communication between moving vehicles and control points is commonplace in road as well as rail transport. Radio links transmit control signals between lineside stations and train-mounted rolling stock, and current developments may replace pulsed, coded signals sent through track-circuits and wheels, or via inductive links, at least in the more advanced train control systems (Hope, 1996). Transmission-based signalling is being considered for modernising the West Coast Main Line between London Euston and Glasgow Central. Following the break-up and privatising of British Rail, this programme was mounted by the Railtrack company, which invited nine companies to bid for a signalling and control system to eliminate lineside signals from most of the 885 route-km some time in the early 21st century. Railtrack determined that the system should eliminate track circuits, which can be replaced by transponders, Doppler radio or other devices. Data concerning position and speed can be continuously supplied to a computerised control centre, from which information and control signals are continuously exchanged with trains and sub-centres. Although automatic driving is possible, it is not stipulated for the WCML modernisation, though automatic train protection will be a central component of a system working at 'level 3' (see below) of the ETCS concept. Railtrack announced that replacement of life-expired signalling along the route by a level 3 system which would eliminate lineside signals was decided by estimates of cost. Savings due to eliminating lineside signals would justify the expense of the 'level 3' signalling system.

The need to get a good economic return caused Railtrack to select a level 3 program rather than gain experience with less technically advanced level 1 and level 2 systems, as other major railways are doing. The level 3 system will be introduced in stages. In the first stage, all communication with trains will be by radio, and telephones, colour light signals and track circuits will be largely eliminated. In future, the lineside equipment most in evidence will be points machines (De Curzon, 1994; Winter, 1994, 1996). In some areas it is probable that conventional signalling will be retained alongside more advanced modes to enable private short-haul operators to work trains over the infrastructure without installing equipment beyond their requirements. The recent 'collapse' of the Railtrack organisation has placed much of the WCML modernisation plan in question. This indicates how British-style privatising of the railway network may hinder technical advance and standardisation. Much of the technology proposed for the WCML project already exists in the USA, and has been demonstrated in the Advanced Train Control Project (ATCS), though Hope (1995) states that ATCS is more a managerial device rather than a train control system. He further comments that the Americans only use radio data to issue movement orders at discrete locations many kilometres apart.

Previous transmission-based signalling systems, such as those on the Paris RER network use an inductive link with continuous track conductors; those on the SNCF (TVM 430) use coded track circuits and beacons, but the proposed WCML system was to use a digital cellular radio network covering Europe and reserved for railway service. However, this network is not yet available. Much remains to be worked out, not least the location of the control centre from which contact with trains would be maintained by voice and radio links. Birmingham has been suggested. This proposed WCML centre would go beyond the practice in the large American centres, which leave movements within complex interlockings to local boxes, and would control all movements outside depots, yards and sidings, with automatic train protection extending inside terminals 'right to the buffers to prevent a driver colliding with them' (Hope, 1995, p. 573) The full program may never be implemented, but 'level 3' or an equivalent will be installed instead of modern colour-light, lineside signals to replace the colour-light signalling installed in the 1960s and 1970s. However, the upgrading of the WCML is facing great delay at the time of writing.

In Australia, recent renovation of signalling and control systems on a large scale makes extensive use of transmission signalling and radio communication (Hands, 1995). In 1995, Queensland Railways took delivery of a radio-based automatic train protection system from Westinghouse Brake & Signal Co., which provides continuous communication by radio between a train and the next signal being approached. Passive transponders provide information for train-mounted computers controlling braking and speed. This system was used because adverse criticism was directed against automatic train protection systems using track loops or beacons, though the application was on long sections of single track and not high-speed main lines or busy rapid-transit railways. Much of Australia's railways outside the city areas is single track, protected by token and ticket which are electrically locked and released on busy routes. Until the 1980s, only major trunk routes between states enjoyed centralised traffic control. Since the late 1980s, modernisation has replaced the token system by train order working which also dispenses with the home signal which controls admission into an interlocking area (Hands, 1995). These orders are sent to the train crew by radio, and are prepared with the aid of computers which prevent conflicting orders being issued. To provide reliable links between the control centre and each locomotive, a combination of land links and communication via global positioning satellites is being installed. This system, based on long-established American dispatching practice, is well suited to Australian conditions which resemble those in the USA where long-distance goods working over single track is of far greater importance than passenger traffic. A four-stage development scheme has been devised to upgrade the system as required from stage 1, which provides two way verbal communication with manual entry of data to computers. Later stages envisage automation, closer integration of management information with train control and automatic issuing of orders. At present, manual operation of points using local ground frames is retained with some push-button control of powered equipment, but later stages of the development programme envisage points setting from the locomotive cab, or automatic setting in accordance with control orders.

20.6 Large-scale integration of systems

In 1995, the president of the Institution of Railway Signal Engineers, E. O. Goddard, reviewed current trends in signalling. He forecast fundamental and extensive transformation as a consequence of engineering innovation in the aerospace and defence industries, and following organisational changes in national and international railway operations. Present day practises which pointed to future change include the use of optical fibre and radio data links to carry both vital ('safety') and non-vital ('non-safety') information.

Elimination of trackside components, and the use of cab signalling to replace lineside signals, are rapidly being accepted as standard practice, at least for the two extremes of our business – high speed lines and metros. How far away is the portable cab, enabling the driver to control the train from any point? Are we also looking at a substantial increase in automatic train operation?

The growth of through international traffic demands that systems be standardised, or interfaced to facilitate efficient operations across national and system boundaries. Goddard suggests that behind a superficial similarity there will be considerable diversity at the subsystems and components level as new design philosophies create equipment for particular circumstances. This is the signalling engineer's version of the locomotive engineer's 'modular' philosophy, in which designing for particular needs is combined with standardisation. Goddard believed that by 2000 the growth of multinational, multidisciplinary companies making and using standard signalling equipment would compel the signalling industry to accept common international telecommunications standards. The need to interface with other railway control networks would enforce standardisation. Already, this is reflected in the systems architectures being defined for the European Train Control System (ETCS) and the American Train Control System (ATCS). Systems using gyroscopes, radar, radio control, and satellite navigation, will enable signalling to merge with control networks. The major train-mounted equipment for controlling power and braking, for status monitoring, and for processing crew and passenger information will then be unified with the 'shorebased' systems for which the signalling engineer is responsible. Replacing the term 'lineside' by 'shorebased' is itself significant, as is the general use of American terminology in signalling literature. The term 'lineside' is felt to be inappropriate because the new techniques are aimed at abolishing the familiar lineside signals, and the control centre need not be near the railway it supervises.

An impressive example of a modern traffic control centre is the James J. Hill Network Operations Center, Forth Worth, Texas, opened in April 1995. This centre was designed to dispatch and manage train movements throughout the whole of the Burlington Northern's 37,350 route-km. Since the centre opened, the Burlington Northern Railroad merged with the Atchison, Topeka and Sante Fe Railway to form a company of 51,500 route-km – the Burlington Northern & Sante Fe. The centre controls the whole system. The Union Switch & Signal Company stated that the Fort Worth centre had sufficient space and computer power to accommodate the Sante Fe dispatchers if these were to move to Fort Worth from the AT&SF control centre at Schaumburg, Illinois. The James J. Hill Center was named after the 19th-century 'Empire Builder' who built many of the railways which became part of the Burlington Northern. It was designed to control all the Burlington Northern's operations from a single centre, from which 30,000 employees, and 65,000 items of rolling stock are managed. It cost 120 million US dollars and represents a major advance on previous examples of modern signalling technology installed by the Union Switch & Signal Company. These include the Dufford Control Centre, in Jacksonville, Florida (Schmidt, 1997), owned by the CSX Transportation Inc.; and the Harriman Dispatch Center in Omaha, Nebraska, owned by the Union Pacific Railroad Corp. Signalling engineers claim that the standards for state-of-the-art technology found in the Fort Worth Center exceed those in any air traffic control centre.

The Fort Worth centre replaced seven dispatching centres when it became fully operational towards the end of 1996. It housed 850 staff, including 470 dispatchers, working 24 hours per day at 92 workstations. The centre takes the form of a single tornado proof building with 7800 sq m floor space. The main control room is a fan-shaped three-storied space with a floor area of 4200 sq m, which contains the teams for governing all operations to move 500 to 600 trains per day through the network. All traction is by diesel-electric locomotive, usually working in multiple. The main displays are on video screens, 5.480 m high arranged in an arc 65.8 m long along the end wall of the fan-shaped room. As reported in *Railway Gazette International* (1995), besides the workstations for dispatchers, there are consoles for managers of train crew, locomotive rostering, terminal operations, which create 'a synergistic relationship between the various operational areas'.

372 *Electric railways 1880–1990*

Figure 20.3 The Folkestone Control Centre showing the Eurotunnel system diagram and the desks from which controllers supervise signalling, train operations, power supply, ventilation, drainage, maintenance and other work. A standby centre at Calais automatically takes over functions if there are failures in the Folkestone Control Centre.

Source: Eurotunnel/QA Photos.

Incorporated within the team staffing the centre are managers responsible for telecommunications, signal maintenance, transport of particular merchandise, intermodal traffic, and through workings by other operators. The centre is powered via two substations, with automatic switchover if one fails. There is a back-up supply for at least three days work. The central feature of the system is the Computer-Aided Dispatching System of the Union Switch & Signal Inc., which expands capacity from simple centralised train control (CTC) to fusion of CTC with computerised management of crews and rolling stock. An unprecedented range and amount of information is available to the dispatchers:

... from CTC on busy main lines, automatic routing, automated meets and passes, and dark territory management via radio and track warrant control to train sheets, track and time permits, speed restrictions and delay reporting. Dispatchers thus work in a largely paperless environment.

In Britain, British Rail Research has developed the 'Control Centre of the Future' (CCF) to achieve the same ends, albeit on a smaller scale. The Folkestone Control Centre (Eurotunnel) represents best practice in the mid-1990s. Following the splitting up of British Rail into private ownership and operating companies, the first such centre at Upminster was planned to enable controllers employed by Railtrack and the Train Operating Unit of the London, Tilbury & Southend route to supervise all operations over 80 route-km. As at Fort Worth, the working environment was intended to be largely paperless. A shift in working philosophy occurred after 1980. Since the 1970s, orthodox practice used computers to monitor track occupancy and to record data concerning delays and modifications to timetables, rather than support decision making (Rawlings, 1994). Controllers needed to collate data supplied through several independent systems before deciding what to do. This took time and the great advances in signalling engineering such as Solid State Interlocking, Automatic Route Setting, and the advent of the Integrated Electronic Control Centre led to a revolution in route control and management. The CCF system is made up of three components which provide a detailed report of real-time operations; a capacity for predicting events (including problems); and methods for improving problem analysis and solution. The provision of decision support tools was the most difficult task. Techniques drawn from artificial intelligence met the requirement of leaving the operator more time to evaluate options and to reach a decision concerning retiming services, rescheduling rolling stock use, rerostering crews and rerouting traffic. The London Tilbury & Southend control centre at Upminster was installed in parallel to new solid state signalling, after experience was gained with the new Integrated Electronic Control Centre at Liverpool Street in 1993. This latter covers the 198 km route from London to Norwich.

It is anticipated that closer integration between operations control and signalling are likely, as Rawlings (1994) relates:

Alterations to the train service generated using the workstation could be passed directly to the IECC's ARS system as an amended timetable. Similarly, in emergencies or long periods of bad weather, a smooth transition to a library-stored emergency timetable could be effected.

The possibility of automatically exchanging data between the major control and management networks concerning rolling stock and crew use, scheduling changes, and civil engineering works was investigated by British Rail.

In a review of electronics railway signalling principles Schmid and Konig (1995) argue that the traditional principles followed in the past have given rise to multi-level train control techniques which actually limit the ability of railway operating departments to respond to change. Many systems have become extremely complicated and reduce reliability and safety. These authors claim that there are weaknesses in the structures of the engineering systems which inter-relate such systems as European Rail Transport Management which optimises traffic flows, and European Train Control which they describe as a multi-level and multi-path secure data communications system which uses standard building blocks and transmission protocols (Tamarit, 1997). They claim that between these two recent systems there is

... an ill-defined domain of infrastructure and vehicle related control and safety systems, designed to country-specific standards and concepts, perhaps with the exception of the emerging ERRI methodology for specifying interlocking interfaces. Classically, these systems would include track occupancy detection and automatic train protection. Not enough effort is being devoted to controlling the critical interaction between vehicle and infrastructure, specifically with elements such as points.

Large-scale projects for upgrading main routes, such as AlpTransit, or the Berlin-Moscow line demanded resolution of these issues. Schmid and Konig fear that instead of radically new concepts being considered, there will result imperfect mixes of older equipment. In previous years signalling relied on messages transmitted from lineside apparatus to the train crew or to automatic driving apparatus. The instructions transmitted elicited a response concerning route and speed. In recent years the trend is towards speed signalling. Whatever the system, there has been route interlocking and in recent times speed signalling. Speed signalling, always the norm in the USA, is used in the European Train Control System which exercises a control function if a driver or automatic train controller contravenes a speed instruction. Complex relationships between system components have resulted from the introduction and integration of new engineering devices for improving signalling and control.

Schmid and Konig argue that the development of solid-state interlocking was 'haphazard, dominated by operator or supplier dogma rather than long term strategies'. They refer to five types of computer based interlocking systems available from European supply groups (GEC-Alsthom; Siemens; Alcatel; Westinghouse-Sasib-Ansaldo; Ericsson). American companies produce their own systems, some of which are marketed through European partners or licensees. Many systems guarantee safety and reliability by using multiple hardware and software, which is expensive, though long established in railway electronics. The GEC-Alsthom Solid State Interlocking with three processors and the Ericsson system using dual hardware are the most complex, whereas the simplest is the Westinghouse single processor with self-testing capacity. Difficulties with the new electronic signal box at Hamburg Altona, Germany, showed current technology to be near the limit of its usefulness, and disruption for several weeks of the InterCity network was caused by dynamic memory problems.

Schmid and Konig suggest that systems designers pay insufficient attention to the connection between complexity and unreliability when designing subsystems like interlockings. Complexity has increased as more functions have been added to deal with local, automatic, remote and centralised control, and multiple aspect signalling, speed signalling, automatic train protection, and automatic train control. Despite this, many traditional features which could have been eliminated have been retained, such as shunting signals. The dominant fail-safe culture is partly responsible because 'any complex system is deemed appropriate, as long as it can be proved that it will always fail in a safe manner'. Systems are thereby designed so that no traffic is free to move once a problem has been detected. Signals then return to danger ('zero-speed aspect'), automatic train protection is activated, and points are locked in the most recently selected position. Some engineers argue that this is too inflexible and a new design strategy is required, especially for interlocking.

The advent of the European Train Control System heralds the end of the lineside signal (Winter, 1994). This system meets the requirements of a European network, carrying international passenger and goods traffic over increasingly standardised railways, and planned to a common strategy. The European Union summit meeting in Corfu in 1994 approved seven major railway projects, scheduled as part of the Trans-European Networks programme, to be supported by Union resources. These include upgrading the railways in East European non-member states, such as Poland and Russia, with modernisation of signalling and control systems as well as rolling stock, and fixed works. It would be prohibitively expensive to re-equip thousands of route-km with classic fixed signals constructed to new standards, and the latest solid state technologies must be used to eliminate lineside signals completely. Economics demands it.

A special working party under the European Union Transport Directorate was set up in 1990 to identify suitable systems, and a range of projects were suggested for concurrent development under the heading European Train Control System. Manufacturers are developing train-mounted equipment ('Eurocab') using open computer architecture to interface with all existing train control apparatus, and the plan of the railway companies themselves is to develop intermittent or quasi-continuous data transmission devices ('Eurobalise' and 'Euroloop') and methods for connecting with continuous data radio links ('Euroradio'). Funding came in equal parts from the European Union and industry. The plan was for the railways to agree on comprehensive specifications to which the manufacturers responded, though conflicts of interest between railways and manufacturers appeared. The railways preferred to employ Eurocab and Eurobalise, whereas industry sought to fit new technologies into the existing infrastructure – a strategy condemned in some quarters as likely to prevent the full realisation of the potential of new technology.

Development work uses a simulator to investigate driver behaviour in a cab fitted with 'Eurocab'. The Eurocab system is intended for routes fitted with ETCS, which can be encountered at three 'levels' At level 1 automatic train protection and speed control complement orthodox lineside signalling, and thereby develop techniques which have been in use for many years, though using the latest communications equipment to activate the ATP apparatus in the rolling stock. As is normal with automatic train protection, the ETCS would not function if a driver acted correctly using visual identification of lineside signal aspects. If errors were made in response to aspects, the automatic train protection would stop the train. This technique goes back to the first years of electric traction in rapid-transit service. At level 2 automatic train control and cab signalling are combined with orthodox train location by lineside stations. A speed display in the cab can complement or replace lineside signals. By integrating the target speed into the control circuitry, automatic driving is possible. As Winter remarks:

In this mode, ETCS is still an overlay complementing conventional signalling, in that train location and the monitoring of train integrity will still be undertaken by fixed equipment such as track circuits or axle counters.

In level 3 the ETCS provides automatic train control with location by apparatus mounted in the rolling stock. Track circuits are replaced by hodometers carried in the train, and signals are transmitted to the rolling stock from lineside telecommunications beacons or masts. This third level system can be used to provide moving block signalling if increased line capacity is required, though the technique of carrying the block with the train is usually found on rapid-transit systems seeking to maximise line capacity. The three-level structure enables the ETCS to fit a particular line and its traffic. It is expected that Eurobalise equipment for the intermittent transmission of data between lineside stations and rolling stock will be used in the near future in conjunction with existing ATP systems. Eurocab is expected to be used with existing ATP and Eurobalise equipment in 1997, and Euroradio should start linking the three component systems in 1999. In 1994, Swiss Federal Railways introduced ETCS level 1 equipment (Eurobalise and Euroloop) in conjunction with Siemens' ATP with continuous speed control, and followed by trials at level 2 to determine the best ways of introducing ETCS into railway technology. The new AlpTransit baseline tunnel, due to open early this century, will use ETCS level 3 with ATP, ATC and lineside equipment reduced to a minimum. French National Railways are investigating transmission-based signalling and control systems in the Paris-Est region to adapt ETCS for continental scale operations, and have developed the 'Astree' real-time train supervisory system to work with levels 1, 2 and 3 (De Curzon, 1994). Experiments with moving block are also being conducted. To serve in these experiments, both locomotives and diesel railcars have been fitted with train-mounted Astree apparatus for trials on the Bondy-Aulnay line.

Work in Germany in 1996 suggests that transmission-based signalling may become the norm for future high-speed railways. The Koln-Frankfurt route, planned with 300 km/h operations in mind, was designed with transmission-based signalling which used train-mounted radio-control systems. No lineside signals were thought necessary, not even as reserve equipment for emergencies. A Siemens-led consortium was awarded a contract in November 1996 for the electrification, power supply, signalling and communications systems for the 135 km central section of the 177 km route. The key component of the signalling and control system is an integrated radio-based ensemble, known as DIBMOF (Dienste integrierende Bahn-Mobilfunk). This exchanges information between the train and the track without the cables laid between the rails which are required in the inductive methods of train control, such as the LZB system introduced in the late 1960s when high-speed services began. Radio-based LZB systems were investigated after the early 1980s, but it was then thought necessary to provide lineside signals as a safety 'fall-back' measure, even where no non-LZB fitted trains operated. The latest radio based transmission signalling is judged to be reliable enough to make lineside signals unnecessary on the new routes. The DIBMOF system also provides links between the train and exterior public communications networks.

20.7 The moving block system

Current trends in signalling point to abolition of the lineside signal; use of transmission based signalling; subsuming of signalling into centralised control; and the shifting of many of the activities associated with the signal box and signals onto a continuously monitored locomotive or train. The result is a signalling 'block' which is carried by the train itself, and is not defined by lineside stations. The extent of this block, and that carried by other trains, is mutually defined through the continuous control and monitoring of all trains in a system. This 'moving block' system has developed in close connection with rapid-transit automatically operated lines – such as the Docklands Light Railway in London. The Docklands Light Railway is noteworthy as a modern British elevated railway, in which driverless operation is usual. It was opened on the 31 August 1987, as a 15 station, 12 km long system which was later considerably extended by the Bank Extension (1.6 km authorised in 1986, and opened in 1991), and the Beckton Extension (7.2 km, authorised in 1989, opened in 1994). A 4.5 km extension to Lewisham on the South Bank of the Thames has been constructed.

It is a LVDC conductor-rail railway, with 750 V DC in the under-running third-rail and is worked by two types of articulated train weighing 39 tonnes tare, 53 tonnes laden. Traction control is by choppers with naturally cooled gate-turn off thyristors (GTO). As opened and operating into the early 1990s, the Docklands Light Railway used Automatic Train Operation to take the train from one station to the next, with Automatic Train Protection ensuring that speed limits were observed and safety was maintained throughout the system. Supervision and control is computerised, and interlocking uses solid-state devices. The supplier of the fixed-block automatic train control system was GEC. The human controller selects the day's timetable from the computer memory, instructs train operators to drive them to a data docking link at the exit from the Operations and Maintenance Centre, after which the central computer which is located in this centre works through a train-mounted computer and takes it through its timetabled duties. Operation by a human driver is provided for and each train is fitted with driving consoles. At each station the train-mounted computer is connected to the central computer via a docking data link which performs safety checks, works doors, times the halt, signals to the driver operator to close the doors, and directs the train-mounted ATO to take the train to the next station. Detection of transpositions of two cables laid along the track keep the train to speed limits. At maximum speed, the train should cross a transposition just less than once a second. Bogie mounted inductive coils pick up the 470 Hz signal sent through the track cables and detect the phase reversal at each transposition. A timer checks the period between the phase reversals and triggers a brake application if speed is too high. Signalling is by codes, sent through the rails, which are detected by induction coils ahead of the leading axle.

Provision is made for manual driving: one mode is supervised by the Automatic Train Protection system up to any safe line speed; the other mode allows unsupervised driving up to 20 km/h. In this, it resembles the principles developed by Dell on London Transport, but using the advantages of modern electronics, computers, and miniaturised equipment which appeared in the years following the opening of the Victoria line and the establishment of Dell's system. The Automatic Train Control on the Docklands Light Railway has three subsystems which offer automatic train protection, automatic train operation and automatic train supervision. The systems were designed to work in railways in which fixed block sections were normal. Block working in railway service dates from the 1840s and for most of its history the blocks have been fixed. That is, they were physically defined as a particular length of track between carefully defined points, protected by lineside signals. Most railways were divided up into fixed blocks, with supervision of each block by a signal box, which controlled the signals and switches for that block, and which communicated with adjacent boxes by telegraph or telephone. As described in earlier chapters, track circuits, centralised control, automatic signalling, and powered operations transformed signalling but fixed blocks remained the norm. Transmission based signalling and centralised control removed the need to divide a route into fixed blocks, and in 1993 the Docklands Light Railway placed a £26.4 million contract with SEL-Alcatel of Canada, an associate of Standard Elektrik Lorentz AG of Stuttgart, to replace the existing fixed-block signalling with moving block and solid-state interlocking. In March 1994, this moving block automatic train control system was brought into operation on the Beckton Extension. The system, which suffered some initial problems, uses a central computer to co-ordinate data about train speeds, position of trains relative to each other, braking performance, etc. Stored information about timetables, speed restrictions, maintenance work, changes to timetable, etc. is used to control the running of the trains – viewed as a group – to optimise system performance. Inductive loop cables laid on the track transmit information between the computer and the train. The loops are transposed every 25 m and are used with train-mounted sensors. Measure of distance-run is got from the number of axle rotations which give the train's location. There are three levels of computer operation. The central computer provides automatic train operation, route setting, and working trains to a timetable which was being regularly modified. At a second level, a Vehicle Control Centre microprocessor keeps the trains a safe distance apart, and operates the interlocking of points under the direction of the central computer. This Vehicle Control Centre micro-processor receives data concerning speed, location and direction of travel from each train, and transmits to the train the most recently calculated safe stopping point; the maximum speed up to that point; and any required information to control braking to keep speed within the predetermined safe profile for each point on the route. These functions were once performed inflexibly by lineside signals and warning signs based on fixed blocks, later supplemented by cab signalling and various automatic train protection and operating devices. In effect, the above system moves imaginary lineside signals and speed restriction signs guarding each train, so that the fixed block becomes a moving one. The third level of computer operation is performed by the train-mounted computer which receives the signals from the Vehicle Control Centre computer and governs braking, acceleration, traction power supply, and transmits data about location, speed, direction and detection of faults in equipment. Standard Elektrik Lorentz AG claim that trains can be worked with a 60 seconds headway at 55 mph, compared with the 90 second headway required for 55 mph trains over a similar railway fitted with fixed block apparatus. This is obviously a considerable increase in capacity over a railway facing a rising demand to carry more passengers within the existing civil engineering structure. Moving block is therefore likely to find first use on electric rapid-transit railways where it is cheaper to use it to increase capacity instead of relaying track, lengthening trains, or rebuilding stations. Proponents of the system claim that it has low inherent maintenance costs, and reduces power consumption because energy-saving coasting of trains can be implemented during peak hours without loss of line capacity. However, the moving block principle demands more effective traffic regulation than fixed-block working.

All major manufacturers of signalling and control apparatus are investigating moving block, and are producing systems of their own. Westinghouse have developed a radio-based moving block signalling system for use on the Jubilee Line Extension (London Underground Limited). This will be the first of its kind on the London Underground network. During a Discussion Meeting on 'Developments in moving block signalling systems' at the IEE, London, on 18 November 1994, Mr M. McGuire of British Rail Research drew attention to the availability of electronics simulators for comparing different signalling systems. The 'vision' system simulates the movement of several kinds of train within different infrastructures, with various signalling systems, to altered timetables. The simulator represents the essential features of alternative signalling and control systems, whether traditional or modern, including moving block. During a 1996 lecture to the IEE, Mr A. J. Annis stressed that developments in engineering, such as computer-based signalling, solid-state interlocking, and automatic route setting, had outstripped the technology used to support train service management (TSM) which aims to keep the train service as close as possible to that planned, and expressed in the timetable. The technology to support operations control decisions is compelled to keep pace with engineering progress in signalling and rolling-stock control. This, and the need to co-ordinate all aspects of decision, control and supervision, has led to the James J. Hill Center at Forth Worth, and the 'Control Centre of the Future' proposed by British Rail (Rawlings, 1994; Schmidt, 1997; Tamarit, 1997).

Senior railway managers are turning to the latest information technology to ensure that movement of rolling stock matches the requirements of the service the railway is supposed to be providing. Analysis of movements throughout a system shows that vehicles are frequently some distance from where the company thought they were. The CSX Transportation corporation recently decided to equip its 2800 main line locomotives with the GPS tracking system which reduces time out of service; locates locomotives more accurately and speeds up turnrounds in depots. In 1999, the CSX system carried out a three month trial of the 'Pinpoint' equipment manufactured by the GE Harris Railway Electronics company, which was installed in 25 'randomly chosen' CW40–8 diesel locomotives (*RGI*, 1999, p. 487). Utilisation of these locomotives was 6 per cent better compared to machines without the equipment, and their mileage was 10 per cent greater. The 'Pinpoint' system automatically notifies power mangers when locomotives have not moved for specified periods of time: 4, 8, 12 and 16 hours. It locates locomotives with improved precision. During the trials it located units which were 'at least 80 km from where CSX thought they were'. The system sends out a signal when a locomotive is 16 km from its destination, greatly reducing the period between actual and expected times of arrival, and thereby shortening turnround time. It monitors fuel consumption continually and identified engines with excessive consumption. The contract which CSX awarded to GE Harris Railway Electronics in July 1999 will result in the first use of the GPS tracking system in North America.

Information technology working through the internet can follow individual shipments through railway networks, national and international. For example, the RailTracker system enables clients, shippers, forwarders, and railway managers to trace a consignment by telephone, internet or websites installed on individual railways. Several information systems, networks and services are in use and are being constantly improved to match shipments to client requirements, pricing policies and market needs. It is becoming increasingly necessary to know contents and destinations of individual consignments to ensure that they are correctly transhipped across national boundaries and from one mode of transport (such as rail) to another (road or waterway). The incompatibility of IT systems within the European rail network is hindering the efficient use of data to win traffic for railways from other transport modes. Information networks are as vital to the running of modern railways as those essential to signalling, control and general communications.

20.8 Automatic control of main line trains

The automatic control of trains has a long history, going back at least to the Post Office Railway. In the 1960s it found a complete expression in rapid-transit systems such as London Transport but it was not used on main lines. Experiments showing how it might be done were carried out in the 1960s, and automatic operations were examined in the 1970s by British Railways in the Derby research centre. Isolated experiments with locomotives were conducted in Germany and elsewhere but no serious attempts were made to introduce automatic working of either passenger or goods trains over main lines. Automatic working of passenger trains which still carried a crewman became common on rapid-transit lines after 1970 and driverless metro trains were introduced in Lille in 1983.

In Canada, driverless working of ore trains on mineral railways began in 1962 using apparatus developed by the General Railway Signal Co. and General Motors for diesel-electric locomotives. This system used a locomotive-mounted antenna which received control signals from the rails ahead of the train and passed them to regulating equipment in the cab. Trials began in 1958 when six diesel electric locomotives owned by the Northern Land Company were fitted with this apparatus for working mineral trains in the push-pull mode, for which a second aerial was fitted to the rear of the train. Five trains were operated over the railway at the same time, without crews (Sergeant, *RGI*, 1996, p. 713). The line was eventually electrified but the control system was retained and worked successfully. It was developed further by General Railway Signal Company. This Canadian application was noteworthy, but it was for a simple system, with one kind of train running to a simple timetable without complex movements. Likewise, the automatic operations, driverless or otherwise, on London Transport and other rapid-transit railways, were simple and regular though more complex in that routes intersected. Automatic working of main line trains was intrinsically more difficult because a particular kind of train had to be controlled as it moved through a network over which many different kinds of train were running at different speeds.

Experiments to make this possible have begun. In Summer 1996 a revenue earning goods train was hauled by an automatically controlled diesel locomotive over the Deutsches Bundesbahn main line. The trial is of great engineering significance. The system used (Automatische Fahrbetrieb) was evolved by the IFS (Institut fur Schienenfahrzeuge) at Aachen Technical University. Research into the system was started by Prof F Frederich, and in September 1994 tests under automatic working were carried out between the main station in Aachen and Aachen West, using a 650 hp three-axled goods locomotive of Class 364, which was later used in the road trial. The basic principle of automatic control assumes that every signal encountered is set to danger, so that the train must halt unless informed to the contrary (Hughes, *RGI*, 1996, p. 563). The information system uses inductive loops, varying in length from 25 to 300 m, and lineside electronic units supplied by Alcatel-SEL. Interface equipment monitors the setting of signals and points, and passes it to the lineside electronic unit (LEU), which processes it, checks with stored data about the route, and transmits signals via the inductive loop to the train. A telegram generator assembles the message passed to the train. The train receives information about signal aspect, distance to next signal, distance to a point of speed change, and location of speed limits. Each loop determines operation to the next loop. Speed, direction of travel, and response to signal aspect is controlled by Alcatel's 'Euroloop' automatic train protection apparatus, and a train mounted computer constantly compares the two channels of the vital safety circuits and the separate control circuits. Any difference between the safety level (ATP) and the control level (ATO) results in emergency brake application. The brakes are applied if speed or wheelslip are excessive. Optical tachogenerators linked to the wheels locate train position, with correction for wheelslip effected with the help of an accelerometer. Speed is calculated using data from the inductive loops and information on train position. If the ATO computer works correctly and controls traction and braking, the ATP system does not apply emergency braking. External intervention to halt the train can be made through radio signals from a signalling control centre, or by a push button on the side of the locomotive. Transceivers along the route provide a radio link with the signalling control centre, and (Hughes, *RGI*, 1996, p. 565):

... telegrams sent by the SST several times a second include the address of the nearest transceiver, and it receives in return a clear signal that also contains the transceiver's address. This is checked on board, together with a number of other security codes. If three consecutive clear signals are invalid or not received, emergency brakes are applied.

On 4 June 1996, a demonstration for SNCF engineers was made. A coach for the German and French engineers was hauled instead of goods, making the train an automatically controlled passenger special running over main line shared with other traffic. The train was driven manually to Brunswick (Braunschweig) marshalling yard after which automatic operation took it to the Volkswagen car engine plant in Salzgitter-Beddingen. The intention is to automate operations over the 53 km route between the VW plant at Salzgitter-Beddingen and the main VW assembly plant in Wolfsburg and work trainloads of components on a 'Run on demand' basis. This revolutionary idea applies the 'just in time' philosophy of supply to rail operations, and is a bold venture which could win back traffic from road haulage, if successful. Instead of running trains to a timetable, the automated goods train will run when the customer needs supply. Equipping other routes for automatic driving will enable industries to enjoy the advantages of 'run on demand' supply by rail.

DB equipped a 1200 hp Bo-Bo locomotive of Class 294 with the automatic apparatus to work heavy trains DB engineers and specialists at Aachen Technical University examined the application of automatic control to individual wagons which could be guided from any point to any other point on the system, a suggestion considered by British railway engineers in the early 1970s. The principle could also be applied to individual passenger carriages, or groups of

Solid-state electronics in signalling, communications and control 383

vehicles, whether diesel powered or electric powered, thus attracting back to the railways low volume goods and passenger traffic running between small centres. Information technology is vital to such developments both for controlling operations and relating individual vehicle movements to customer needs. Many rapid transit and underground systems include one or several lines with completely automatic operation, for example the metro lines in Lille, Toulouse, Lyon and Rennes which use MATRA-VAL (now Siemens) equipment. These systems derive from those installed by Westinghouse in the automatic railway which provided the "Skybus" airport link service in Pittsburg in 1965.

Chapter 21
The electric railway 1965–95

21.1 The need for increased performance 1965–95

In 1965, the single-phase AC electric main line railway linked London Euston to Crewe, Manchester and Liverpool. This first phase of the WCML electrification marked a culmination and a beginning. It was sometimes described as the former steam railway improved by electric traction and dieselisation of non-electric operations. Steam traction was surpassed in every way, but the railway's commercial and economic functions dated from the pre-electrification period. The reports of Dr R. Beeching and others revealed the need for a new kind of railway to serve a rapidly changing economy. Railways throughout industrialised nations concentrated on providing the services they could do best, and were reshaped within limits set by state policy and funding. Unremunerative operations were reduced or eliminated. These included pick-up and mixed goods traffic; shunting and marshalling; branch-line services and stopping trains. Concentration was on long-distance passenger trains, and unit goods trains. Other services survived, including 'cross-country' trains and urban rapid-transit trains, but these were sometimes dependent on subsidies from the state or local authority. Staff was reduced and maximum utilisation made of standardised rolling stock. These steps transformed the railways, including the WCML.

The WCML became a new kind of railway when the extensions to Glasgow were completed, and the high-powered locomotives of Class 86 and 87 were introduced. Traction performance far exceeded that of previous modes and raised the question of how much speeds might be increased within the existing loading gauge. This matter was important in all countries. In the 1960s rising prosperity enabled people to use relatively cheap air transport over medium- and long-distance routes formerly served by trains only. American railways were badly affected and long-distance railway passenger traffic declined to a small residue operated by Amtrak with federal aid. Within European states, the air network rivalled the train for distances above 400 miles or 600 km. The aeroplane also threatened the express train over shorter journeys, for example between Manchester and London. At the same time, motorways spread to form a network which enabled cars to travel much faster than they did over the old roads. The public turned to the modern, cheap, mass-produced car – usually bought through hire-purchase or got second-hand – and rail services over all distances declined when road improvements enabled the car to match the train in journey time. Rapid-transit rail traffic in and around great cities should be judged a special case, defined by social need and inadequacy of road transport.

Motorways and the high-capacity lorry drew away from the railways goods traffic that was once regarded as transportable only by rail. The light, powerful diesel engine, new suspensions and power-assisted steering created fast articulated lorries able to move tonnages far in excess of those hauled by the biggest vehicle of the 1940s. The use of containers, trailers and roll-on, roll-off ferries further threatened the railways' share of goods traffic. The railway planners sought a considerable increase in the speed of services to check a general decline in all kinds of traffic. A marked increase in train speed was deemed essential to win back passengers from airlines and motorways over distances between 400 and 600 km. The question was asked – how fast could services be operated within the standard loading gauges without a radical departure from the main forms of railway technology? A broader question concerned the forms, orthodox or unorthodox, which guided land transport might take in future. Three general strategies were considered. One was to run trains over existing railway routes within the existing loading gauge at markedly higher speeds. Very fast services could then be run without the great cost of building new railways. The Advanced Passenger Train project of British Rail was of this kind. There were drawbacks to this concept because of the difficulties in pathing very high speed trains over tracks occupied by trains running at much lower speeds. These difficulties were reduced by eliminating low-speed goods trains and stopping trains; and working unit goods trains by night. Improvements to signalling and control made the project practicable. The APT concept was particularly attractive in Britain where building new railways through industrial areas would be extremely costly. Tilting trains would be needed to preserve passenger comfort through curves at high speed. Electric traction could provide the high powers required, though gas-turbine power was used in the experimental stage. Conventional electric and diesel electric trains, with occasional use of gas-turbine sets, provided better services when worked over routes where track had been improved, and were able to check or reverse the loss of passengers to airlines and roads. The success of the diesel-electric High Speed Train over British main lines showed what conventional railway engineering at its best could do.

The second solution, pursued by the Japanese, the French, the Germans, the Italians and others was to construct new railways segregated for very high speeds, over which standard trains of new design worked to standard timetables. This eliminated the difficulties of pathing fast services through much slower ones and left older lines free to take goods traffic and less-rapid passenger services. On segregated routes, trains worked through curves at a single, defined speed rather than running within the range of speeds taken by the different kinds of train passing through the curves on general-purpose railways. The curves on segregated lines were given the correct rail superelevation to minimise passenger discomfort at the stipulated line speed, and tilting trains were not required. The new lines were built to interconnect with existing networks, so that the latter could be used to gain access to stations in major cities. Most of the new lines were of standard gauge. There were disadvantages to this scheme. Some important towns were bypassed, or served by expensive loops, to enable the new routes to be aligned satisfactorily. The high-speed lines were costly to build and were opposed by farmers and landowners. State funding or EC support was needed to build them, and the growing European network of segregated high-speed lines and interfaced, improved older lines is partly the result of socio-political policy to bind nations and community into a closer-knit whole. The segregated high-speed network is best if money is available for construction, and, despite the success of high-speed trains designed to work over upgraded existing routes, it is the only proven means of achieving service speeds of 300 km/h. However, very large sums of money are needed to build them, and political disputes over funding can interfere with strategic planning of high-speed networks. Line construction takes years, during which funding policy might change following an election. In 1999 the German Transport Minister announced that construction of the 192-km Nurnberg-Erfurt Neubaustrecke through the Thuringer Wald had been suspended because it was not likely to be financially viable. Availability of funds remains the most important limiting factor in all railway engineering developments, setting a limit to design and planning ambitions.

The third approach to improving guided land transport assumed that conventional railways had reached their limit so that much higher operational speeds could only be obtained with radically new systems, many of which could hardly be termed railways. It is significant that not one of the many systems which were tried approached the segregated high-speed electric railway in performance. Most fail to equal the modern conventional general-purpose railway when worked by electric or diesel-electric locomotives. Radically different systems have been proposed from the early years of railways, and are usually advanced as the best means for gaining greater speeds in passenger traffic. Some systems were never intended for general-purpose traffic, but were designed for military or agricultural application or for use in unusual locations. The 19th- and early 20th-century Brennan Monorail; Lartigue Monorail; and Langen System fall into this category. The latter survives still as the electric suspended monorail railway between Barmen and Elberfeld in the Wuppertal, Germany, where it follows the line of the river down a valley, lined by towns. A cable-suspended version of the Langen System evolved into a common form of mountain cableway. Several systems were intended for widespread application. Examples are the Kearney and Bennie railcars of the period 1920–50. These never progressed beyond demonstration units, and failed to become established.

When very high speed conventional trains were being developed, several projects were begun to introduce new guided transport systems for fast, long-distance communication. They were not intended to replace the general-purpose railway, with its goods and slow passenger traffic, but were intended to provide medium- and long-distance passenger services much faster than any orthodox railway. Their proposers failed to anticipate the improvements to conventional traction brought about in the next 10 to 15 years. The British Guided or Tracked Hovercraft (Bailey, 1994); the French 'Aerotrain'; and the current German and Japanese 'Maglev' projects are well-known examples. The Aerotrain achieved line speeds exceeding 400 km/h on test near Orleans in the 1970s. At present, only the 'Maglev' alternative is being pursued with serious commitment and full-scale test facilities. The Japanese and German work does not suggest that 'Maglev' is an economic substitute for existing TGV, ICE or 'shinkansen' systems, or is likely to be better than the segregated electric railways of the foreseeable future.

21.2 The origins of very high speed train operations

The high-speed electric railway dates from the trials conducted in 1955 by the SNCF (French National Railways) and directed by F. Nouvion, the deputy head of electric traction research. The tests studied behaviour of rolling stock and permanent way at very high speeds and identified problems needing solutions if such speeds were to be normal. High speeds had been attained during trials on previous occasions in several countries. The tests carried out between 1901 and 1903 with three-phase traction on the Marienfeld-Zossen military railway witnessed speeds up to 210.2 km/h (130.6 mph), attained by a locomotive and railcars from Siemens & Halske and AEG (Lasche, 1901). These trials revealed the potential of electric traction but did not establish it in regular service. In Germany between 1931 and 1939, Kruckenberg and colleagues demonstrated high speeds with petrol-engined and oil-engined vehicles, and speeds exceeding those of the contemporary American lightweight diesel trains were reached. The famous Kruckenberg Schienenzeppelin ('Rail Zeppelin') was driven by an airscrew and had an early 'steerable' axle, actuated by a hand-wheel. It was used to investigate vehicle behaviours at speeds up to 230.2 km/h (143 mph), which remained the world rail speed record until 1954. German and American diesel sets raised the speed record for diesel traction from 165 km/h (102 mph) set by the VT877 railcar between Berlin and Hamburg in December 1932, to the 181 km/h (112.5 mph) reached by the 'Pioneer Zephyr' between Denver and Lincoln in May 1934, and the 205 km/h (127.4 mph) reached by the Leipzig railcar between Ludwigslust and Wittenberge in February 1936. In June 1939, the speed record for diesel traction was set at 215 km/h (133.6 mph) by the Kruckenberg train between Hamburg and Berlin. In Italy, in July 1939, an electric set ETR 200 reached 203 km/h (126.1 mph) between Bologna and Milan. During the 1930s, steam traction had reached 200 km/h on two authenticated occasions: 200.4 km/h (124.5 mph) in Germany in May 1936 by 05. class engine no. 002, and 202.8 km/h (126 mph) in Britain, on Stoke Bank, in July 1938 by A4 class engine no. 4468. The high-speeds trials with electric, petrol and diesel traction bore fruit during the period of economic recovery after the 1939–45 war when high-speed diesel trains were introduced in Belgium, the Netherlands and Luxemburg. These 'Benelux' operations and the Trans European Express services owed much to the pioneer work of Kruckenberg.

The records set by steam traction remained isolated examples of what a technology might achieve as it approached obsolescence. There were designs for steam locomotives able to work at 250 km/h (155 mph). Project Lubeck (Glasers Annalen, 1943) proposed two 4–8–4 five-cylinder compound expansion engines coupled back-to-back with a condensing tender between them. A Chapelon designed high-speed steam locomotives in the late 1940s, but these were not developed. A railway in which very high speed was the norm required electric or internal combustion engined traction to provide rolling stock with a sufficiently great power to weight ratio.

The first speed trials which were to lead to the modern high-performance long-distance railway took place over five days in February 1954 along the Paris-Lyons main line, which had been electrified in 1952 using the 1500 V DC system. The tests used locomotives and carriages, not sets. The locomotive was CC type no. 7121 (CC7121), which reached 243 km/h (151 mph) with a three-coach train between Dijon and Beaune. The engine was not subjected to any major modification before making the trials which broke all previous speed records on rail, including the highest set by the Kruckenberg car in 1931. In 1955, speed tests were resumed, on the Bordeaux-Hendaye main line between Morcenx and Lamothe where there was a long, flat, straight section. This time, the rolling stock and supply system were modified to increase performance (Tassin, Nouvion and Woimant, 1986). Two mobile substations increased the traction supply voltage from 1500 V to 1900 V DC, and the two test locomotives were regeared and given monobloc wheels and new pantographs redesigned for high speed. One engine was a BB type with two four-axled bogies; the other a CC type with two six-axled bogies. The test train was made up of three vehicles fitted with rubber fairings to improve air flow and the last car had a rounded tail section added to reduce drag. In March 1955, the four-axled locomotive hauled the test train at 276 km/h, and the Alsthom-built six-axled machine no. CC7107 became the first locomotive anywhere to exceed 300 km/h when it reached 326 km/h. The highest speed was gained by the MTE-built BB9004, which set a world record for conventional railway rolling stock of 331 km/h, which stood until 1981 when it was broken by the SNCF TGV (371 km/h).

The tests showed what must be done to introduce regular services at greatly increased speeds. Both locomotives had suffered from pantograph collapse caused by frictional heat, and the fragments were hot enough to set fire to lineside trees. The most severe problem concerned the riding of the vehicles at high speeds. The bogies hunted so violently that the wheels dealt blows which distorted rails and disturbed sleepers and ballast. SNCF photographs of damaged track suggest that the trains came close to derailment. The slipstream and wake around and behind the train raised dust, disturbed ballast, and generated sparks and noise. These difficulties faced engineers planning a conventional 'wheel-on-rail' system for high speed, and led to proposals for guided transport involving magnetic-levitation, or air-cushion hover-trains. In the end, systematic research and development of conventional railway vehicles removed these obstacles, provided funding and political support were adequate (Hughes, 1988; Potter, 1987).

Japan was the first nation to construct a new electric railway. The Tokyo-Osaka Shinkansen or New Trunk Line was opened on 1 October 1964, which is a date of considerable importance in the history of the post-war regeneration of the world's railways. The new trunk line was segregated from the existing Japanese system, and was built to standard (1435 mm) gauge rather than the narrow (1067 mm) gauge of the older network. It provided greatly increased passenger-carrying capacity over the 515 km route between Tokyo and Osaka. During the economic recovery of Japan in the 1950s, this corridor witnessed the rapid growth of cities and an increase in demand for rapid communication, especially with Tokyo. The former Tokyo-Osaka main line formed 3 per cent of the Japanese National Railways' route mileage, yet carried 25 per cent of the passenger and goods traffic. In 1959, the best time between the two cities was 6 hours 40 minutes, giving an average speed of 83 km/h. The route was electrified between Tokyo and Kyoto in 1956. Part was four- or even six-track, and quadrupling all of it was considered, but judged an inadequate solution.

In 1958, it was decided to build an entirely new line to standard gauge, which would reduce travelling time to 3 hours 10 minutes, giving an average speed of 162 km/h (101 mph). A standard gauge railway would permit higher speeds and vehicles with a larger carrying capacity than would a narrow gauge line. The system was projected to work with trains running at 260 km/h, but the World Bank, which was funding construction, reduced this to 210 km/h. The electric system was 25 kV, 60 Hz rather than the HVDC (1500 V) used throughout much of Japan. Contemporary French developments and Japanese trials with 20 kV single-phase AC systems on a 29 km section of the Senzan line influenced the decision to choose single-phase AC at Japanese industrial frequency. An initial plan to work container-goods trains over the line was abandoned when passenger traffic increased so much that it demanded all the daytime carrying capacity. Night was reserved for maintenance of track and fixed works, which were subject to severe wear even at the reduced speed of 210 km/h imposed by the World Bank. The new line eliminated level crossings and permitted high speeds. About one third of the route was on viaducts. The slab track generated noise and considerable works were undertaken to shield nearby residential areas following strict nuisance-suppression legislation in 1975. All new high-speed lines, in France, Germany and elsewhere faced strict legislation compelling noise suppression along routes passing close by residential or commercial areas. The trains were multiple-unit sets rather than locomotive-hauled rakes. Sixteen car ensembles drew current from the overhead catenary at 25 kV through eight pantographs to supply motors on each of the 64 axles of the train. This distribution of motors gave smooth, high acceleration. Total traction power was of the order 11,840 kW (15,880 hp). Multiple-unit sets enjoyed advantages over locomotive-hauled rakes for very high speed passenger services (Hughes, 1988):

... such as rapid turn rounds at termini. (They) permitted the electrical equipment needed for the 8900 kW continuous power rating to be distributed along the length of the train, so ensuring that the weight was evenly spread and that axle-loads did not exceed 16 tonnes.

It was essential to keep axle loads down to reduce the destructive forces on permanent way. Research into the interplay between high-speed vehicles and track was undertaken in those countries that developed the new railway technology for home use or export. The first Shinkansen sets had a traction motor on each axle and used rheostatic braking to reduce wear on disk brakes which were employed below 50 km/h. The brakes could stop a train from 200 km/h in 3.6 km, which was less than the 5.3 km required by the cab signalling system. Fast acceleration and rapid deceleration were needed, as there were 10 intermediate stops along the 515-km route. The line carried only one kind of train working the same service, though there were two timetables, and so the signalling and communications systems did not have to accommodate different braking distances and speeds. Lineside signals were therefore not used and trains were directed by speed signalling installed in the cab, working from pulse-modulated audio-frequency track circuits, developed from similar equipment found on rapid-transit networks. Automatic driving was recognised as a possibility, but the driver was expected to respond to speed signals, subject to automatic braking if instructions were ignored. Some trains stopped at all intermediate stations and others called at only two. The average journey times differed, and the faster trains needed to overtake the slower trains. This was accommodated by using automatic route-setting at intermediate stations where overtaking would occur.

Once authorisation for the first Shinkansen was given in 1958, the project advanced rapidly. Construction began in April 1959 and the first trains were working when the Olympic Games were hosted by Japan in Tokyo in 1964. The official opening was 1 October 1964, though a limited service had begun in August. Journey time between Tokyo and Osaka was reduced from 6.5 hours over the narrow-gauge line to 4 hours over the new. Performance was improved further once the track and works had settled in, and in October 1965 the time between Tokyo and Osaka was reduced to 3 hours 10 minutes, with an average speed of 162.6 km/h. This was the fastest scheduled rail service in the world and the first to provide a start-to-stop average of over 161 km/h (100 mph). Since then further Shinkansen routes have been built with rolling stock developed for much higher speeds.

The spread of the Shinkansen network was not a story of unopposed progress, and it met with opposition from lineside residents, environmentalists, politicians and economists. It took place while Japan went through crises over public spending levels, and when a drastic reorganisation of the Japanese National Railways was undertaken, leading to privatisation. Despite these difficulties, the Shinkansen network was extended, encouragement coming from the success and profitability of the first line. Later projects were questioned on economic grounds. Costs were very high because of the need to build noise suppression works with huge concrete screens and deflecting walls. A high percentage of the routes was in tunnel, reaching 40 per cent on the 1982 Sanyo line. All this increased costs and the 1980s extensions to the Shinkansen network were opposed as certain loss-makers. Nevertheless, the system was extended, and the 1982 Tohuku Line (465 km) and the Joetsu Line (270 km) were engineered for speeds of 260 km/h (160 mph). These lines were worked by new rolling stock, with aluminium bodies and an installed power of 11,040 kW (14,800 hp) in 12 and 16 car sets. The 12-car units were capable of 260 km/h, though limited at first to 210 km/h (Williams, 1985; Hughes, 1988; Machefert-Tassin, Nouvion and Woimant, 1986). Whatever the economic justification of the later Shinkansen lines, there is no questioning their engineering success and the promise of higher speeds from the latest generation of trains which will run faster than 300 km/h (Hughes and Masatka Kimata, 1994).

21.3 The development of the SNCF train à grande vitesse

The high-speed trials of 1954 and 1955 caused the SNCF to develop rolling stock for the first regular passenger services in Europe to run at 200 km/h (125 mph). The first service was provided by the 'Capitole' express between Paris and Toulouse in 1967. Before the early 1970s the French envisaged using high-speed locomotives and lightweight coaches, drawing supply from overhead contact wires (HVDC or HVAC). A most important objective was to run these trains over existing railways, using the stations, viaducts and fixed works already in place, and to path them between other trains, fast and slow. The construction of new, segregated lines reserved entirely for fast passenger trains was not envisaged. Before introducing regular passenger services at 200 km/h over existing railways, the SNCF ran speed trials south of Bordeaux to explore train behaviour between 200 and 250 km/h. There were over 400 runs with conventional four-axled and six-axled electric locomotives, and observations were made of the effects of high speed on rolling stock, permanent way, lineside structures, other trains and bridges. The noise and dust nuisances were investigated, and assessments made of the ability of the signalling, control and communications networks to deal with the proposed new services.

In 1967, the 'Capitole' express entered service between Paris and Toulouse, and other high-speed trains called the 'Aquitaine' and the 'Etendard' were introduced between Paris (Austerlitz) and Bordeaux. This distance of 581 km (361 miles) was covered in 3 hours 50 minutes, giving an average speed of 152 km/h (94 mph). Improvements to track and fixed works were carried out in the 1970s which permitted running at 200 km/h, and by the mid-1980s 400 km of the Paris-Bordeaux route allowed trains at this speed. Services on other French routes were also upgraded.

This new generation of trains required remarkable locomotives. One noteworthy design was the CC-40100 type, introduced in 1964 to pull the Trans European Expresses (TEE) and international services between France, Belgium, the Netherlands and Germany. This six-axled locomotive could operate from four current supplies: 1500 V DC; 3000 V DC; 15 kV, 16.66 Hz single-phase AC; and 25 kV, 50 Hz single-phase AC. There were four separate pantographs, one for each supply, and safety devices protected electrical systems from damage if a pantograph contacted an unsuitable supply. An AC supply was fed into a fixed ratio transformer and converted to DC in a silicon bridge rectifier. The DC was controlled by motor grouping and rheostats. These locomotives were Alsthom products and had lightweight steel bodies. They met all European loading gauge requirements. Each bogie had one double-armature series-wound Alsthom direct-current traction motor, fully suspended with Alsthom flexible drive. Control was rheostatic, and the electric brake was rheostatic. Weight in working order was 108 tonnes. Ten of these engines were built, in two varieties. Engines CC-40101 to CC-40104 had alternative gear ratios which could be engaged when the locomotive was in motion. These gave two maximum speeds of 240 km/h (150 mph) and 160 km/h (100 mph). Continuous power output was 3670 kW (4920 hp). The second group of engines, CC-40105 to CC-40110, were given gear ratios for 220 km/h (137 mph) and 180 km/h (112 mph), but these could only be changed in the workshops (Harris, 1981). Continuous output for these machines was 4480 kW (6005 hp). The class was successful, and in 1973 six of the second type were built for SNCB (Belgium National Railways). The French locomotives usually worked from La Chapelle depot in Paris on national and international services through the Nord region. Equally noteworthy were the CC-6500 locomotives introduced in 1969, some of which were geared for high speed and were used to work the 'Aquitaine' and 'Etendard' expresses. These were an Alsthom-MTE product, and their mechanical components had much in common with the 1968 CC-7200 diesel-electric class – namely the three-axle single-motor bogie designed to reduce destructive track working by having fully suspended motors. The motors were of the Alsthom double-armature DC series-wound type, fully suspended with cardan-shaft drive. The class worked as high-speed express passenger locomotives and as heavy-duty express goods locomotives because two gear ratios were available, though these could only be changed when the locomotive was stationary. Top speed in the fast gear was 220 km/h (135 mph) and in slower gear it was 100 km/h (62 mph). The current supply system was 1500 V DC, with overhead contact wire and pantograph collection, though a subsection of the class was equipped for third-rail collection over the Chambery-Modane line in the Savoy, which was not provided with overhead distribution until 1976. The CC-6500 class comprised 74 locomotives rated at a continuous power of 5900 kW (7905 hp) with maximum output of 6500 kW (8700 hp). The type proved successful, and lower powered derivatives modified to work from 3000 V DC distribution were exported to Morocco and Yugoslavia (Harris, 1981). The CC-6500 class took over operation of the 'Aquitaine' and 'Etendard' trains between Paris, Austerlitz and Bordeaux, and these averaged 152 km/h over the 581-km journey.

The 12-year test programme gave the SNCF a 200 km/h (125 mph) service, which in the 1970s was superior to any express passenger service working outside Japan. Experience stressed the need to reduce axle loads and to increase power to weight ratio if faster regular services were to be achieved. Because of this, engineers thought that locomotive-hauled trains would be abandoned, and replaced by the Japanese model of lightweight sets with the motors distributed along the train. However, the locomotive has survived as a feature of the very fastest trains in France and Germany, as both the latest TGV and ICE systems testify, though the third generation of ICE trains has motors distributed throughout the train. It proved possible to reduce axle load, and to raise power to weight ratio, without distributing power, traction and control equipment throughout the train, and replacing the locomotive by the multiple-unit form. In the TGV, ICE and 'Eurostar' trains, locomotive vehicles are evident, though the multiple-unit form of high-speed train remains an alternative employed in Japan, Italy and Germany.

In the 1960s, as the 200 km/h locomotive-hauled express programme was developing, the SNCF began systematic research into very fast, light trains using an experimental gas-turbine set with hydro-mechanical transmission converted from a two-car diesel unit. The aviation-type gas-turbine was rated at 1050 kW (1400 hp) and gave a power to weight ratio of 17 kW (22.8 hp) per ton for the two-car set, which still retained its 330 kW diesel engine in the power car as a source of extra power. The axle load was 11.8 tons compared to the 19.2 tons of the CC-6500 locomotive. Trials with this pioneer unit encouraged the SNCF in 1968 to order ten four-car sets, capable of 180 km/h (112 mph) using aviation gas-turbines. Introduced in 1970, these ETG sets ran between Paris, Caen and Cherbourg with an average speed of 125 km/h over the 372 km, reducing the previous best journey time by 50 minutes. These were followed by more powerful five-car sets intended to run at 200 km/h (125 mph) on services between Lyon and Strasbourg, Nantes and Bordeaux, and the role of the gas-turbine in high-speed, light-vehicle propulsion seemed established.

In 1969, the SNCF investigation into high-speed trains was advanced by the order for a newly constructed unit, made up of three 'trailers' between two end power cars. The train was articulated, and adjacent vehicles were carried on four axled bogies, which, with the outer bogies under each power car, gave 12 axles in total. Each axle was powered. There was a gas-turbine-driven alternator in each power car which provided 1880 kW for traction power and 180 kw for auxiliaries and air conditioning. The train set weighed 197 tons and the ratio of total traction power to weight was 18.2 kW/ton. Gas-turbine sets of 2400 kW were installed later and the power to weight ratio was then increased to 23.3 kW/ton. This remarkable train was the TGV-001, the first of the French TGV trains, which represented railway technology at its most advanced. The experimental train began trials in April 1972 and on the 8 December of that year reached a speed of 318 km/h (198 mph). It reached 250 km/h (155 mph) in over 1000 trips, and during four years of trials covered 320,000 km, whereas (Williams, 1985) the APT-E reached 241 km/h on only three occasions and covered 37,800 km. Design changes reduced unsprung mass. The traction motors were suspended from the bodies of the vehicles to drive the wheels through cardan shafts and axle-mounted gearboxes. Flexible helical springs formed the secondary suspension, rather than air suspension. Both these features became widespread in high-speed rolling stock.

In the 1960s and 1970s, several railways considered using tilting coaches to minimise passenger discomfort on curves where cant deficiency was high, but the SNCF rejected such proposals because they were complex and required costly and lengthy experiments to perfect. The SNCF plan was to introduce new trains, using gas-turbine power, to run over conventional main lines which would be improved for high-speed operations. At the time, these lines were carrying trains of the latest coaching stock hauled by locomotives of the CC-6500 class. The initial route considered for the first TGV service was the old PLM main line from Paris to Lyon via Dijon, and on to Marseille. Development of this line faced problems similar to those encountered by the Japanese when planning to increase capacity between Tokyo and Osaka. The PLM line was operating near its limit and it would prove difficult to path very fast trains through slower goods and passenger trains. Installing extra track would only defer the need to provide greater capacity at the still higher speeds which research and experiment suggested were possible. The French therefore followed the Japanese example and constructed a new railway from the outskirts of Paris to the suburbs north of Lyon – the Ligne à Grande Vitesse (LGV, or High Speed Line). Unlike the Japanese new trunk routes, this route connected with the existing SNCF network and was constructed to the same track and loading gauge. The new trains could use ordinary railways to gain access to city centres and to complete journeys between places where a new line was not justified. Electric traction was selected for the LGV and for all future extensions to the programme after an initial proposal to employ gas-turbine power was changed. The problems of picking up power from overhead contact wire at very high speeds were resolved, and much of the route which the TGV would share with ordinary trains was already electrified. Gas turbine traction proved less economic than expected when oil prices increased after the 'oil crisis' of 1973–74, and the long-term prospects for getting high performance with reliability and low maintenance were better with electric traction. An all-electric traction system promised better integration of all components of the railway – power supply, signalling and control – than did one in which the rolling stock was powered by gas-turbines. However, gas-turbine and diesel-powered TGV sets were not and are not ruled out for lightly loaded LGV.

The LGV was electrified at 25 kV, 50 Hz single-phase supply, which was by then the French standard, though the PLM system was electrified using the earlier standard of 1500 V DC. In 1978, two electric test units were constructed by Alsthom as dual-voltage sets for pre-production trials. The power cars in these trains were not articulated to the trailers, though the latter were articulated with each other in an eight-car rake between outer power units. The power cars carried separate DC and AC pantographs, one of each on each car. Both DC pantographs were used when under the 1500 V DC overhead contact wire, but only one AC pantograph was raised when running over the LGV with HVAC supply. This was due to trouble with the trailing AC pantograph at very high speed – a matter later resolved. AC supply for the two power cars used the leading pantograph which fed a 25 kV bus carried along the roof of the vehicles. This arrangement was variously approved and condemned by different authorities. Through-train omnibus-lines for carrying power supply were first mooted in the 1890s, and have never gained general acceptance because of the fire risk. In some countries they are avoided, if not illegal, and the refusal to use them has restricted the design of high-speed electric train sets.

The pre-production trains were rated at a continuous power output of 6450 kW (8650 hp) which was sufficient to move a 420-ton set at the then-maximum of 270 km/h (168 mph). The arrangement of traction motors was followed in later developments in France. Instead of all axles being powered, motors were fitted to the eight axles of the two power cars, which had two four-wheeled bogies each. Motors were also mounted on the axles of the two four-wheeled bogies which carried the outer end of the articulated rake of trailers. This provided 12 powered axles in all and perhaps justifies referring to the power car as a locomotive. The pre-production sets were successful and ushered in a development in long-distance high-speed railway services which continues to this day. Construction of the new line started in December 1976 and the route, which involved 410 km of new railway, carried the first fare-paying passengers at 260 km/h on 27 September 1981. Curves of minimum radius 4.2 km and pointwork designed for 160 km/h helped to reduce the schedule between Paris and Lyon to two hours. The 97 production trains were similar to the pre-production sets.

Cab-signalling was installed and there were no lineside signals, apart from marker boards at the start of each block. In this, the French system resembled the Japanese Shinkansen. Signal aspects indicated at least five blocks ahead and set speed limits for each section approached. If five sections ahead were clear, a 'way clear' sign (VL – Voie Libre) was displayed. Four sections clear corresponded with a cab signal of 270, meaning that the next marker board could be passed at the line maximum of 270 km/h with readiness to reduce to 220 km/h in the succeeding section. Indications following the 270 were 220, 160 and 000. With a clear section ahead, a display of 000 meant that the train must stop at the next block marker board. The signalling practice came from modern rapid-transit systems where automatic operation had been tried and where brakes were applied if speed tolerances were exceeded. A triple braking system was fitted. Above 80 km/h rheostatic braking was used on the powered axles. On the non-powered axles, disk brakes operated through all speed ranges. All wheels, whether powered or not, were fitted with tread brakes which acted between 160 km/h and halt.

The new trains established new standards of service. In 1983, the Paris to Lyons start-to-stop average speed was 213.6 km/h. The Paris-Geneva journey time was reduced from 5.5 hours to 3.5 hours using seven triple-supply sets modified to work under the Swiss 15 kV, 16.66 Hz single-phase AC system. In February 1981, the world rail speed record was first raised to 371 km/h, and then increased to 380.4 km/h by electric TGV sets. Since then, it has been progressively raised to the present world record for wheeled railway vehicles of

515 km/h held by the TGV, and regular services operate at 300 km/h (186 mph). Expansion of the LGV network was supported by the state. In September 1981, the French President Mitterand announced that studies were being undertaken for a second line to serve western France. In September 1989, the western branch of the TGV-Atlantique line was opened for service with a maximum speed for commercial operations of 300 km/h, and the construction of the TGV-Nord line was started. The LGV-Atlantique became the fastest railway in the world. On 5 December 1989, a TGV set a world speed record of 482.4 km/h, and on 18 May 1990, the present world speed record of 515.3 km/h was set.

In 1993, the LGV-Nord was completed to link Paris with the Calais and the Channel Tunnel portal area. Since 1986, construction of the Channel Tunnel had proceeded with great success. The French and British service tunnels had met on 1 December 1990; and in October 1991 the main tunnels joined. On 28 October 1992, the SNCF, SNCB (Belgian National Railways) and British Rail displayed the 'Eurostar' trains intended to work the services through the Channel Tunnel between London, Paris and Brussels. These trains differed in important details from the earlier TGV sets, though the influence of the latter on Eurostar design was obvious, and the 'Eurostar' trains were commonly referred to as 'TGV Transmanche'. The Eurostar services through the Channel Tunnel began on 6 May 1994. Speed in the tunnel was restricted to 160 km/h and 300 km/h was permitted elsewhere.

The SNCF showed that segregated lines worked by trains of high power could meet the requirements of a modern industrial state for fast long-distance passenger transport on land. This policy proved more successful than projects to integrate very high speed trains within existing systems or to build new guided systems like 'Tracked Hovertrain', 'Alweg Monorail', 'Maglev' or 'Jetrain'.

The new lines were expensive, but in France excessive tunnelling and earthworks were avoided because steep gradients of 1 in 29 were worked at speed by electric traction. In France, unlike Japan, there were no high mountain ranges, deep gorges, and mountainous coastlines to compel engineers to put much of the route in tunnel, and the LGVs were built without too much urban disruption. Despite this, there were delays in construction arising from political and social objections, though the LGVs were generally supported by the French state and public. The network continues to grow and is being linked with high-speed routes in Belgium, the Netherlands, Germany, Spain and Italy. The success of these LGVs in providing regular services at speeds far in excess of previous maxima has discredited those proposals for high-speed land transport systems which are radically different from the orthodox railway, and which were projected on the assumption that wheel-on-rail systems had reached their limit of performance.

Figure 21.1 The Eurostar train represents the modern TGV designed for international services via the Channel Tunnel. It operates from LVDC conductor rail at 750 V; from HVDC overhead supply at 3000 V DC; and from HVAC overhead supply at 25 kV, 50 Hz. It is here shown working from 25 kV, 50 Hz supply.

Source: GEC-Marconi Research Centre.

21.4 The British Rail Advanced Passenger Train project

The British Rail Advanced Passenger Train project sought greatly increased speeds within the railway engineering infrastructure of the 1970s. It would avoid the expense of building new lines: hence the cost of developing a highly innovative train would be recovered. This was a different philosophy from that pursued in France, where an advanced but much less innovative train was used over an expensive new infrastructure of conventional form. In Britain, there was no possibility of the latter course being taken. Funds were lacking; there was no dedication by the state to build new railways, and great opposition to railway construction through densely populated areas was anticipated.

The project owed much to research into vehicle and rail interaction pioneered by British Rail at the Railway Technical Centre established in Derby in 1964. Before the high speeds era begun by the Shinkansen and TGV trains, there were problems in working fast passenger trains over routes occupied by slow goods trains. For any train, there is a critical speed above which lateral oscillations become dangerous, and this speed is related to the profile of wheel tread and rail head. The wheel tread profile is crucial. It needs to be coned to self-steer round curves, but the action is never precisely in equilibrium and the contact point between wheel and rail oscillates along the rail surface. The oscillation depends on rail and wheel tread wear; vehicle suspension and inter-coupling; rail and track structure; and condition of ballast and track bed. In the 1960s, it generally became dangerous above 160 km/h (100 mph) for passenger trains and above 70 km/h (44 mph) for four-wheeled goods wagons. Goods trains shared the track with express trains over the majority of the system and their lower speeds markedly reduced line capacity, but attempts to run them faster caused derailment due to hunting and misalignment of four-wheeled vehicles on the track. One objective of the research into vehicle-rail interaction was to develop a high-speed goods wagon able to run at express speeds. Continuous braking and the introduction of bogie wagons for containers or unit loads would enable fast good services to be worked if hunting and destructive rail working were eliminated. The research suggested ways of increasing the speed at which the oscillations became dangerous and destructive of track.

British Railways Research pioneered this investigation. Potter (1987, p. 50) claims that British Railways was the first in Europe to use computers for scientific and engineering applications, having bought a digital computer in 1957, and he quotes Dr Jones, head of research in 1964, as stating that electronics enabled the advanced research to be carried out. Electronic measuring and recording devices plus computer modelling provided the insight needed to raise the critical speeds at which destructive hunting arose for various kind of vehicle. The model developed by the Railway Technical Centre at Derby was particularly valuable. As a result, a four-wheeled goods wagon was developed that was tested on a rolling test bed in Derby at 225 km/h (140 mph) and tried on the railway at 160 km/h (100 mph). A ten-unit rake of such goods wagons was tested over 192,000 km (120,000 m) at high speeds without dangerous hunting, vehicle misalignment with track and other vehicles, or excessive wear. This work and similar research done in other countries raised the critical speeds at which trains could run without danger of derailment due to hunting. It provided the solution to the major problem revealed by the SNCF tests in the 1950s.

The critical speed for bogie passenger vehicles was always much higher than for the traditional goods wagon and for regular services remained at about 160 km/h (100 mph) since speeds in this region were first attained in the early 20th century. In the 1950s and early 1960s it was rare for a regular service train to reach 160 km/h (100 mph) when pulled by a steam locomotive, and speed restrictions generally prevented this. The work on vehicle-rail interaction in the 1960s made it possible to run trains at 200 km/h (124 mph) in the near future, and to reach 250 km/h (155 mph) a few years later. Speeds of 320 km/h and even 400 km/h were shown to be possible on railway track with steel-wheeled vehicles. Subsequent research raised this to over 550 km/h, and 515 km/h was achieved by a TGV in 1990. Dr S Jones persuaded the British Railways Board to support the APT project rather than unusual guided schemes by indicating the unrealised potential in steel-wheel-on-steel-rail systems. In 1968, a year before the TGV-001 was ordered by SNCF, the British Railways Board and the Ministry of Transport agreed to fund the APT project. Potter (1987) has described the role of research, the dispute over priorities, the rivalries between departments within British Rail, and the support or opposition to the APT. It was obvious from the start that the scale of funding enjoyed by the Shinkansen and LGV could never be expected by the APT project, which was supported as a cheap method of getting very fast railway services. Potter (p. 58) indicates just how cheap it was.

From its very beginning in 1967 to 1982, the APT project cost £43 million, under £3m a year. Over the same period expenditure on new motorways and other trunk roads was £5,300m . . . and the Concorde project alone cost £2,000m to develop from 1962 to 1976 (£140m a year). The Paris-Lyon TGV line and stock cost about £1,000 (million). At 1985 prices, the APT's development costs would be around £100m.

The research and development work is told from the engineer's viewpoint by Williams (1985). Special research and testing facilities were built at the RTC in Derby and a design team was recruited. Many members came from outside the railway industry. A disused railway line became a test track for experimental vehicles. The objectives of the APT project as declared in January 1969 included reaching an operational speed of 250 km/h. This demanded maximum speeds 50 per cent higher than were reached by ordinary express trains working the WCML. The APT needed to pass round curves 40 per cent faster than existing expresses, and was to operate on existing routes within the limits of existing signalling. This latter requirement was a severe one. Passenger comfort standards were to be maintained during high-speed passage round curves. Another severe demand was to prevent track repair costs due to the APT from exceeding those of conventional trains. Energy consumption and noise levels were to be kept down. The cost per seat-km was to be roughly equal to that of existing express trains. Most of these requirements could be met by following the practice which the Japanese and French were evolving for high-speed trains. Traction motors could be suspended from the body to drive the wheels through shafts and lightweight gearing. Gas turbines could be body mounted and the drive taken to the wheels through hydro-mechanical, mechanical, or electric transmission. These techniques, which kept unsprung weight to a minimum, combined with methods for suppressing hunting, would keep track damage down. Good aerodynamical design would reduce energy consumption and reduce noise and dust cloud generation.

The requirement that dominated the design concerned passenger comfort during high-speed passage round curves. A passenger in conventionally arranged transverse seating will experience no sustained sideways force in a train on straight track. If the train rounds a non-superelevated curve on the flat, there will be a sideways force which could cause discomfort when the bend is severe and speed high. There would also be a shift in train weight from the inner to the outer rail, unequal rail and tyre wear, and an unreduced overturning moment. If the outer rail is raised or superelevated by canting the plane of the rails through an angle with respect to the horizontal, the discomfiting sideways force can be reduced. It can be eliminated completely if the train rounds a curve, with a particular cant angle, at a speed which causes the resultant force on the passenger to be perpendicular to the plane of the seat or train floor. For any curve, there will be one speed when this occurs and when rail and tyre wear are equalised. On segregated lines, such as the Shinkansen and LGV, each curve can be laid with this cant angle, calculated for the known speed at which all trains will pass. However, the APT was designed to run over routes where several kinds of train worked at different speeds. These included slow goods trains and non-express passenger trains. The track cant angle at any curve could not be set for one kind of train: it had to be a compromise for all. A cant angle which was exactly correct for the APT would give an inwardly directed force for passengers in slow trains. It had to be less than the theoretical requirement for the APT and be nearer the requirement for ordinary expresses, whilst not exceeding by too much the angle for slow goods trains and stopping trains. There was a necessary 'cant deficiency' between actual cant and the cant required to eliminate sideways force for the planned APT cornering speed. There would be some outwardly directed centrifugal force. Observation suggested that the maximum cant deficiency should be about 4.25 degrees, corresponding to a centrifugal force on typical curves of about 0.07g. If sideways forces were higher, walking along the train would prove difficult and objects could slide off tables. On British main lines the cant angle was kept below 6 degrees to accommodate the majority of ordinary trains, but this was not sufficient to keep sideways forces below 0.07g when the APT went round curves at 160 km/h or more. Some of the cant deficiency had to be made up by tilting the body of the train at high speed on curves. Without this, speeds would need to be reduced on curves and the purpose of the APT thwarted. It was calculated that if the APT vehicle bodies were tilted up to 9 degrees, they could take curves (already canted at up to 6 degrees) with total cant of 15 degrees. This would keep discomfiting centrifugal forces within limits whilst permitting high speed and equalising rail and tyre wear. Tilting trains were not new. Passive systems in which the coach body swings like a pendulum were tried in the 1930s and the idea is older. Pendular systems with a necessarily high pivot could not fit into the British loading gauge unless the body was made too narrow for a commercial train. The APT was therefore designed with a low pivot near the centre of mass. Tilting needed to be powered by mechanical action, because the passive pendulum action could not be used. This system was labelled 'active'. Active and passive systems were being investigated in several countries at the time: Canada, Italy, Spain, Sweden and Switzerland.

The APT active tilting mechanism comprised hydraulic tilting jacks which responded to electronic sensors of lateral acceleration. The jacks were the tilting bolster on which the coach body was air-suspended. Tests investigated the time that elapsed between the sensors measuring lateral acceleration and the jacks tilting the body through the required angle. This was found to be too long when the sensors were in the car being tilted, and the sensors for a particular tilt mechanism were shifted to the vehicle in front. The active-tilt requirement dominated the design of the train. The bodywork of the carriages had to slope inwards towards the top to avoid fouling the loading gauge when tilting at 9 degrees with respect to the track. The links between the non-tilting bogies and the tilting body had to be flexible, and the pantographs needed compensation mechanisms to maintain contact with the overhead supply conductor on curves. Tilting ruled out use of a through-train omnibus line for linking end power cars to allow the lead pantograph only to be used Through-train traction power buses have generally been avoided on safety grounds in Britain. This affected train configuration.

The project was a great engineering success and a remarkable achievement, considering the innovative design and funding that was minimal compared to that which the motor industry would invest in developing a new model. As in France, a fast experimental train was built with gas-turbine power units to try full-scale vehicles on the Old Dalby test track. This was the APT-E (Experimental), completed in 1972. Specially constructed test vehicles were used in preliminary investigations of tilt mechanism, bogie design, suspensions, and train-track interaction over several rail formations. One special train was the POP unit, which first ran at the Old Dalby track in September 1971. This train was so called because its initial configuration was two power car body shells sharing an APT-E articulation bogie with no trailer between: hence 'Power-O-Power'. A trailer car was inserted later. The body shells were open frameworks, fitted with small test cabins to shelter crew and equipment. They are described and illustrated by Williams (1985). The POP train was powered by conventional locomotives and coupled to test cars to investigate bogies and tilting mechanisms at normal speeds before they were tried at high speeds using the APT-E.

The APT-E became known as the 'Aerospace Train' and was described in an article of that name by B. Gunston in the 19 October 1972 issue of *Flight International*. It was heralded as a new kind of train which relied on aeroindustry techniques and expertise. This gas-turbine-powered train was first tested in July

1972. The APT-E was built as a four-car articulated set, and in the words of Williams (p. 24):

... it was designed as a suspension test bed and only four cars were needed to conduct the various experiments and carry the necessary instrumentation. It was not intended to be a prototype passenger carrying train which would normally run with up to 12 trailer cars depending on service and performance requirements. Four traction turbine alternator sets and their associated control equipment were installed in each power car. They consisted of Leyland gas turbines driving Houchin 400 Hz, 3000 rpm alternators. The turbine was a two shaft design with the free power turbine being coupled via reduction gearing to the alternator. A fifth turbine alternator set was mounted at the rear of the power car providing a 415 V, three phase, 50 Hz auxiliary power supply.

The turbines were fuelled by standard railway diesel locomotive fuel. The APT-E was originally planned to have mechanical transmission from body-mounted electric motors to a bogie-mounted gearbox. To reduce design time, orthodox axle-suspended DC traction motors were substituted and geared to give a track speed of 312 km/h (195 mph). Unsprung mass was high and power bogie design was 'compromised' but the arrangement served to test suspension, tilt-mechanism, and other vital components. The prototype train, APT-P was constructed with body-mounted motors and shaft drive, with reduced unsprung mass.

APT-E had rheostatic braking in the speed range 72–250 km/h (45–155 mph). The train first ran under power on 28 June 1972. Between then and 1976 it was subjected to trials on the Old Dalby track and in the Derby area. After modifications there were more tests. In three-car form it began runs over the former Midland main line between Derby and Leicester on 20 August 1973. Many problems were encountered and overcome, and the set was modified in 1974 and subjected to further trials on the Old Dalby line and on main line. Speeds of over 210 km/h were attained, and during August 1975 new British records were set when APT-E reached 243.5 km/h (151.3 mph), and then 245.1 km/h (152.3 mph) between Uffington and Goring during driver training sessions on the Great Western Region. These were to stand as British records until broken by APT-P in December 1979. The APT-E demonstrated the great promise of the APT project. Technical problems could be overcome, though there were ominous signs of labour disputes concerning proposed single-driver operation. Support for the APT was growing in the British Railways Board, which wished to upgrade passenger services on all main lines, and authority for constructing three prototype trains classified as APT-P was given in September 1974. At this stage, two general types were envisaged: a gas-turbine-powered train for use over non-electrified routes, and an electric version. In the event, the high-speed diesel-electric train (HST-125) took over fast services on non-electrified routes and worked the fastest services in Britain. It set the then British record of 210.8 km/h (131 mph) in June 1973 between York and Darlington, and set the then diesel speed record of 230.5 km/h (143.2 mph) between Northallerton and Thirsk in the same month. After the advent of this very successful train, the APT project became more closely associated with electric traction.

The final test run with the APT-E was made on 2 April 1976, and the set was sent to the National Railway Museum in York in June. Responsibility for the APT-P project lay with the Chief Mechanical and Electrical Engineer's Department, which began design work in October 1973. Experience with APT-E suggested that major modifications would be necessary. The APT-E was a test bed for suspension and braking systems and not a pre-prototype for APT-P. The Research Department supplied to the designers the data obtained from testing the APT-E. The WCML was the most suitable route for the APT train, where the greatest commercial success was to be won, and therefore electric traction was chosen rather than gas-turbines. The traction drive had been well proven by French, German and later British experience. The traction motors were body-mounted and drive was taken through cardan shafts to lightweight gearboxes on the axles. Suspension, bogie design and drive owed much to contemporary research in Germany, Switzerland and France. The drive, articulated bogie design and suspension reduced the unsprung mass per wheelset to 1500 kg, compared to 3300 kg for conventional express trains hauled by the 'Deltic' diesel electric locomotive or Class 87 electric locomotives. Tests showed that the APT on straight track at 200 km/h exerted half the forces on the track as did the Class 87 at 160 km/h, and both trains exerted the same forces on the track on curves though the APT was going faster. The braking system included hydrokinetic brakes built into the conical axle, which dissipated kinetic energy as heat which was rejected to the atmosphere through radiators. This provided a light, compact brake which reduced unsprung weight, but it gave trouble. It was intended to function above 80 km/h, and wheel tread brakes acted at lower speeds to rest. The target maximum speed for the APT-P was 250 km/h. There were difficulties with signalling because the APT-P would operate over lines already signalled by lineside colour lights, but at faster speeds. The signalling system in place had four aspects, whereas French experience showed that five aspects at least were needed for the speeds envisaged with the APT. The LGV system was a newly built line and was constructed with the necessary system described above, without lineside signals. This option was not open to the APT. It had to fit in with a signalling system which the majority of trains would continue to use. As Potter remarks (p. 67):

To operate APT safely, a fifth aspect would have to be provided to give a driver warning to slow down to 200 km/h where conventional signals showed green one block before a double amber. This gives a total stopping distance of 3,075m. Totally resignalling the routes to be operated by the APT by installing a fifth aspect . . . was seriously considered by British rail, but the costs were seen as prohibitive if conventional signals were to be used. In addition, there were doubts as to whether it would be safe to add a fifth aspect to visual signals, since in marginal situations it could confuse drivers.

Likewise, lengthening the block sections to increase braking distances would also be prohibitively expensive. Potter concludes:

Quite remarkably, this problem was simply shelved. Except for special tests, when the track was cleared ahead of it, the APT was restricted to 200 km/h. This 'fifth aspect' question was put to one side.

A control and information system was installed known as Control-APT, which used passive transponders on the track, powered by a radio beam from an aerial under the train, to prevent drivers from becoming confused when operating APTs and conventional trains over the same tracks. A precoded signal from the transponder set up a speed limit sign in the cab, which progressively changed as the train moved to the next transponder. This speed signalling system applied the brakes automatically if audible warning signals were ignored. The system also sealed air intakes to the coaches to prevent hurtful air pressure rise in tunnels.

New manufacturing techniques were required for the APT-P, which was built with large aluminium extrusions in place of the riveted stress skins of the APT-E. The general configuration of the train was constrained. The 1973 scheme proposed a 12-coach train, seating 534 passengers, and a possible 14-coach, 653-seat version which was the longest platforms could accommodate. Such trains would need 6000 kW output, which at that time meant two power-cars of four axles or one with six axles, unless the French practice was followed of powering the end bogies of the trailers set. It was decided to use two power cars per rake and reduce buffing forces in the pushing mode of operation and keep down the weight of the bogie. The power cars could be placed at the outer ends of the rake, as with the TGV, but disturbances in the overhead supply set up by the leading pantograph might prevent the trailing pantograph from being used. A train traction-power bus was ruled out by the need to tilt the coaches, so it was decided to place the two power cars in the middle. This split the train into two as far as passengers were concerned because only a staff corridor gave access through the power cars.

Several different configurations were considered at different stages of the project, including reduced formation, single end power cars and a power car at each end; these plans were never realised, though they influenced the IC225 project which resulted in the trains now working the ECML. The first power car was built at Derby Locomotive Works and ran in September 1977. A special test train was assembled for the units of the APT-P which included a laboratory coach and a generator coach to provide traction power for moving the train when not on electrified lines. An HST diesel-electric power car was used for this purpose. Some equipment from the APT-E was reused. Labour problems delayed the project but a three-car set established a British railway speed record in December 1979 by reaching 261 km/h. By March 1980 the three trains were being tested, one of them undergoing endurance trials between Glasgow, Preston and Euston. Hydrokinetic brake failure and parting of the conical axle derailed the associated wheel-set at 200 km/h (125 mph) but the train stopped safely without injury to personnel. Tests of the three trains stopped but resumed in the summer of 1980. In that year, BR included the APT-P in the public timetable and one full-length train of 12 passenger coaches and two power cars was scheduled to make three return trips per week between Glasgow and London in 4 hours 15 minutes for the 640 km. This gave a start-to-stop average of 150 km/h, with some 200 km/h running. The first train did not run in this service until 7 December 1981, and the first return trip was delayed by minor defects. The second and third trips were interrupted by extremely bad blizzards when components froze and travel was hindered by drifts. The APT-P was then withdrawn, supposedly until May 1982, but it never re-entered service. In September 1982, it was announced that the APT would not be used in regular passenger service, though it might be used as a relief train. Trials continued with BR staff serving as passengers, but in 1986 the APT concept was abandoned and the APT-P trains were sold for scrap to a Rotheram scrap merchant.

Many factors were identified as responsible for the failure of the project, including underfunding and defective management and organisation. Design defects in the tilting mechanism and the hydrokinetic brakes would have been overcome, but there was evidence of poor-quality manufacture and assembly which suggest that (Potter, 1987, p. 130)

... the design of the APT was unrealistic given the manufacturing methods and maintenance facilities of the rail industry.

Hughes (1988, p. 58) remarks:

Perhaps the biggest problem was too much innovation on one train, but there is no doubt that weak management contributed to the woes. On top of that there was bad engineering, bad public relations, appalling industrial relations, and sheer bad luck.

In the 1980s, traffic on the WCML was declining and the confident expectations, which followed electrification north of Crewe to Glasgow, that there would be a marked increase in passenger numbers was not justified when much London to Glasgow business traffic was lost to the air shuttle service. The expense of resignalling railways to exploit the full speed potential of the APT was prohibitive and, without resignalling, speed was limited to little better than the HST. Glasgow declined as a major industrial and commercial centre. The ECML was not then electrified, and there was no gas-turbine or diesel-powered APT in prospect. The strategy adopted for high-speed services was to rely on the diesel-electric HST over non-electrified routes and to develop a modern, conventional high-speed electric train for the WCML. The design would make use of lessons learned from the APT project and would draw on Continental experience with very fast locomotive-operated push-pull non-tilting trains. The result was the successful IC225 train, which works the fastest British electric express services at the time of writing (2001). Tilting trains have found some use on existing routes where they can speed up the schedule considerably, without necessarily involving very high speeds. The Japanese National Railways Class 381 train sets of 1973 were built to run below 120 km/h (75 mph), but they can round curves at 96 km/h (60 mph) where ordinary trains are limited to 80 km/h (50 mph). On the Japanese narrow-gauge system of 1067 mm (3 ft 6 in), this provides worthwhile reductions in journey time. They are built to work from both 1500 V DC and 25 kV, 50 Hz/60 Hz AC supplies, and each train is equipped with six motor cars, each having four 100 kW motors, or 2400 kW in all. The Italian ETR 401 Pendolino four-car train set of 1976 is much faster, and was built for standard gauge express service. It was financed by Fiat. Each set has two 250 kW motors on each of the four cars, giving 2000 kW in all. Each of the eight motors is body-mounted, driving one of the axles through longitudinal cardan shafts and gears, as is common for high-speed units. Supply is taken from 3000 V DC overhead distribution. Maximum speed is 250 km/h (156 mph). Maximum axle load is limited to 10.5 tons. Used between Rome and Ancona (298 km) they reduced journey time by some 23 minutes, or approximately 12.5 per cent, though it was calculated that a reduction of 45 minutes was possible. Similar trains were later supplied to the RENFE (Spanish National Railways). Derivative sets (ETR 450) were ordered in 1987–88 by the Italian Railways for 250 km/h services in the period before a true 'TAV' (Treno ad Alta Velocita) was developed and introduced. Tilting trains are under order for those Anglo-Scottish express services on the WCML which after privatisation have been provided by the Branson franchise-holders. A review of recent rolling stock is provided by Kaller and Allenbach (1995).

21.5 Developments subsequent to the TGV

The success of the French TGV and the Japanese Shinkansen gave rise to very high speed trains running over segregated high-speed routes in other countries, notably Germany and Italy. Experiments with electric, gas-turbine and diesel-electric fast trains in Russia, the USA and Canada did not lead to regular very high speed services in those nations. High-speed services on segregated routes after the French model were introduced in Germany, and the Channel Tunnel 'Eurostar' services owe much to the example of the French TGV network, which is now being extended into the Low Countries and Germany (Semmens and Machefert-Tassin, 1994).

In the 1960s, the German Federal Railways in West Germany conducted a series of high-speed runs with passenger trains hauled by Class E03 Co-Co electric locomotives. The E03 electric locomotive ranks among the most remarkable in the world and is recognised as a classic design. The prototypes were introduced in 1965 and series construction dates from 1970. The design was a joint one, by Rheinstahl-Henschel, Siemens and Deutches Bundesbahn's Central Design Office in Munich. Four prototype locomotives were built in 1965 and were demonstrated during the Munich International Transport Exhibition, when 200 km/h (125 mph) was regularly reached on test. They pulled the 'Blauer Enzian' over the Munich-Augsburg section, which was the first European express service scheduled at this speed. The Co-Co locomotives drew supply from overhead distribution 15 kV, 16.66 Hz single-phase AC. They were extremely powerful. The continuous output was 5950 kW (7975 hp) for the prototypes, 7440 kW (9970 hp) for the series type. One-hour ratings were 6420 kW (8600 hp) for the prototypes, 7780 kW (10,425 hp) for the series types. The prototypes had a ten-minute rating of 9000 kW (12,000 hp), and the series type had 10,400 kW (13,950 hp). Mechanical parts were made by Henschel, Krauss-Maffei and Krupp, and electrical equipment came from Siemens,

AEG-Telefunken and Brown-Boveri. They were fitted with six single-phase commutator motors, which were frame-suspended and drove through flexible cardan shafts. Two of the prototypes had Henschel spring drive. Rheostatic brakes were fitted. Maximum speed was rated at 200 km/h (125 mph), and one locomotive was geared for 250 km/h (155 mph). Introduced as Class E03, they were later reclassified as 103.

These locomotives successfully demonstrated high-speed running in Germany in the 1960s. The prototype could accelerate a train of eight coaches from rest to 200 km/h in 3 minutes, and during the 1965 Munich Transport Exhibition 222 runs at 200 km/h were made. These 1965 runs and other high-speed trials revealed the limitations of existing signalling systems which were designed for a shorter braking distance. The equivalent of extra aspects was required. This was obtained by laying an inductive cable along the track which transmitted information about signal aspects and speed restrictions well ahead of the train, using principles well tried in rapid-transit service. The cab signalling apparatus was advanced for a locomotive in the mid-1960s and incorporated a computer to calculate the maximum safe speed compatible with stopping before the next signal set at halt. A cab display showed both actual speed and target speed. Brake application was automatic if actual speed exceeded the set target. This system, the LZB (Linienzugbeinflussung), became mandatory in DB trains intended for 200 km/h operation. The tests with automatic control and signalling suggested the practicability of automatic (driverless) locomotive working, and this was demonstrated using the E03 locomotive. The driverless experiments were successful, like those on the London Underground lines, but the practice did not become the norm.

Working very high speed trains by locomotives wore the track badly, and between 1967 and 1977 the DB gradually abandoned 200 km/h running. Investigations took place into locomotive-track interaction; high-speed bogie and wheelset design; suspensions and integrated complete drives. Destructive loading of track by locomotives and multiple-unit trains was greatly reduced. Permanent way was upgraded, and after 1977 200 km/h running was allowed again. In the 1970s, the German Federal Railways prepared plans for modernising the system. These involved upgrading routes and constructing new segregated lines over which very high speed German 'TGV' trains would run. These plans have been progressively updated, with considerable strategic amendment following the union of the former West and East Germany into one country, with increased emphasis on east-west traffic rather than north-south. The modernisation programme of 1970 was followed by important state plans, including the Federal Transport Infrastructure plans of 1980 and 1985 which authorised the construction of a New Lines network and work to raise speed on existing routes. As Rahn (1986) relates:

The technology of the New Lines is determined by their full integration in the existing network and the demand for mixed traffic with the new Intercity Express high-speed multiple units at up to 250 km/h and fast freights at up to 120 km/h.

The New Lines were built with maximum gradients limited to 1.25 per cent (1 in 40) and standard curves of 7000 m radius (minimum radius 5100 m). The signalling was developed from the proven LZB system of continuous automatic train control with cab signalling, radio control and electronic signal boxes. The Re250 overhead line was specially developed for 250 km/h services and installed. The ICE (InterCity Express) system was planned so that it would integrate with the existing network and with any more advanced railway network likely to be developed for very high speeds. Radio control of the trains was envisaged, in conjunction with integrated centralised traffic and train control. Train position was continuously transmitted to signal boxes and computerised train running control centres, which in the 1980s were represented by the one at Stuttgart Central. As the ICE programme developed, the possibilities were examined of incorporating high-speed goods and postal services over selected routes at night.

The German programme, like the French, involved a lengthy research schedule. Rahn (1986) states that it was organised to investigate six major sectors of railway engineering, which were (i) vehicle/track interaction; (ii) vehicle technology (brakes, controls, aerodynamics, etc.); (iii) track and fixed works design; (iv) control systems, and the integration of control and communications systems into one unity; (v) power supply, including pantograph/contact wire interaction; and (vi) environmental issues such as noise, vibrations, air disturbance, electrical interference, and human response to ICE. The interaction of the rolling stock with the track was the key issue and attempts were made to make

all the physical-technical effects amenable to a mathematical description, so that optimum structural solutions can be calculated in advance by means of appropriate assured mathematical methods.

Many experiments were carried out to check the validity of mathematical models and to investigate effects which could not be expressed mathematically. Vehicle behaviour up to speeds of the order of 500 km/h was modelled and results showed the great potential for development inherent in the wheel-on-rail system. Full-scale prototype vehicles were tested under controlled, scientific conditions on a roller test rig constructed in stages between 1974 and 1980 in Munchen-Freimann. It was ready for two-axle ensemble tests in 1977 and for four-axle trials in 1980, when complete vehicles could be investigated on it. It was possible to vary a large number of parameters affecting the total behaviour of vehicles on track and the test bed was used to check the accuracy of the mathematical models as they evolved. Trials on the roller test rig guided the design of new wheelsets, high-speed bogies, three-phase drive systems and components suitable for 350 km/h operations. Components for still higher speeds were proposed.

By 1982, it was possible to specify the requirements for prototype Intercity Experimental trains, which were to be 'the all-embracing example of the railway technological possibilities realized'. The German ICE project, like the French TGV and the Japanese Shinkansen, was used to build up an expertise and manufacturing ability directed towards international consultancy and export of

German railway engineering. As early as 1978, the DB outlined a high-powered unit for 250 km/h maximum speed, drawn up by a joint design office through which the railway engineering industry and the state railway co-operated. The design was for a multiple-unit ensemble, with one E120 locomotive at each end providing 5.6 MW continuous power, and streamlined unpowered carriages with 600 seats. The ICE concept was defined by work carried out after 1979 at the behest of the Federal Ministry of Research and Technology at the experimental railway installation at Rheine-Spelle-Freren. The ICE concept was pursued and is still being developed.

It was stated clearly that the ICE must strengthen the competitiveness of German industry in the world railway engineering market. In 1985 the first ICE was handed over to DB and given its public debut. The first power car 410 001 was accepted on 19 March 1985, and the first central unpowered trailer vehicles were handed over on 31 July. The first complete train was marshalled in September 1985, and public demonstrations were begun in November. A German speed record of 317 km/h was set on 26 November between Bielefeld and Essen. The power unit was of the Bo-Bo type, and intended for a maximum speed of 350 km/h. Maximum output was 4200 kW with 3640 kW continuous. Total weight was 78 tonnes. Length was 20.8 m. Unsprung mass was reduced to 7.5 tonnes. Since the successful debut, the ICE has been subjected to an intensive development programme, which has resulted in a formidable rival to the Japanese Shinkansen trains, and the French TGV.

21.6 Current developments in main line electric traction

The high-speed railway was established in Europe and Japan, and spread to the USA, South Korea and other lands. In July 1996, the DB placed contracts for the 177-km Koln-Frankfurt line, scheduled for opening in May 2000, but recent German projects were delayed after resistance from 'environmentalist' groups. Upgrading the Erfurt-Halle route to permit 250 km/h running, with 300 km/h long term, was approved, in order to reduce Munchen-Berlin timings by half to 4 hours. Unfortunately, recent cut-backs in state funding threaten the completion of these schemes, though there is support to improve the international railways between Germany and Poland. In 1996 there were no services at 300 km/h in Germany. In late 1996 only three railways ran trains at this speed, over LGV lines in France, Spain and Belgium. The successful trials of the Series 500 train on the Japanese Railways West between Shin Osaka and Hakata encouraged JRW to plan for 300 km/h services from spring 1997 with start-to-stop average speeds between Hiroshima and Kokura (192 km) of 253 km/h, which will be surpassed only by the SNCF average of 258.7 km/h over the 207 km between St Pierre des Corps and Massy TGV, introduced in summer 1996. However, objections to further development of LGV networks, in France, Germany and elsewhere, are being mounted, mainly on the grounds of high construction costs and anticipated low returns. The TGV 'Master Plan' for 4700 km of new lines costing Fr210 bn at 1989 prices was approved by the then French government but was rejected as unrealistic by the succeeding Secretary of State for Transport. Finance ministers in several countries are objecting to high costs and likely low returns on investment if more routes are built. Pressure is being brought on LGV planners to reduce costs, and tilting TGV sets have been proposed in more than one country as a way of raising route speed without constructing new lines or upgrading old ones.

In the USA, the demand for improved passenger services increased, particularly in the North East Corridor linking Boston, New York, Washington and Philadelphia. Financial aid came from Congress in the mid-1990s to keep threatened passenger trains running and to fund improvements (*RGI*, November 1996). This enabled the National Railroad Passenger Corporation ('Amtrak') to plan both the introduction of TGV trains ('The American Flyer') and new high-speed locomotives based on French practice. In August 1996, the US Federal Railroad Administration published its report 'High Speed Ground Transportation for America', which considered three main options (*RGI*, October 1996). There was the 'Accelerail' scheme for upgrading existing routes to take 240 km/h tilting trains. There were European-derived LGV systems with segregated lines and TGV. Third, there was a Maglev system with speeds of 480 km/h. Granted the status of railway passenger traffic throughout North America, the 'Accelerail' scheme is the most likely. One wonders why ambitious projects for Maglev routes continue to be proposed when no major Maglev railway has been operated anywhere and the TGV runs faster.

The USA plans concentrate on the routes in the North East Corridor which have always been important and served by the fastest trains in the USA. In the 1930s, on the Pennsylvania Railroad electrified lines in this area, the GG1 locomotives pulled heavy express passenger trains and surpassed the fastest steam-worked timetables. These 'Corridor' lines saw repeated attempts in the post-steam traction era to raise the speeds of fast operations. In the 1960s, the Metroliner electric sets were tried, and in 1970 the experimental tilting 'Turbotrain' built by United Aircraft was tested on the New Haven-Boston service. In recent years, Federal aid has allowed civil engineering works to be strengthened and track to be realigned and speed limits increased. In 1999, funds became available for purchasing new rolling stock to revolutionise Amtrak's express services within the North East Corridor. There is a major project to electrify the New Haven-Boston line with the 25 kV, 60 Hz system.

The first rail services in North America at LGV speeds will use trains derived from the French TGV Duplex sets. A consortium of Bombardier and GEC Alsthom is constructing a fleet of advanced, tilting trains to be called 'American Flyers', which meet specifications laid down by the National Railroad Passenger Corporation and the Federal Railroad Administration (*RGI*, 1996). The power cars use asynchronous motors and can work with the three kinds of power supply in the North East Corridor: 11 kV, 25 Hz; 12.5 kV, 25 Hz; and 25 kV, 60 Hz. There will be two power cars per train, with six passenger trailers between. The trailers are not articulated because of the need to tilt each vehicle. Each

Figure 21.2   High-speed service between Washington D.C. and New York City worked by AEM-7 locomotive, derived from Swedish ASEA design, and introduced into the USA in 1980. Built by Electro-Motive (General Motors) at La Grange, Illinois.

Source: Amtrak.

power car is 4600 kW. The asynchronous motors will be suspended from the bogie frame, to drive through flexible transmission, with one motor per axle. The entire train is based on the GEC Alsthom units developed for the SNCF. The original plan was to introduce these trains before 2000 but the trains had not run by January 2001.

Amtrak ordered 18 'American Flyer' trains, and 15 electric locomotives for 200 km/h working based on GEC Alsthom's BB36000 'Asytrit' (asynchronous tricurrent) design for the SNCF. This was in turn derived from the BB26000 'Sybic' (synchronous bicurrent) locomotives. These will give 6000 kW continuously at rail, which is a practically standard power rating for high-performance general-purpose locomotives intended for commuter, intercity and long-distance trains. They will work under the three supply systems found in the North East Corridor. French-dominated companies lead in the export of TGV systems to other countries, and it remains to be seen if the Germans and Japanese will threaten this position.

In Japan, very high speed runs were made on the Central Japan Railway, which was established after the reorganisation of the former Japanese National Railway. This railway operates the latest generation of electric sets, the 300X Series (Hughes and Masatka Kimata, 1994). The 300X began trials in January 1995 and in the autumn of that year was tested at 350 km/h. In July 1996, the Series 300X set of the Central Japan Railway reached 443 km/h and became the second fastest train in the world. Continuous power rating is 45 kW per tonne. The previous Japanese record was set on the East Japan Railway by the STAR 21 electric unit in 1993. The Central Japan Railway announced that the 300X train was the key element in a programme to develop a high-performance railway system which integrated rolling stock, fixed works, civil engineering, methods of operation and control. Joint development programmes are planned with the other Japanese railways which are now autonomous after reorganisation of the former JNR. The Central and West companies intend to co-operate and introduce the next generation of Shinkansen trains, the Series N300, on the Tokaido and Sanyo lines. The Series 300 project unites the development philosophies expressed in the advanced trains of both railways: the Central's Series 300X and the West's Series 500 trains. The first 16-car prototype of the Series N300, intended for 270 km/h operation, was scheduled for autumn 1997 and was used to raise operational speeds at later dates.

The tilting train, accepted for the 'American Flyer' programme, is being considered elsewhere. In September 1996, the joint operation of ETR tilting train sets was begun between Milan and Geneva and between Milan and Basle. These trains are operated by Cisalpino AG, which has been formed by Italian Railways, Swiss Federal Railways and the Bern-Loetschberg-Simplon Railway. The trains are dual-voltage and work through the 3 kV DC system of Italian Railways and the 15 kV, 16.66 Hz system of Switzerland. The sets are able to run at 250 km/h, though 200 km/h is the maximum possible in Switzerland. The trains reduced the Milan to Geneva schedule time to a shortest of 3 hours 40 minutes, saving 24 to 30 minutes; and reduced the Milan to Basle time to 4 hours

23 minutes, saving 47 minutes. Nine ETR470 sets were running in summer 1997, and it is hoped to extend these services to Venice. Chinese Railways are developing trains based on the Swedish-built X2000 trainset, similar to the unit which was subjected to three months of tests between Oslo and Kristiansand where journey time was reduced from 4 hours 26 minutes to 3 hours 46 minutes over 353 km. Norwegian State Railways intends to make use of the X2000 train on the Oslo-Bergen, Oslo-Trondheim and other routes.

In Japan, a second generation of tilting trains was introduced by the JR West system in 1996 to work the 'Ocean Arrow' or 'Super Kurishio' services. The new trains of Series 283 have active tilting apparatus developed by Hitachi which is used on other Japanese tilting trains such as the JR Kyushu 'Sonic 883 'Wonderland Express" sets. These trains raise speeds over the old 1067 mm gauge lines which have frequent, severe curves in the hills and coastal regions of Japan. Overhead supply is 1.5 kV DC which is inverted on board the train to provide variable voltage, variable frequency supply for the three-phase asynchronous traction motors. The maximum speed is 130 km/h. Six-car and three-car units are run in six- or nine-car formations.

In Britain, tilting trains have been ordered for the West Coast Main Line to reduce the journey time between London and Glasgow by up to one hour; and between London and Manchester by 30 minutes. The first units have been built in Italy. In Europe, the German high-speed network continues to be extended, though there have been setbacks. Plans were made in 1996 to add 400 km between Koln-Frankurt and Nurnburg-Erfurt, in the face of opposition from 'environmentalist' pressure groups and some political disquiet over costs. However, work on the Nurnburg-Erfurt line was halted because the state feared it would prove uneconomic. This is one example of changes in funding policy interrupting long-term railway projects several years after start. Repeated changes of strategy at ministerial level make it difficult to carry through comprehensive plans for modernising infrastructure which take many years to complete.

In South Korea, the first 300 km/h operation outside Europe and Japan was planned to begin between Seoul and Taejon in 2000 using French GEC Alsthom equipment derived from the latest TGV and Eurostar trains and LGV (Berton, *RGI*, 1996, pp. 735–38). The power car of the Korean train was completed at the Belfort factory of GEC Alsthom. It was designed for 300 km/h maximum service speed with power taken from 25 kV, 60 Hz supply. Formation will be a power car and powered trailer at each end with 16 trailers between, providing 935 seats. Continuous power for the train is 13.2 MW from 12 synchronous motors. A total of 46 trains have been ordered, 12 to be built in France, the others to be built in Korea. The whole enterprise is an exercise in technology transfer. Line construction and power supply are the responsibility of Korean construction companies, with French advice. Rolling stock, catenary, signalling and control equipment are to be provided through a Korean TGV Consortium which has 12 French and Korean member companies. A 'Core System Group' will oversee the integration of components into a unified system of rolling stock,

Figure 21.3 Example of modern high-power locomotive design. The Russian VL85 twin-unit locomotive was introduced in series production in 1985 for working from 25kV, 50 Hz supply. Each unit has three two-axle bogies and is rated up to 5000 kW. Solid-state rectifiers and DC motors are used. The VL86 development of 1985 used asynchronous motors. The VL15 was developed with parts in common for working under 3000 V DC supply.

Source: Sergey Dovgvillo.

catenary, Automatic Train Protection, Solid State Interlocking and Centralised Traffic Control. The anticipated traffic is in excess of any carried by the French home LGVs. It is 120 million passengers a year compared to 18 million on the Paris-Sud-Est route. The power supply system reflects this. Transformer output to the line between Seoul and Pusan is to be 960 MVA passed through eight substations of 120 MVA each to catenary over the double-tracked main route with auto-transformers every 10 km. A headway of 3 minutes between trains is planned. The training of Korean engineers in France and a major French presence in South Korea has been an essential part of the technology transfer.

Rapid improvements continue to be made to locomotive design, paralleling those made to electric multiple-unit sets. The Adtranz Italia plant in Vado Ligure, Italy (Adtranz is the name given to ABB-Daimler Benz) constructed 20 Bo-Bo locomotives of 87 tonnes for Italian State Railways able to work at 200 km/h which provide 6 MW at rail. This E412 Class is designed to work under three supply systems: 1.5 kV DC; 3 kV DC, and 15 kV, 16.66 Hz. Maximum power of 6 MW is obtained between 95 and 200 km/h drawing supply at 3 kV DC. Because no special control and conversion equipment for 1.5 kV DC is carried, the maximum power falls to 2700 kW when working from that supply. When working from the 15 kV, 16.66 Hz supply, the continuous power rating is 5500 kW. The four traction motors are asynchronous and are rated at 1530 kW each. The multiple-use of basic components of the power electronics systems contributed to the excellent power to weight ratio of the locomotive.

Recent Adtranz activities include co-operating with General Electric Transportation Systems of the USA to produce a prototype lightweight diesel-electric locomotive with asynchronous motors, aimed at the world market. Pakistan Railways have ordered 30 units. The General Electric diesel engines are rated at 2240 kW, and each AC motor is supplied through its own inverter. Adtranz contributions include the Henschel 'Flexifloat' bogie and the mechanical components. This 'Blue Tiger' series of locomotives is planned to include variants using different General Electric engines with power outputs from 1640 to 3280 kW (*RGI*, 1996, p. 779). This co-operation between American and European industrial combines in the diesel-electric field may be succeeded by co-operation in the electric locomotive field if the American combines build up an expertise in electric traction to match that of the European-based builders. Co-operation between manufacturers of signalling and communications equipment has long been global. The near future will probably see a combination between one of the major North American manufacturers of locomotives (General Electric or Electro-Motive) and one of the European combines.

Increase in the speed of goods trains has not kept pace with the improvements to passenger services, but research has been devoted to that end. The union of several programmes for developing high-speed goods wagons, integral trains and automatic control of goods trains should revolutionise freight working in the not-too-distant future. In October 1996, German Railways began testing a diesel-powered multiple-unit goods train. The prototype is made up of five flat wagons for carrying containers with a driving cab at each end. Four underfloor

The electric railway 1965–95 417

Figure 21.4 The power-operated swing-nose crossing represents advances in permanent way which enable the full potential of modern traction to be realised.
Source: Railtrack.

Volvo automotive diesel engines provide power for 120 km/h. Driverless operation is being considered. Trains made up of such sets would upgrade goods services. Self-propelled units could be broken into or out of longer trains, to serve selected locations. Diesel and electric versions have been suggested. Rapid-transit and express passenger operations were the sources of much of the innovation that has created the modern railway. The need to revolutionise goods working in the face of road competition may stimulate innovations in freight operations which are applied with benefit to passenger and rapid-transit systems.

Bibliography

The sources referred to throughout the text, using the 'Harvard' system of reference, are included in this bibliography. Many engineering papers and booklets are anonymous, especially those dating from before 1950. These are listed under the journal in which they appeared, or the company that issued the document, for example *Railway Gazette*, or Westinghouse Electric.

ABBOTT, R. A. S.: 'Vertical Boiler Locomotives and Railmotors built in Great Britain' (Oakwood Press, 1989)

ACLAND, H. A. D.: 'The Kitson-Still locomotive', *Trans. Inst. Eng. & Ship. Scot.*, vol. LXXIII, 1930, pp. 316–401

ACWORTH, W. M.: 'The railways of England' (John Murray, London, 1900, 5th edn)

ACWORTH, W. M.: 'Elements of railway economics' (Oxford, 1911)

ADAMS, E. D.: 'Niagara Power: History of the Niagara Falls Power Company 1886–1918' (privately published Niagara Falls, NY, 1927), 2 vols

ADAMS, W., and PETTIGREW, W. F.: 'Trials of an express locomotive', *Proc ICE*, 1896, vol. CXXV, pp. 282–95

AEG DAIMLER BENZ (Anon): '12X launches modular locomotive family', *RGI*, Aug. 1994, p. 497.

AGNEW, W. A.: 'Electric trains: their equipment and operation, including notes on electric locomotives, electropneumatic brakes, regenerative braking, etc.', (Virtue, London, 1937). 2 vols; vol. 1 is devoted almost entirely to the Sprague exemplar

AHRONS, E. L.: 'The British Steam Railway Locomotive 1825–1925' (Locomotive Publishing Co., 1927; reprinted Ian Allan, 1966)

ALFORD, L. P.: 'Henry Lawrence Gantt: Leader in Industry' (Harper and Brothers, NY, 1934)

ALLEN, C. J.: 'British Pacific locomotives' (Ian Allan, 1975)

ANDERSON, A. F.: 'Robert Davidson – Father of the Electric Locomotive' *Proceedings of History of Electrical Engineering Summer Weekend, Proc. IEE*, 1975, pp. 8/1–17

ANDREWS, H. H.: 'Electricity in transport' (English Electric Co., 1951). Review of stock built by EE

ANDREWS, H. I.: 'Railway Traction: The principles of mechanical and electric railway traction' (Elsevier, 1986)

ARMAND, L.: 'Motive power trends on European Railways', *Proc. IMechE,* 12 June 1947, pp. 239–45

ARMSTRONG, J. H.: 'The Railroad: What it is; What it does. The Introduction to Railroading' (Simmons-Boardman, Omaha, USA, 1979)

ARMYTAGE, W. H. G.: 'A social history of engineering' (Faber and Faber, 1961). Outline history of engineering from Stone Age to present

ASHE, S., and KEILEY, J. W.: 'Electric Railways' (Constable, 1905)

ATWELL and BASTON: 'Electrical equipment for the Chesapeake and Ohio Railway Co. steam turbine-electric locomotives', *Trans. AIEE,* vol. 68, 1949, pp. 145–48

AUSTIN, E.: 'Single phase railways' (Constable, 1915)

AUTOCAR: 'The tram and road cars of the future', *Autocar,* 7 Jan. 1905, p. 25. Precis of paper by A. Clark delivered to Royal Scottish Society of Arts, Edinburgh

BACHELLERY, M.: 'Chemins de fer electriques' (Balliere, Paris/Simmons Bordman, NY, 1925)

BAGWELL, P. S.: 'The transport revolution from 1770' (Batsford, 1974)

BAILEY, C. (Ed.): 'European Railway Signalling' (Institution of Railway Signal Engineers, A & C Black, 1995)

BAILEY, M. R.: 'The Tracked Hovercraft', *Trans. Newcomen Soc.* (London), Vol. 65, 1993–94, pp 129–145

BAILEY, SMITH and DICKEY: 'Steamotive: A complete steam-generating unit. Its development and test', *Mech. Eng. (USA),* vol. 58, pt. 12, 1936, pp. 771–80

BAKER, C.: 'The Metropolitan Railway' (Oakwood, 1951)

BARBILLON, L., and GRIFFITH, G. F.: 'Traite pratique de traction electrique' (Bernard et Cie, Paris, 1903)

BARBILLON, L.: 'La traction electrique a courant continu' (Albin Michel, 1923)

BARKER, T. C., and ROBBINS, M.: 'History of London Transport' (George Allen and Unwin, vol. 1, 1963; vol. 2, 1974)

BARNETT, C.: 'The Audit of War' (MacMillan, 1986)

BARWELL, F. T.: 'Some speculations on the future of railway mechanical engineering', *Proc. IMechE,* vol. 176, no. 3, 1962, pp. 61–106

BAYLISS, D. A.: 'The Post Office Railway' (Turntable Productions, 1978)

BEECROFT, G. D., FREW, I. D. O., HOLMEWOOD, A., RAYNER, B., and STEVENSON, B.: 'London Transport railways handbook' (Foxley Press, Chelmsford, 1983). Lists dates by which lines and stations were opened

BELL, A. M.: 'Locomotives' (Virtue, London, u.d. but circa 1946, 2 vols). Illustrated review of contemporary designs, chiefly British and mainly steam locomotives

BELL, A. R., CHALMERS, W., CLELAND, W., *et al.*: 'Railway mechanical engineering' (Gresham, London, 1923, 2 vols). Series of papers covering contemporary mechanical and electrical engineering by leading company engineers and consultants

BENEST, K. R.: 'Metropolitan electric locomotives' (London Underground Railway Soc./Electric Railway Soc., 1963)

BERTON, F.: 'South Korea buys TGV technology package', *RGI,* Nov. 1996, pp. 735–737

BEYER-PEACOCK QUARTERLY REVIEW: 'The turbine-condenser locomotive', *BPQR*, vol. 1, no. 2, April 1927; July 1927, pp. 2–4; Oct. 1927, pp. 3–8; Jan 1928, pp. 4–13

BEZILLA, M.: 'Electric traction on the Pennsylvania Railroad, 1895–1968' (Pennsylvania State University Press, 1980)

BINNEY, E. A.: 'Electric traction engineering' (Cleaver Hume, 1955)

BLACK, R. M.: 'The history of electric wires and cables' (Peter Peregrinus, 1983)

BLONDEL, A., and DUBOIS, F. P.: 'La traction electrique sur voie ferree' (Baudry, Paris, 1898)

BONAVIA, M. R.: 'The organisation of British Railways' (Ian Allan, 1971)

BONAVIA, M. R.: 'The history of the Southern Railway' (Unwin Hyman, 1987)

BOND, R. C.: 'A commentary on the change from steam traction on Britain's railways and some thoughts on the future', *Proc. IMechE*, 1963–64, vol. 178, pt. 1, pp. 1–26

BOWEN-BOWERS, B.: 'Edison and early electrical engineering in Britain', *History of Technology*, annual vol. 13, pp. 168–180

BOWERS, B.: 'A history of electric light and power' (Peter Peregrinus/IEE, London, 1982)

BRIGHT, A. A.: 'The electric light industry: technological change and economic development from 1800 to 1947' (Arno Press–New York Times Co., Arno NY, 1972)

BRILLIE, M. E.: 'Application du moteur a hydrocarbures a la traction sur voies ferrees' (Genie Civil, 1923), pp. 272–414

BRITISH RAILWAYS BOARD: 'The reshaping of British Railways' (BRB, Marylebone, London, 1963). The 'Beeching' report

BRITISH RAILWAYS BOARD: 'The development of the major trunk routes' (BRB, Marylebone, London, 1965). Review of 1964 conditions and an estimate of 1984 traffic

BRITISH RAILWAYS BOARD: 'Report on Organisation' (BRB/HMSO, London, Dec. 1969)

BRITISH TRANSPORT COMMISSION: 'Performance and efficiency test bulletin no. 15: British Railways Standard Class 8 3-cyl. 4–6–2 express passenger steam locomotive no. 71000' (BTC, London, 1957)

BRITISH TRANSPORT COMMISSION: 'Performance and efficiency test bulletin no. 16: 1-Co-Co-1 2000 HP main line diesel-electric locomotive 10203' (BTC, London. Undated, but tests 1955)

BRITISH TRANSPORT COMMISSION: 'Performance and efficiency test bulletin no. 19: English Electric 'Deltic' 3300 HP Co-Co diesel-electric locomotive' (BTC, London, Sept. 1956)

BROWN, F. A. S.: 'Nigel Gresley' (Ian Allan, 1961, 1975)

BROWN, H. F.: 'Economic results of diesel electric motive power on the railways of the United States of America', *Proc. IMechE*, vol. 175, no. 5, 1961, pp. 257–317. Brown criticised dominance of US market by makers and favoured increased use of electric traction

BRUCE, J. G.: 'Tube trains under London' (London Transport Publ., 1968)

BRUCE, J. G.: 'Steam to silver' (Capital Transport, 1983, London Transport, 1970). Illustrated history of London Transport electric railways

BRUCE, J. G.: 'A hundred years of development of electric traction' (The Electric Railway Society, 1985). The John Prigmore Memorial Lecture

BULLEID, O. V. S., HAWKSWORTH, F. W., IVATT, H. G., and PEPPERCORN, A. H.: 'Railway power plant in Great Britain', *Proc. IMechE*, 12 June 1947, pp. 235–39

BURGH, E. P.: 'Electric traction' (McGraw-Hill, 1911)

BURN, W. S.: 'Diesel engine flexibility', *Trans. NE Coast Eng. & Ship.*, vol. XXXVIII, 1921–22, pp. 241–316

BURNS, D.: 'Electrical practice in collieries' (Griffin, 1920)

BURTT, P.: 'The principal factors in freight train operating' (George Allen and Unwin, 1923). Compares British and American techniques, organisation, operations, equipment, waggon loads, locomotives

BURTT, P.: 'Railway electrification and traffic problems' (Pitman, 1929)

CALISCH, L.: 'Electric traction' (Locomotive Publishing Co., London, 1913)

CAMBOURNAC, M. L.: 'Progres recents dans les installations et materiel de la SNCF', *Memoires de la Societe des Ingenieurs Civils de France*, Fascicules 1 et 2, Jan/Fev. 1949, pp. 6–12

CAMDEN, A. C.: 'The diesel engine in railway service', The Railway Engineer, August 1927, pp. 290–92, 303

CANTLIE, K.: 'Improving power of steam locomotives', *Rail Engineering Int.*, vol. 3, pt. 2, 1973, p. 71

CARDWELL, D. S. L.: 'From Watt to Clausius' (Heinemann, 1971). History of thermodynamics and heat engines practice from origins

CARDWELL, D. S. L.: 'On Michael Faraday, Henry Wilde and the Dynamo', *Annals of Science*, 49, 1992, pp. 479–87

CARLING, D. R.: 'Locomotive testing stations, Part 1 & 2', *Trans. Newcomen Soc.*, 1972, pp. 105–82

CARLSON, F. G. & PETERS, A. J.: 'Electronic actuation improves freight train braking', *RGI*, Sept. 1996, pp. 583–586

CARLSON, S. P., and SCHNEIDER, F. W.: 'PCC: The car that fought back' (Interurban Special no 64, Interurban Press, Glendale, Ca., USA, 1980)

CARTER, F. W.: 'Railway electric traction' (Arnold, 1922)

CASSERLEY, H. C.: 'Steam locomotives of British Railways' (Hamlyn, London, 1975, 1961). Photographic record of all types taken over by British Railways in 1948, plus standard classes built by BR

CASSON, H. N.: 'Significance of the World Power Conference', *Industrial Management*, Nov. 1924, 68, p. 256

CERA (CENTRAL ELECTRIC RAILFANS' ASSOCIATION): 'An interurban goes modern', Reprinted Bulletins 20–34 describing updating of equipment and operations on US interurbans. (CERA, PO Box 503, Chicago, Ill., USA, 1977, 1942, 1941)

CERA: 'Indiana railroad system', *Bulletin 91*. Review of the very extensive Indiana interurban system. (CERA, Chicago, 1980)

CERA: 'Interurban to Milwaukee', *Bulletin 106*. History of the North Shore Line interurban railway before 1926. (CERA, Chicago, 1962)

CERA: 'Route of the Electroliners', *Bulletin 107*. History of the North Shore Line after 1926. (CERA, Chicago, 1963)

CERA: 'Chicago's rapid transit, vol. 1: rolling stock, 1892–1947', *Bulletin 113* (CERA, PO Box 503, Chicago, Ill, USA, 1973). Vol. II, *Bulletin 115* (CERA,

Chicago, 1976) describes rolling stock developments after the formation of the Chicago Rapid Transit Co. in 1947

CERA: 'Electrification by General Electric', *Bulletin 116*. Collection of General Electric Co. papers of 1923, 1927 and 1929. (CERA, PO Box 503, Chicago, Ill., USA, 1976)

CERA: 'Westinghouse Electric Railway Transportation', *Bulletin 118*. Collection of Westinghouse papers of 1915, 1917, 1922, 1924, 1929 and 1936. (CERA, PO Box 503, Chicago, Ill., USA, 1979)

CHALKLEY, A. P.: 'Diesel engines for land and marine work' (Constable, 1914)

CHAPELON, A.: 'La locomotive 242-A-1 de la SNCF', *Rev. Gen. des Chemins de Fer*, Dec. 1947, no. 12, pp. 397–444. Description of ultimate development of the compound expansion, high-performance steam locomotive

CHAPELON, A: 'Le centenaire de la locomotive Crampton et un siecle de progres en traction a vapeur', *Memoires de la Societe des Ingenieurs Civils de France*, Fascicules 1 et 2, Jan./Fev/ 1949, pp. 172–264. Outline history of steam traction from origins

CHAPELON, A.: 'La locomotive a vapeur' (Balliere et Fils, Paris, 1952)

CHAPMAN, S. D., and CHAMBERS, J. D.: 'The beginnings of industrial Britain' (University Tutorial Press, 1970). Origins of factory system

CLARK, W. J.: 'Electric railways in America from a business standpoint', *Cassiers Magazine*, 16, Aug. 1899, pp. 520–21

CLEVELAND-STEVENS, E.: 'English railways: their development and their relation to the state', (Routledge, 1915)

CLOUSTON, R. W. M.: 'The development of the Babcock boiler in Britain up to 1939', *Trans. Newcomen Soc.*, vol. 58, 1986–87, pp. 75–87

COLLECTIF D'AUTEURS: CHAPELON, A., MACHEFERT-TASSIN, Y. M., et al: 'Histoire des Chemins de Fer en France' (Paris, 1963)

COLLECTIF D'AUTEURS: 'Electricite et Chemins de Fer', *AIIICF-AIIE, 10 Colloque 1995*, PUF, Paris, 1996

COLLINS, H.: 'Steam turbines' (McGraw Hill, 1909). Practical advice on operating the major contemporary types

CONDIT, C. W.: 'The pioneer stage of railroad electrification', *Trans. American Philos. Soc.*, vol. 67, pt. 7, 1977, issued as separate monograph Nov. 1977

CONDIT, C. W.: 'The port of New York: a history of the rail and terminal system from the beginnings to Pennsylvania Station' (University of Chicago Press, 1980). Contains account of Pennsylvania Railroad electrification of the NY terminal, river tunnels and lines in New York area

COOK, R. J.: 'Super-power steam locomotives' (Golden West Books, USA, 1966). Illustrated history of the Lima Locomotive Works and its Woodard-inspired 'super-power' products

COOPER, B. K.: 'Elcctric trains in Britain' (Ian Allan, 1978)

COOPER, B. K.: 'BR motive power since 1948' (Ian Allan, 1985)

COURSE, E.: 'London Railways' (Batsford, London, 1962). General history and geography of London railways, mainly surface lines

COX, E. S.: 'British Standard Steam Locomotives' (Ian Allan, 1973)

COX, E. S.: 'Locomotive panorama, Vol.1' (Ian Allan, 1974). Brief description of British steam-electric project based on GE 'Steamotive'

COX, H. R.: 'Some fuel and power projects', *Proc. IMechE*, vol. 164, 1951, pp. 407–24.

CROSBY, O. T., and BELL, L.: 'The electric railway', *The Electrician*, London, 1892

CUDAHY, B. J.: 'Under the sidewalk of New York' (Stephen Greene, USA, 1979). Development of the three underground railways that formed the core of the present NY Metropolitan Transit Authority

DAVIES, W. J. K.: 'Diesel rail traction' (Almark Publications, London, 1973). Brief illustrated history

DAWSON, P.: 'Electric traction', *Electrician*, 1909

DAWSON, P.: 'Electric railway contact systems', *JIEE*, vol. 58, 1920, pp. 838–57

DAWSON, P.: 'Electric traction on railways', Railway Mechanical Engineering (Gresham, London, 1923, 2 vols) pp. 121–258

DAY, J. R.: 'The story of London Underground' (London Transport Publ., 1974, 1963)

DE CURZON: 'Astree real-time train supervision will complement ETCS development', *RGI*, Sept. 1994, pp. 567–570

DIAMOND, E. L.: 'Recent improvements in the efficiency of the steam locomotive', *Proc. IMechE*, Jan.1925, vol. 1, pp. 53–68

DIAMOND, E. L.: 'Investigation into the cylinder losses in a compound locomotive', *Proc. IMechE*, May 1927, pp. 465–517

DICKINSON, R. E.: 'Electric trains' (Arnold, London, 1927)

DICKSON, W. K. L., and DICKSON, A.: 'The life and inventions of Thomas Alva Edison' (New York, 1894)

DOVER, A. T.: 'Electric traction' (Pitman, London, 1954, 1929, 1918, 1917)

DUFFY, M. C.: 'Mechanics, thermodynamics and locomotive design: the machine-ensemble and the development of industrial thermodynamics', *History & Technology*, 1983, **1** (1), pp. 45–78

DUFFY, M. C.: 'Mechanics, thermodynamics and locomotive design: the turbine condenser locomotive', *History and Technology*, vol. 3, 1986, pp. 87–122

DUFFY, M. C.: 'The electric power industry and exemplary techniques', *IEE Proc. A*, 1986, **133** (3), pp. 159–72

DUFFY, M. C.: 'Andre Chapelon, thermodynamics and the steam locomotive', *Trans. Newcomen Soc.*, vol. 58, 1986–87, pp. 11–26

DUFFY, M. C.: 'Mechanics, thermodynamics and locomotive design: the high-pressure steam locomotive', *History and Technology*, 1987, vol. 3, pp. 155–192

DUFFY, M. C.: 'The Still engine and railway traction', *Trans. Newcomen Soc.*, vol. 59, 1987–88, pp. 31–59

DUFFY, M. C.: 'Mainline electrification and locomotive-electric systems', *IEE Proc. A*, 1989, **136** (6), pp. 279–89

DUFFY, M. C.: 'The American Steam-Turbine-Electric Locomotive', *Trans. Newcomen Soc.*, vol. 57, 1985–86, pp. 79–99

DUFFY, M. C.: 'The Schmidt high-pressure locomotive and its influence on American and British locomotive design', *Trans. Newcomen Soc.*, vol. 63, 1991–92, pp. 103–32

DUFFY, M. C.: 'Three-phase motor in railway traction', *IEE Proc. A*, 1992, **139** (6), pp. 329–37

DUFFY, M. C.: 'The Metadyne in Railway Traction', *Trans. Newcomen Soc.*, vol. 72, no. 2, 2000–01, pp. 235–68

DUMMELOW, J.: '1899–1949' (Metropolitan Vickers Electrical Co. Ltd, Manchester, 1949) History of British Westinghouse and its development to become part of Metropolitan Vickers.
DUNLOP, J.: 'Internal combustion locomotives', *Trans. Inst. Eng. & Ship. Scot.*, vol. LXVIII, 1925, pp. 234–78
DUNSHEATH, P.: 'A history of electrical engineering' (Faber, London, 1962)
DU RICHE-PRELLER, C. S.: 'A hundred-ton electrical locomotive', *Engineering*, 1893, **55** (1), pp. 772–74, 794–99, 806–7, 834–36
DURTNALL, W. P.: 'Correspondence on the internal-combustion locomotive', *Trans. NE Coast Eng. and Ship.*, vol. XLI, 1925, pp. 197–99
DURTNALL, W. P.: 'Electric locomotives for mainline railways: improved English models' (Rep. Modern Transport, London, 1926)
DUTTON, S. T.: 'Railway Signalling Theory & Practice' (Railway Engineer, 1928)
DYER, F. L., and MARTIN, T. C.: 'Edison: his life and inventions' (New York, 1910)
DYMOND, A. W. J.: 'Operating experiences with two gas turbine locomotives', *Jnl. Inst. Loco. Eng.*, vol. 43, pt. 2, no. 232, 1953, pp. 268–336
ELECTRICIAN: 'Turbo-electric-locomotives', *Electrician*, 7 Jan. 1910, p. 509
ELECTRICIAN: vol. CI, 1928, pp. 121–23. Brief, anon. note on electric locomotive for Great Indian Peninsula Railway.
ELEKTROTECHNISCHE ZEITSCHRIFT: 'Neue Elektrische Lokomotive System Heilmann', *EZ*, vol. 18, no. 15, 1897, p. 223
ELLIS, C. H.: 'British railway history 1830–1876' (George Allen and Unwin, 1954)
ELLIS, C. H.: 'British railway history 1877–1947' (George Allen and Unwin, 1959)
ENGINEER.: 'Trials of the Heilmann electric locomotive', *Engineer*, vol. 84, 1897, pt. 2, p. 505
ENGINEER: 'The Ramsay condensing turbine-electric locomotive', *Engineer*, 24 March 1922, pp. 327–28
ENGINEER.: 'An Italian double-ended diesel-electric locomotive', *Engineer*, 27 March 1925, p. 358
ENGINEER.: 'Lord Kelvin and the Diesel engine', *Engineer*, 7 March 1947, pp. 186–87; 14 March, pp. 210–11
ENGINEERING: 'A turbo-electric locomotive', *Engineering*, vol. 88, 1909, pt. 1, p. 613
ENGINEERING: 'The Reid-Ramsay Electro-turbo-locomotive', *Engineering*, vol. 90, 1910, pt. 2, 8 July, p. 54
ENGINEERING: 'Coal-burning gas-turbine locomotives', *Engineering*, vol. 178, 1954, p. 351
ENGLISH ELECTRIC Co.: 'An introduction to alternating current traction equipment' (EE, 1960)
FELL, L. F. R.: 'The compression ignition engine and its applicability to British railways', *Proc. IMechE*, vol. 124, Jan.–Jun. 1933, pp. 1–66
FERGUSON, T.: 'Electric railway engineering' (Macdonald and Evans, London, 1955)
FITT, W.C. (Ed.): 'Union Pacific FEF3 Class 4-8-4 Locomotive Drawings' (Wildwood Publications, Michigan, 1975)

FLANAGAN, T.: 'Northern Electric Railway' (Rohrbeck, USA, 1980). Brief account of typical interurban, opened in NE Pennsylvania in 1907
FLOHIC: 'Le Patrimoine de la SNCF et des Chemins de Fer en France', Editions FLOHIC (2nd edition), Paris, 1999
FORD, H & CROWTHER, S.: 'My Life and Work' (Garden City Publishing, 1922, 1926)
FOWLER, H. (Chairman of discussion): 'Electrification of English main line railways', Proc. IMechE, Jan. 1922, pp. 317–30
FREW, I. (Ed.): 'Britain's electric railways today', Electric Railway Society/ Southern Electric Group, 1983. Review of electrified sections of British railways, identifying systems used and dates of installation
FRY, L. H.: 'Some experimental results from a three-cylinder compound locomotive', Proc. IMechE, Dec. 1927, pp. 923–1024
GAHAN, J. W. 'The Line beneath the Liners' (Countyvise, Birkenhead, 1983)
GARREAU, M.: 'L'etat actuel de l'electrification de Chemins de Fer', Revue Generale des Chemins de Fer, August, 1938
GEE, B.: 'Electromagnetic engines: pre-technology and development immediately following Faraday's discovery of electromagnetic rotations', History of Technology, Annual vol. 13, 1991, pp. 41–72
GEIPEL: 'Discussion of Swinburne and Cooper', JIEE, 1902, p. 1040
GENERAL ELECTRIC (USA): 'The electric divisions of the Chicago, Milwaukee and St. Paul Rly', GE, Nov. 1927 (reprinted in CERA Bulletin 116)
GENERAL ELECTRIC (USA): 'The New York Central Electrification', GE, Jan. 1929 (reprinted in CERA Bulletin 116)
GERARD, E.: 'Traite complet d'electro-traction' (Weissenbruch, Brussels, 1897)
GIBBS, G.: 'The New York tunnel extension of the Pennsylvania RR: station construction, road, track, yard equipment, electric traction and locomotives', Trans. Am. Soc. Civil. Eng., 68, Oct. 1910, pp. 238–55
GIBBS, W. J.: 'Electrical machine analysis using tensors' (Pitman, 1967)
GLASERS ANNALEN (GUNTHER), 'Locomotives for High Speed' (In German), Glasers Annalen, Juli 1943, Heft 13/14, pp. 210 et seq.
GLASSPOOLE, W. F.: 'Some thoughts on gas turbine locomotives', Jnl. Inst. Loco. Eng., vol. 47, pt.5, 1957, pp. 521–43
GOODWARD, W. & GANNON, M. J.: 'Virtual Reality offers 3-D simulation in design and planning', RGI, Nov. 1994, pp. 735–738
GORDON, H. H.: 'Some aspects of metropolitan road and rail transit' (ICE, London, 1919). Booklet of 25 Nov. 1919 meeting. Comparison of road, tramway and rail traffic in London
GOSS, W. F. M.: 'Report to American railway master mechanics: assoc. committee on front ends' (Purdue, 1905)
GOSS, W. F. M.: 'Locomotive performance' (John Wiley, NY, 1907)
GRESLEY, H. N.: 'High pressure steam locomotives', Proc. IMechE, vol 120, 1931, pp. 101–206
GRESLEY, H. N.: 'Locomotive experimental stations', Proc. IMechE, vol. 121, July-Dec 1931, pp. 23–53
GRIME, T.: 'Possibilities of increased efficiency in railway locomotives', Trans. NE Coast Inst. of Engineers and Shipbuilders, vol. XXXIV, 1922–23, pp. 592–665

GUILLEMIN, A. (revised by S. P. THOMPSON): 'Electricity and Magnetism' (Macmillan 1891). Theory and practice of electrical science in 1890s
GUY, H. L.: 'The economic value of increased steam pressure', *Proc. IMechE*, 1927, pp. 99–213
HAMILTON, J. A. B.: 'Britain's railways in World War 1' (George Allen and Unwin, 1967). Includes account of circumstances leading to grouping of the railways into four companies in 1922
HANDS: 'Australia and radio control of trains', *RGI*, Sept. 1995, p. 575
HANNAH, L.: 'Electricity before nationalisation' (Macmillan, 1979)
HARDING, F., and EWING, D. D.: 'Electric railway engineering' (McGraw Hill, 1911, 1916, 1926)
HARDY, B.: 'Standard tube stock, Part 1, 1922–1945' (London Underground Railway Society, 1986)
HARDY, B.: 'Standard tube stock, Part 2, 1945 onwards' (London Underground Railway Society, 1987)
HARDY, B., and CONNOR, P.: 'The 1935 experimental tube stock' (London Underground Railway Society, 1982)
HARDY, B., FREW, I. D. O., and WILLSON, R.: 'A chronology of the electric railways of Great Britain and Ireland', Electric Railway Society Monograph, 1981
HARESNAPE, B.: 'Electric locomotives', British Rail Fleet Survey no. 6 (Ian Allan, 1983)
HARESNAPE, B.: 'Diesel shunters', British Rail Fleet Survey no. 7 (Ian Allan, 1984)
HARESNAPE, B.: 'Early prototype and pilot scheme diesel-electrics', British Rail Fleet Survey no. 1 (Ian Allan, 1981, 1984)
HARRIS, K.: 'World electric locomotives' (Janes, 1981)
HAUT, F. J. G.: 'The history of the electric locomotive' (George Allen and Unwin, 1969)
HAWKES: 'Review of railway signalling and communications' (Kempe's Engineers' Yearbook, 1986)
HAY, W. W.: 'Railroad engineering' (Wiley Interscience, 1982, 2nd edn). Much revised and reprinted standard text used throughout the 20th century.
HEELER, C. L. (Ed.): 'British railway track: design, construction and maintenance' (Permanent Way Institution, London, 1979, 5th edn). Review of modern permanent way structures
HENNESSEY, R. A. S.: 'The electric railway that never was: York–Newcastle 1919' (Oriel Press, Newcastle, 1970)
HENNESSEY, R. A. S.: 'The Electric Revolution' (Oriel Press, Newcastle, 1972)
HILLS, R., and PATRICK, D.: 'Beyer Peacock: Locomotive builders to the world' (Transport Publ. Co., 1982)
HILTON, G. W., and DUE, J. F.: 'The electric interurban railway in America' (Stanford UP, CVa., USA, 1960)
HINDE, D. W.: 'Electric and diesel locomotives' (Macmillan, 1948)
HIS MAJESTY'S STATIONERY OFFICE (HMSO): 'Report of the Committee on Main Line Electrification' (HMSO, London, 1931). The 'Weir Report'.
HOBSON, J. W.: 'The internal-combustion locomotive', *Trans. NE Coast Eng. Ship.*, vol. XLI, 1925, pp. 145–242

HOLT, E. K.: 'Transponders and coded track circuits underpin Northeast Corridor resignalling', *RGI*, Sept. 1994, pp. 575–578

HOOLE, K.: 'North East Railway buses, lorries and autocars' (Nidd Valley Narrow Gauge Railways Ltd, Knaresborough, 1969)

HOOLE, K.: 'The North East Electrics' (Oakwood, 1987)

HOOLE, K.: 'The Electric Locomotives of the North Eastern Railway' (Oakwood, 1988)

HOPE, R.: 'EIRENE spans Europe's radio frontiers', *RGI*, Aug. 1996, pp. 493–494; also Letter (Aurelius) *RGI*, Oct. 1996, p. 621; ATCL & WCML, *RGI*, Sept. 1995, p. 571

HOPKINSON, C.: 'Electric Tramways', *Proc. ICE*, Vol. 151, 1902–03, Pt. 1, pp 39–141.

HORNE, M.A.C. 'The Northern Line' (1987), 'The Central Line' (1987), 'The Victoria Line' (1988), 'The Bakerloo Line' (1990) (Rose/Nebulus Books)

HOWSON, F. HENRY, 'World's Underground Railways' (Ian Allan, 1964)

HOWSON, H. F.: 'London Underground' (Ian Allan, 1951, 1986)

HUGHES, M., 'Rail 150' (Ian Allan, 1988)

HUGHES, M.: 'Driverless trials open up prospect of freight on demand', *RGI*, Sept. 1996, pp. 563–566

HUGHES, T. P.: 'Thomas Edison: professional inventor' (London, 1976)

HUGHES, T. P.: 'Networks of power: electrification in Western society' (Baltimore, 1983)

HUGHES, M. & KIMATA, M.: 'Series 300X refines the shinkansen package', *RGI*, Jan. 1994, pp. 29–30

HURCOMB, C.: 'Report on types of motive power (Railway Executive)' (British Transport Commission: British Railways, 1948)

JACKSON-STEVENS, E.: 'British electric tramways' (David & Charles, 1971)

JARDIM, A.: 'The first Henry Ford: a study in personality and business leadership' (MIT Press, Cambridge, 1970)

JARVIS, A., 'The Liverpool Overhead Railway' (Ian Allan, 1996)

JENKIN, C. F., 'Single-Phase Electric Traction', *Proc. ICE* (London), Vol. 167, Nov. 1906, pp. 28–101

JOHNSON, R. P.: 'The steam locomotive' (Simmons-Boardman, 1942, 1981). Contains economic comparison between steam and diesel traction in USA

JOHNSON, J., and LONG, R. A.: 'British Railways Engineering 1948–1980' (Mechanical Engineering Publications, 1981)

JOSEPHSON, M.: 'Edison: A Biography' (New York, 1959)

JOY, S.: 'The train that ran away' (Ian Allan, 1973). Financial history of British Railways, with comment on traction policy and track costing by former chief accountant

KALLER, R. & ALLENBACH, J.-M.: 'Traction Electrique' (2 vols.) (Presses Polytechniques et Universitaires Romandes,Lausanne, 1995). Review of current railway and tramway practice

KEARNEY, C.: 'Erone' (The Eaton Press, 1943, 1950). A novel written by the inventor of the Kearney high-speed electric railway, who imagines a world transformed by engineering, and organised along rational, 'technocratic' lines

KEILTY, E.: 'Interurbans without wires' (Interurbans special, no. 66, Interurbans Press, Glendale, Ca., USA, 1979). Illustrated history of the railmotor car in the USA, from the McKeen cars of 1908 to 1951 trials with a rail bus

KENNEDY, R.: 'The book of modern engines and power generators' (Caxton, u.d but circa 1905, 6 vols). Thorough, very well illustrated review of contemporary wind, water, steam, oil and gas engines, and electric power equipment

KENNEDY, R.: 'The book of electrical installations' (Caxton, u.d. but circa 1912, 3 vols)

KENNEDY, R.: 'Steam turbines: their design and construction' (Whittaker, 1910)

KERR, C.: 'Why a turbine-electric steam locomotive', *Westinghouse Engineer*, Sept. 1947, pp. 130–132

KERR, D. M.: 'Signalling and Communications' (Section O3, Kempe's Engineers' Year Book, Miller Freeman, 2000), pp. 2427–2459

KICHENSIDE, G. M. & WILLIAMS, A., 'British Railway Signalling' (Ian Allan, 1980, 4th edn)

KICHENSIDE, G. M. & WILLIAMS, A., 'Two Centuries of Railway Signalling' (Oxford Publ. Co., 1998)

KIEFER, P. W.: 'A practical evaluation of railroad locomotive power', (Simmons-Boardman, NY, 1942, 1945)

KIEFER, P. W.: 'Railway power plant from the United States' point of view', *Proc. IMechE* (London), 12 June 1947, pp. 245–51

KITSON CLARK, E.: 'An internal combustion locomotive', *Proc. IMechE*, April 1927, pp. 333–98. The Kitson-Still locomotive

KRANZBERG, M. & PURSELL, C.: 'Technology in Western Civilization', 2 Vols. (Oxford, 1967)

KRON, G.: 'Diakoptics: the piecewise solution of large scale systems' (Macdonald, London, 1963)

LAMB, D. R.: 'Modern railway operation' (Pitman, London, 1941, 1927). Organisation of passenger and goods stations, marshalling yards, goods handling, motive power, rolling stock, signalling, traffic administration and railway-owned road services. Useful record of British steam railway between the wars.

LAMMING, C.: 'Cinquante ans de traction á la SNCF,' *CNRS Editions*, Paris, 1997

LANGDON, W.: 'On the supersession of the steam by the electric locomotive', *Jnl. IEE*, vol. XXX, 1900–01, pp. 124–231

LASCELLES, T. S.: 'The City and South London Railway' (Oakwood, 1955, 1987)

LASCHE, O.: 'High speed railway car of the AEG, Berlin', *J. Inst. Elect. Eng.*, 1901–02, 31, pp. 24–76

LEE, C. E.: 'The Metropolitan District Railway' (Oakwood, 1956)

LEE, T. R.: 'Turbines Westward' (T. Lee Publ., AG Press, Kansas, 1976). Photographic history of Union Pacific Railroad gas-turbine locomotives.

LEONARD, H. W.: 'A new system of electric propulsion', *Trans. Amer. Inst. Elect. Engrs.*, 1892, 9, pp. 566–77

LEONARD, H. W.: 'Discussion Contributions', *Trans. Amer. Inst. Elect. Engrs.*, 1892, 9, pp. 761–93. The discussion of the Ward Leonard paper 'A new system of electrical propulsion'

LEONARD, H. W.: 'How shall we operate an electric railway extending 100 miles from the power station?', *Trans. Amer. Inst. Elect. Engrs.*, 1894, 11, pp. 78–107

LEONARD, H. W.: 'Note on recent electrical engineering developments in France and England', *Trans. Amer. Inst. Elect. Engrs.*, 1895, 12, pp. 36–51

LEONARD, H. W.: 'Volts. v. ohms – speed regulation of electric motors', *Trans. Amer. Inst. Elect. Engrs.*, 1896, 13, pp. 377–86

LESLIE, S. W.: 'Boss Kettering: Wizard of General Motors' (New York, 1983)

LEUPP, F. E.: 'George Westinghouse. His life and achievements' (Murray, 1919)

LEWIS, L. P.: 'Railway Signal Engineering, (Mechanical)' (Railway Engineer, 1912. 3rd edn (revised by J. H. Fraser), 1932)

LEWIS, M. J. T.: 'Early wooden railways' (Routledge and Kegan Paul, 1970)

LILJEGREN, C. O.: 'Coal, oil or wind?', *Trans. Inst. Eng. & Ship. Scotland*, 1920–21, 64, pp. 242–309

LIND, A. R.: 'From horsecars to streamliners: an illustrated history of the St. Louis Car Co.' (Transport History Press, USA, 1978). Illustrated history of the railway equipment made by this famous carbuilder

LIND, A. R.: 'Chicago surface lines: an illustrated history' (Transport History Press, USA, 1974, 1979). History of the largest street railway system in the world and the feeder bus lines. 'Second edition supplement' (THP, 1979)

LINECAR, H. W. A.: 'British electric trains' (Ian Allan, 1947, 1949)

LOCOMOTIVE, CARRIAGE & WAGGON REVIEW: 'Ramsay condensing turbo-electric locomotive', *LCWRev.*, vol. 28, no. 356, 1922, pp. 92–93

LOCOMOTIVE, CARRIAGE & WAGGON REVIEW: '4-8-0 four-cylinder triple-expansion locomotive, Delaware & Hudson Railway', *LCWRev.*, 15 Aug. 1933, pp. 227–29

LOCOMOTIVE, CARRIAGE & WAGGON REVIEW: 'Recent developments in French steam locomotives', *LCWRev.*, 14 Aug. 1937, pp. 238–42

LOCOMOTIVE CYCLOPEDIA (USA): 'Train Shed Cyclopedia No. 20: diesel-electric locomotives 1925–1938', reprinted extracts from Locomotive Cyclopedia, Simmons Boardman, USA. (Newton K. Gregg, Novato, Ca., USA, June 1974)

LOMONOSSOFF, G. V.: 'Problems of railway mechanics', *Proc. IMechE.*, 1931, 120, pp. 643–59

LOMONOSSOFF, G. V.: 'Diesel traction', *Proc. IMechE*, 1933, pp. 537–613

LOMONOSSOFF, G. V.: 'Introduction to railway mechanics' (Oxford University Press, 1933)

LOMONOSSOFF, G.V., and LOMONOSSOFF, G.: 'Condensing locomotives', *Proc. IMechE*, London, vol. 152, 1945, pp. 275–303

LYDALL, F.: 'Electrification of the Pietermaritzburg-Glencoe section of the SAR', Reprint in booklet form of paper presented to IEE, 29 March 1928

MACHEFERT-TASSIN, Y. M.: 'La Traction Electrique a courant alternatif monophase a frequence industrielle et la material moteur a redresseurs,' *Revue S. W.*, 1958 (Reprint 1960), **12**

MACHEFERT-TASSIN, Y. M.: 'Les Automotrices et Locomotives a redresseurs au silicium S. W.'. (Ed. Le Materiel Electrique S. W., Schneider et Maury, Levallois, 1960)

MACHEFERT-TASSIN, Y. M., NOUVION, F., and WOIMANT, J.: 'Histoire de la traction electrique: Des origines a 1940' (La Vie du Rail, Paris, 1980), vol. 1

MACHEFERT-TASSIN, Y. M., NOUVION, F., and WOIMANT, J.: 'Histoire de la traction electrique: De 1940 a nos jours' (La Vie du Rail, Paris, 1986), vol. 2

MACLAREN, M.: 'The rise of the electrical industry during the nineteenth century' (Princeton, 1943)
MACLEOD, J.: 'The steam turbine locomotive', *Trans. Inst. Eng. & Shipbuilders in Scotland*, LXXIII, pt.II, Nov. 1929, pp. 49–96
MAJUMDAR, J.: 'Economics of railway traction' (Gower, 1985)
MANSON, A. J.: 'Railroad electrification and the electric locomotive' (Simmons Boardman, NY, 1925)
MARECHAL, H.: 'Le chemin de fer electriques' (Beranger, Paris and Liege, 1904)
MARSDEN, C. J.: 'A pictorial record of the diesel shunter' (Oxford Publ. Co., 1981)
MARSDEN, C. J.: '100 years of electric traction' (Oxford Publ. Co., 1985). Pictorial history of electric traction
MARTIN, H.: 'Production et distribution d'energie pour la traction electrique' (Beranger, Paris and Liege, 1902)
MASON, E.: 'The Lancashire and Yorkshire Railway in the 20th Century' (Ian Allan, 1954, 1974)
McGRAW ELECTRIC RAILWAY DIRECTORY, 1924: (Trolley Talk, USA, 1980). Reprint of 1924 directory of milages, routes, power house locations, repair shop sites, gauges, rolling stock and ownership for North America, Mexico and West Indies
McGUIGAN, J. H.: 'The Giant's Causeway tramway' (Oakwood, 1964)
MECHANICAL ENGINEERING (USA): 'Coal-fired steam-turbine-electric locomotive', *Mech. Eng. (USA)*, 1955, 77, pt.7, pp. 588–95
MELLANBY, A. L., and KERR, W.: 'The use and economy of high-pressure steam plants', *Proc. IMechE*, 1927, vol. 1, pp. 53–98
MERZ, C. H., and McLELLAN, W.: 'The use of electricity on the NER and upon Tyneside', *Brit. Assoc. Adv. Sci. (Cambridge)*, 22 Aug 1904, reprinted as booklet, Engineer/BAAS, 1904
MERZ, C. H., McLELLAN, W., and LIVESEY, SON and HENDERSON: 'Electric traction on the Central Argentine Railway', reprint of articles from *Railway Gazette*, Oct. and Nov. 1918
MERZ, C. H., McLELLAN, W., and LIVESEY, SON and HENDERSON: 'Electric traction on the Buenos Ayres Western Railway', reprint from *Railway Engineer*, Westminster, undated but circa 1924
MERZ, C. H., and McLELLAN, W., and VICTORIAN GOVERNMENT RAILWAYS: 'Electrification of the Melbourne suburban system', reprint in booklet form of articles taken from *Railway Gazette*, *Railway Engineer* and *Power User*. Undated, but circa 1926
MIDDLETON, W. D.: 'The interurban era' (Kalmbach, USA, 1961)
MIDDLETON, W. D.: 'The time of the trolley' (Kalmbach, USA, 1967)
MIDDLETON, W. D.: 'When the steam railroads electrified' (Kalmbach, USA, 1974)
MIDDLETON, W. D.: 'Grand Central' (Golden West, USA, 1977). History of Grand Central termini, including the electrification of the 'new' station using the Sprague system.
MILSTER (*Note*): *Jnl. of the International Stationary Steam Engine Society*, vol. 17, no. 3, 1995, p. 23
MOLLISON, J.: 'Historical references to the progress in the use of high pressure steam', *Trans. Inst. Eng. and Ship. Scot.*, vol. LXX, 1926, pp. 783–803

MOODY, G. T.: 'London's electrifications 1890–1923' (Electric Railway Soc., 1961)
MOODY, G. T.: 'London suburban railways' (Electric Railway Soc., 1963)
MOODY, G. T.: 'Southern Electric 1909–1979' (Ian Allan, 1979)
MORDEY, W. M. & JENKIN, B. M. 'Electric Traction on Railways', *Proc. ICE*, vol. 149, 1901–02, Pt. 3, pp. 40–199
MORGAN, D. P.: 'Diesel traction in North America' (Concise encyclopedia of world railway locomotives, Hutchinson, 1955), pp. 107–42
MOULTROP, I. E., and ENGLE, M. D.: 'Higher steam pressures for power generation', *Trans. Inst. Eng. and Ship. Scot.*, vol. LXXIII, 1930, pp. 402–55. Reviews American-lead in this subject
MUMFORD, L.: 'Technics and Civilisation' (Harcourt Brace, NY, 1934). A general history which accentuates the creative role of technology, and presents engineering as a cultural activity. Mumford introduces a classification system for history of technics based on the age of a particular practice
MUMFORD, L.: 'Pentagon of Power' (Secker & Warburg, 1971)
NEVINS, A., and HILL, F.: 'Ford: the times, the man, the company', vol. 1 (NY, 1954); 'Ford: expansion and challenge, 1915–1933', vol. 2 (NY, 1957)
NOCK, O. S.: 'The Locomotives of Sir Nigel Gresley' (Railway Publ. Co./Longmans Green, 1945, 1946)
NOCK, O. S.: 'Fifty Years of Railway Signalling' (Inst. Railway Signalling Eng./Ian Allan, 1962)
NOCK, O. S.: 'British Railway Signalling' (George Allan & Unwin, 1969)
NOCK, O. S.: 'The Gresley Pacifics' (Guild Publishing, London, 1985)
NOCK, O. S.: 'The Gresley Pacifics' (Guild/David and Charles, 1986, 1982). Application to British express locomotive design of findings of Goss, Fry and Woodard
NORTH, S. H. (Ed.): 'Britain's fuel problems' (HMSO, 1927). Collection of papers from *Fuel Economist*
O'BRIEN: 'Application of the electric locomotive to main line traction on railways', *JIEE*, vol. 58, 1920, pp. 858–69. Commentary in vol. 59, pp. 339–42.
ODUM, H. T. & ODUM, C. E.: 'Energy Basis for Man and Nature' (McGraw-Hill, 1976)
OGILVIE, J.: 'Should the third rail electrify the world's trains?', *New Scientist*, 26 Oct. 1978, pp. 273–75
PARODI, H.: 'Le Development Actuel de la Traction Electrique sur les Grands Reseaux de Chemins de Fer', *Revue Generale des Chemins de Fer*, January, 1920
PARSONS, R. H.: 'The development of the Parsons steam turbine' (Constable, London, 1936)
PARSONS, R. H.: 'The early days of the power station industry' (Cambridge University Press, 1939)
PARSONS, W. B.: 'New York Rapid Transit Subway', *Proc. ICE*, vol. 173, 1907–08, pp. 83–213
PARSHALL, H. F. & HOBART, H. M.: 'Electric Railway Engineering' (Constable 1907)
PASSER, H. C.: 'The electrical manufacturers' (Cambridge/Harvard, 1953)
PATCHELL, W. H.: 'Application of electricity to mines and heavy industries' (Constable, 1913)
PEARSON, F. K.: 'Isle of Man tramways' (David and Charles, 1970)

PERRY, J.: 'The steam engine and gas and oil engines' (Macmillan & Co., 1899, 1902, 1920)
PHILLIPSON, E.: 'The steam locomotive in traffic' (Locomotive Publishing Co., London, 1949) Running shed equipment, practice and organisation
PIERCE, J. R.: 'Signals' (Freeman, San Francisco, 1981). Growth of electrical and electronics communications technology. Contains brief history of Strowger switch
PINKEPANK, J. A.: 'Diesel spotters guide' (Kalmbach, USA, 1967)
PINKEPANK, J. A.: 'Second diesel spotters guide' (Kalmbach, USA, 1973) Comprehensive photographic catalogue of American diesel locomotive practice arranged by manufacturer: EMD, GE, ALCO, BLW, FM, LIMA and smaller makers
POLE, W.: 'Life of Sir William Siemens' (Murray, London, 1888)
POLITICAL AND ECONOMIC PLANNING PRESS: 'Locomotives: A report on the industry', (PEP, London, 1951). State of British locomotive building industry after World War Two
POLLINS, H.: 'Britain's railways: an industrial history' (David and Charles, 1971). Labour, financial, political and engineering aspects. A useful background to technical developments
POST, R. C.: 'The Page locomotive: federal sponsorship and invention in mid-nineteenth century America', *Tech. & Cult.*, 13, 2 April 1972, pp. 140–69
POTTER, S.: 'On the right lines? The limits to technological innovation' (Francis Pinter, London, 1987). Discusses innovation, successes and failures, within context of high-speed rail traction since 1930
POULTNEY, E. C.: 'Two remarkable locomotives', *Engineer*, 23 Oct. 1925, pp. 439–41; 10 June 1927, pp. 621–22. Woodard's class A1 and 8000 described, with test results.
POULTNEY, E. C.: 'Two large American locomotives', *Locomotive (LCWRev)*, 15 March 1929, pp. 98–100
POULTNEY, E. C.: 'Modern American express locomotives', *Locomotive*, 15 June 1929, pp. 196–200; 15 July, pp. 231–33; 14 Sept., pp. 291–95. Evolution of the post-Goss, Woodard-influenced type in the USA
PRIGMORE, B. J.: 'London Transport tube stock till 1939' (Electric Railway Society, 1960)
PURDAY, H. P. F.: 'Diesel engine design' (Constable, 1948)
PUTZ and BASTON: 'C&O turbine-electric locomotives', *Rly. Age*, vol. 123, no. 12, 20 Sept. 1947, pp. 48–51
PUTZ and BASTON: 'Power plants for the C&O locomotives', *Westinghouse Engineer*, Sept. 1947, pp. 132–35
QUIGLEY, H.: 'Electrical power and national progress' (George Allen and Unwin, 1925)
RAHN, T. (Ed.): 'ICE High-Tech. on rails' (Hestra-Verlag, Darmstadt, 1986). Description of German very-high-speed train and its electronics equipment
RAILWAY CORRESPONDENCE & TRAVEL SOCIETY: 'Locomotives of the LNER Part 10B: railcars and electric stock' (RCTS, 1990)
RAILWAY GAZETTE: 'A new steam turbine electric locomotive', *RG*, vol. 13, no. 3, 1910, pt. 2, pp. 72–74, 78
RAILWAY GAZETTE: 'The Shildon & Newport electrification' (article reprinted in booklet form) *RG*, undated but circa 1915

RAILWAY GAZETTE: 'Automatic signalling on the Central London Railway' (article reprinted in booklet form) *RG*, 12 June 1914

RAILWAY GAZETTE: 'The last word in steam operated suburban train services' (article reprinted in booklet form) *RG*, 1 Oct. 1920. Reprinted by Great Eastern Railway Society, London, 1984

RAILWAY GAZETTE: 'The Ramsay electro-turbo-locomotive', *RG*, vol. 39, no. 12, 1923, pt. 2, pp. 362–66

RAILWAY GAZETTE: 'Mercury arc rectifiers', *Railway Gazette Electric Traction Supplement*, 27 July 1934, pp. 172–76

RAILWAY GAZETTE: 'The fastest electric service in the world', *Railway Gazette Electric Traction Supplement*, 11 Jan. 1935a, pp. 70–75. Review of intensive American interurban operations with express passenger trains and container trolley freight

RAILWAY GAZETTE: 'Rectifier locomotives', *Railway Gazette Electric Traction Supplement*, 11 Jan. 1935b, pp. 78–80

RAILWAY GAZETTE: 'International Railway Congress Number', *RG*, London, 1954

RAILWAY GAZETTE INTERNATIONAL, *Review Issue*, October 1984. Reviews mid-1980s impact of computer and electronics on signalling, dispatching and control

RAILWAY LOCOMOTIVE (USA): 'N&W test turbine locomotive', *Railway Loco.*, vol 129, pt. 1, 1955, pp. 56–60. Norfolk & Western tests of steam-turbine-electric locomotive

RANDELL, W. L.: 'S. Z. de Ferranti' (Longmans Green, 1943)

RANSOME-WALLIS, P. (Ed.): 'The concise encyclopedia of world railway locomotives' (Hutchinson, 1955). Review of steam, oil-engined, and electric traction in the 1950s. The survey of the last generation of steam locomotives is excellent

RAPIER, R. C.: 'On the fixed signals of railways', *Proc. ICE*, London, 1874, vol. XXXVIII, Session 1873–74, part II, pp. 142–247. Reviews development of signalling and interlocking after 1830

RAVEN, V.: 'Railway electrification', *Trans. NE Coast Eng. & Shipbuilders*, Dec. 1921–22, vol. XXXVIII, pp. 173–240

RAVEN, V.: 'Electric locomotives', *Proc. IMechE* (Paris meeting 1922), pp. 735–81; 1057–82

RAVEN, V., and WATSON, H. A.: 'Railway electrification in the USA', Report submitted to board of North Eastern Railway, October 1919

RAWLINGS, D.: 'Control Centre of the Future gives operators the means to regulate effectively', *RGI*, Sept. 1994, pp. 583–588

RAYNER, B.: 'Southern electrics: a picture survey' (Bradford Barton, 1975)

READER, W. J.: 'Architect of air power: the life of the first Viscount Weir' (Collins, 1968). Very little on Weir's work on setting up the electricity grid, and advising on railway electrification. Concentrates on his role in developing RAF strategic bombing theory, and his role in Ministry of Munitions

RECK, F. M.: 'The Dilworth Story' (McGraw Hill, 1954). Biography of pioneer developer of diesel-electric locomotives

REED, R. C.: 'The streamline era' (Golden West Books, Ca., 1978)

REICH, L. S.: 'The making of American industrial research: science and business at General Electric and Bell, 1876–1926' (Cambridge, 1985)

REID, T.: 'Some early traction history', *Cassiers Mag.*, 16 August 1899, pp. 357–70
RENNIE, A.: 'The Still engine for marine propulsion', *Trans. Inst. Eng. Ship. Scot.*, vol. LXV, 1922, pp. 412–521
REPP, S.: 'Super Chief: train of the stars' (Golden West Books, USA, 1980). Contains description and illustration of first commercial passenger diesels and first Pullmann rolling stock designed for diesel haulage
REVUE GENERALE DES CHEMINS DE FER, numero 42f, Dunod, Paris, Septembre 1981. Special issue devoted to Paris-Lyon segregated route for TGV and its equipment
REVUE GENERALE DES CHEMINS DE FER: 'Le tunnel sous la manche', 12th December, 1993, 2nd February, 1994. Special issues devoted to the Channel Tunnel
RICHARDSON, A.: 'Evolution of the Parson's steam turbine' (Engineering, London, 1911)
ROBERTSON, K.: 'Leader: Steam's last chance' (Sutton, 1988)
ROBERTSON, K.: 'The Great Western Railway Gas Turbines' (Sutton, 1989) British design in 1870s.
ROGERS, H. C. B.: 'Chapelon' (Ian Allan, 1972)
ROGERS, H. C. B.: 'The last steam locomotive engineer: R.A. Riddles' (George Allen and Unwin, 1970)
ROHRBECK, B.: 'Johnstown Traction Company' (Rohrbeck, Pa., USA, 1976, 1980)
ROSTOW, W. W.: 'The Stages of Economic Growth' (Cambridge University Press, 2nd edn, 1971)
RUDORFF, D. W.: 'Steam generators' (Charles Griffin, 1938)
RUFFELL, B.: 'Track Circuits', *Jnl. of the Signalling Record Society*, No. 60, Nov./Dec. 1996, pp. 204–207
RUSSELL, J. H.: 'Great Western diesel railcars' (Wild Swan Publ., 1985)
SANKEY, H. R.: 'The thermal efficiency of steam engines', *Proc. Inst. Civil Eng.*, London, 1896, CXXV, pp. 182–242
SANKEY, H. R.: 'The thermal efficiency of steam engines', *Proc. Inst. Civil Eng.*, London, 1898, CXXXIV, pp. 278–312
SANKEY, H. R.: 'Energy chart: practical application to reciprocating steam engines' (Ruffell (Sig. Rec., D,
SCHMID, F. & KONIG, N.: 'Review of train mounted signalling and communication systems', *RGI*, Sept. 1995, p. 561; (Letters) June 1996, p. 331
SCHMIDT, T.: 'Dispatching advances will boost efficiency', *RGI*, Dec. 1997, pp. 863–865. Describes CSX Jacksonville operations centre
SCHRAMM, J. E., and HENNING, W.,H.: 'Detroit's street railways, vol. I, 1863–1922' (CERA Bulletin 117, Chicago, 1978)
SCHRAMM, J. E., HENNING, W. H., and DWORMAN, T. J.: 'Detroit's street railways', vol. II, (CERA Bulletin 120, Chicago, 1980). Illustrated history of Detroit electric street railways from 1922 to conversion to buses in 1956, including goods working and suburban extensions
SCHULTZ, R.: 'A Milwaukee transport era – the trackless trolley years' (Interurbans Special no. 74, Interurban Press, Glendale, Ca., USA, 1980). Illustrated history of Milwaukee trolleybus system

SCOTT, E. K.: 'Notes on electric traction by three-phase alternating currents', *J. Inst. Elect. Eng.*, 1899, 28, pp. 108–19

SEDDON, N.: 'The elements of electric transmission for diesel locomotives: a review of the fundamental considerations', *Jnl. Inst. Loco. Eng.*, vol. 47, pt. 5, 1957, pp. 492–520

SEMMENS, P.: 'A history of the Great Western Railway, Part 2: The Thirties' (Guild Publ. Co., London, 1985). Electrification project, pp. 81–87

SEMMENS, P & MACHEFERT-TASSIN, Y.: 'Channel Tunnel Trains: Channel Tunnel Rolling Stock and the Eurotunnel System' (Eurotunnel, 1994)

SERGEANT, W.: 'Letter: Crewless in Canada', *RGI*, Nov. 1996, p. 713

SHAW, J.: 'Equipment & Working: Results of the Mersey Railway under Steam and under Electric Traction', *Proc. I.C.E.*, Vol. 177, 1909–10, Pt.1, pp. 19–46

SIGNALLING STUDY GROUP: 'The Signal Box: A Pictorial History and Guide to Designs' (Oxford Publishing Co., 1986)

SIMMONS-BOARDMAN: 'Car and Locomotive Cyclopedia' (Simmons-Boardman, USA, 1980, 1985). Review of current North American rolling stock and list of manufacturers and suppliers

SIMMONS-BOARDMAN: 'Early diesel-electric and electric locomotives' (Rail Heritage Publ./Simmons-Boardman, 1983). Photographic record of main types in operation before the mid-1950s

SLOAN, D. B.: 'George Gibbs and E. Rowland Hill' (Newcomen Society of America, 1957). Biographies of major figures in the New York railway terminals electrification schemes

SMITH, R. H.: 'Electric traction' (Harper and Bros., London and NY, 1900, 1892)

SNELL, A. P.: 'Present and future development of the electricity supply', President's Address, Engineering Section, Annual Report, British Association for the Advancement of Science, Oxford Meeting, 1926, pp. 161–66

SNELL, J. F. C.: 'Distribution of Electrical Energy', *Proc. I.C.E.*, Vol. 159, 1904–5, pp. 143–254

SONE, S.: 'Ceaseless march of electronics transforms Japan's traction technology', *RGI*, March 1995, pp. 143–150

SPRAGUE, F. J.: 'The solution of the municipal rapid transit problem', *Trans. Amer. Inst. Elect. Engrs.*, 1888, 5, pp. 358–98

SPRAGUE, F. J.: 'The multiple-unit system for electrical railways', *Cassiers Mag.*, 16, 4, Aug. 1899, pp. 460–68

SPRAGUE, F. J.: 'Correspondence and discussion', *Jnl. IEE*, XXX, 1900–1, pp. 175–77. Sprague suggested that electric traction could not generally displace steam traction, and would be limited to multiple-unit operation

SPROUGHT, A. J.: 'Mail Rail: The Post Office Underground Railway', *Proc. Evolution of Modern Traction Meeting*, London, 14 Nov. 1992, pp. 134–70.

STAGNER, L.: 'The Ultimate Development', *Trains (USA)*, Aug. 1975, pp. 23–55

STAMP, T., and STAMP, C.: 'William Scoresby: Arctic Scientist', (Caedmon of Whitby Press, 1975), pp. 175–85. Early attempts to develop electromagnetic engines

STANGL, P. E.: 'High speed rail comes to the North East Corridor', *RGI*, Nov. 1996, pp. 726–728

STANIER, W. A.: 'Position of the Locomotive in Mechanical Engineering', *Proc. IMechE*, London, vol. 146, 1941, pp. 50–61

STEINHEIMER, R.: 'The electric way across the mountains' (Carbarn Press, Tiburon, Ca., 1980). Photographic record of the Chicago, Milwaukee, St Paul and Pacific RR HVDC electrification

STEPHENSON (CARBUILDERS) CO.: 'Electric railway cars and trucks, 1905'. Reprint of the John Stephenson Co. carbuilder's catalogue for 1905. Introduction by G. W. HILTON (Glenwood, USA, 1972)

STILLWELL, L. B.: 'The electrical transmission of power from Niagara Falls' (American Institution of Electrical Engineers, 23 August 1901), pp. 444–527

STOFFELS, W.: 'Lokomotivbau und Dampftechnik' (Birkhauser, 1976)

STOVER, J. F.: 'American Railroads' (University of Chicago, 1961). General economic history of US railways

SWETT, I. L.: 'Cars of Pacific Electric: Vol. 1: city and suburban' (Interurban special no. 28, Interurban Press, Glendale, Ca., USA, 1975). Illustrated record of all Pacific Electric city and suburban cars

SWETT, I. L.: 'Cars of Pacific Electric: Vol. 3: combos, locomotives & non-revenue cars', (Interurban special no. 37, Interurban Press, Glendale, Ca., USA, 1979). Locomotives and goods fleet

SWINBURNE, J.: 'Some limits in heavy electrical engineering', *Electrician*, 1901–02, L, pp. 274–76, 315–17, 344–46, 394–97. Led to controversy about entropy concept

SWINBURNE, J., and COOPER, W. R.: 'Problems of electric railways', *Jnl. IEE*, March/April 1902, pp. 972–1041

TAKEMURA, K.: 'JR East updates shinkansen control', *RGI*, Aug. 1996 pp. 499–501

TALBOT, F. A.: 'Electrical wonders of the world' (Waverley, undated but circa 1920, 3 vols). Popular, but accurate, informative and very well illustrated review of contemporary practice based on manufacturers' press releases and official photographs

TALBOT, F. A.: 'Railways of the world' (Cassel/Waverley, undated but circa 1920, 3 vols). As above

TAYLER, A. T. H.: '600/750V DC Electric & Electro-Diesel Locomotives of the Southern Railway and its Successors', *Trans. Newcomen Soc.*, vol. 68, 1996–97, pp. 231–64

TAYLOR, A. M.: 'Central station supply economics', *JIEE*, vol. 39, 1907, pp. 364–413

TEICH, W.: 'System technology: traction vehicle technics for all applications' (ABB Henschel, Mannheim, 1994)

TESSIER, M.: 'Traction electrique et thermo-electrique' (Editions scientifiques Riber, 1978)

THOMPSON, F. R., 'Electric transportation' (Int. Textbook Co., Scranton, USA, 1940)

THOMPSON, S. P.: 'Polyphase electric currents' (London, 1900)

THOMPSON, S. P.: 'Discussion of Swinburne and Cooper', *JIEE*, 1902, pp. 1028–31

TOUGH, J. M.: 'Passenger conveyors: an innovatory form of communal transport' (Ian Allan, 1971). Review of moving platforms, escalators, guided carriages

TREWMAN, H. F.: 'Railway electrification' (Pitman, London, 1924)
TROTTER, A. P.: 'Overhead Transmission Lines', *Proc. ICE*, vol. 169, 1906–7, pp. 183–267
TUFNELL, R. M.: 'The diesel impact on British Rail' (Mech. Eng. Publ., 1979)
TUFNELL, R. M.: 'The British railcar' (David and Charles, 1984)
TUFNELL, R. M.: 'Prototype locomotives' (David and Charles, 1985)
TWISS, B.: 'Managing technological innovation' (Longman, 1980)
URQUHART, J. W.: 'Dynamo construction; a practical handbook for the use of engineer-constructors and electricians in charge' (Crosby Lockwood, 1895, 2nd edn). Good review of contemporary practice
URWICK, L. & BRECH, E. F. L.: 'The Making of Scientific Management', (Pitman, London, vol. 1, 1951, vol. 2, 1953)
VICKERS, R. L.: 'DC electric trains and locomotives in the British Isles' (David and Charles, 1986)
VON URBANITZKY (translated from German 1885 edition, revised and enlarged by R. WORMELL and R. MULLINEUX WALMSLEY): 'Electricity in the service of man' (Cassell and Co., 1896, 1895, 1888, 1885). Well-illustrated review of electrical engineering equipment and its use
WEBB, B.: 'The British internal-combustion locomotive 1894–1940' (David and Charles, 1970)
WEBB, B.: 'English Electric main line diesel locomotives' (David and Charles, 1978)
WEBB, B.: 'The Deltic locomotives of British Rail' (David and Charles, 1982)
WEBB, B., and DUNCAN, J.: 'AC Electric locomotives of British Rail' (David and Charles, Locomotive Studies, 1979)
WEIR, J. G.: 'Some limiting conditions of external-combustion engine efficiency', *Trans. Inst. Eng. and Ship. Scot.*, 1923–24, vol. LXVII, pp. 317–58
WELLINGTON, A. M.: 'The economic theory of railway location' (John Wiley/ Chapman Hall, 1889, 1887). One of the greatest classics in engineering and economics which influenced all serious attempts to model the machine-ensemble and set it against a broader background of commercial and other activities
WESTING, F.: 'Penn. Station, its tunnels and side-rodders' (Superior Publ. Co., Seattle, 1978). Photographic record of PRR terminal electrification and the DD1 locomotives
WESTINGHOUSE BRAKE & SIGNAL COMPANY: 'A Century of Signalling' (Westinghouse, 1956)
WESTINGHOUSE ELECTRIC (USA): 'A history of the development of the single-phase system', *Westinghouse*, March 1929 (reprinted in *CERA Bulletin* **118**, 1979)
WESTINGHOUSE, G. W.: 'The electrification of railways: an imperative need for the selection of a system for universal use', Joint meeting of IMechE (London) and ASME, 29 July 1910, London. Summary in *Engineering*, 12 Aug. 1910, pp. 244–49
WESTWOOD, J (Ed.): 'Trains: Complete Book of Trains & Railways' (Octopus Books, Singapore, 1979), pp. 183–84
WHIPPLE, F. H.: 'The electric railway' (Detroit, 1889, reprinted by Orange Empire Railway Museum, Ca., USA, 1980)

WHITE, R. D.: 'Current Development in Traction Power Electronic Converters', *Proc. Evolution of Modern Traction seminar*, London, 19 Nov. 1994, pp. 16–37
WHYTE, A. G.: 'Electricity in locomotion' (Cambridge University Press, 1911)
WILKE, G.: 'Modern battery railcars', *Jnl. Inst. Loco. Eng.*, vol. 47, pt. 5, 1957, pp. 455–91
WILLIAMS, C. W., and CLARK, D. K.: 'Fuel – its combustion and economy' (Crosby Lockwood, 1891)
WILLIAMS, H.: 'APT: a promise unfulfilled' (Ian Allan, 1985)
WILSON, E., and LYDALL, F.: 'Electric traction' (E. Arnold, 1907 and 1908, 2 vols)
WILSON, E. H.: 'William Adams, 1823–1904', *Trans. Newcomen Society*, vol. 57, 1985–86, pp. 125–148. Developed the 4-4-0 suburban tank engine as used on the North London Railway, and the Circle lines, and the 4-4-0 express engine on the LSWR
WILSON, H. RAYNAR: 'Mechanical Railway Signalling' (Railway Engineer, 1900). Very thorough technical review of mechanical signalling: cabins, interlocking, levers, linkages and signals
WILSON, H. RAYNAR: 'Power Railway Signalling' (Railway Engineer, 1908). Detailed review of signalling instruments, automatic signalling and power frames for electro-pneumatic, low pressure pneumatic, electric, electro-mechanical and hydraulic systems
WINTER, P.: 'ETCS will end the day of the lineside signal', *RGI*, Sept. 1994, pp. 561–564; 'Pilot applications will prove ETCS concept', *RGI*, Aug 1996, pp. 487–490
WISE, G.: 'Willis R. Whitney, General Electric, and the origins of US industrial research' (New York, 1985)
WOODWARD, A. J., and CAIN, B. S.,: 'Design of the Union Pacific steam-electric locomotive, Parts 1 and 2', Mechanical Engineering (USA), 1939, 61, pt. 10, pp. 709–14; pt. 11, pp. 817–21
WOODWARD, G.: 'The Liverpool Overhead Railway: Innovation in Engineering', *Proc. of Merseyside Maritime Museum seminar*, 13 Nov. 1993, pp. 57–83
YEOMANS, K. A.: 'Ward Leonard drives: 75 years of development', *Electronics and Power*, April 1968, pp. 144–48

Index

AC power supply
  advantages 115
  HVAC *see* high-voltage alternating current
  main line electrification 321–38
  rectifiers 169
Advanced Passenger Train (APT) 300, 364–5, 386, 394, 399–407
AEG
  mercury-arc rectifiers 182, 248
  single-phase AC 123, 126–7, 139
  three-phase system 116, 119–20, 351
Alsthom Company
  GEC-Alsthom 305, 374, 411, 413
  locomotives 256, 389, 395
  mercury-arc rectifiers 256
  SLM flexible drives 326
American Locomotive Company (Alco)
  Alco-General Electric 241
  Alco-Westinghouse 136
  locomotives 131, 227–8, 236, 242
Amtrak
  'American Flyer' 411, 413
  North-East Corridor 363–4, 411
arc-rectifiers *see* mercury-arc rectifiers
Armstrong-Whitworth 155–7, 159, 211
Associated Electrical Industries (AEI)
  air-blast circuit breakers 326
  electric traction 215, 327–8, 333
Association of American Railroads 217, 288, 367
automatic electric railways
  Post Office Railway 185, 188, 381
  rapid-transit railways 185, 381
automatic electric railways *cont.*
  United States 293, 296–7
  Victoria Line 293–6, 362
automatic signalling
  Baker Street & Waterloo Railway 99
  City & South London Railway 96–8, 185
  District Railway 67–8, 99, 104, 190
  Great Northern & City Railway 69, 99

  innovation 98–105
  Lancashire & Yorkshire Railway 106
  Liverpool Overhead Railway 8, 77, 97–8, 105, 185
  London & South Western Railway (LSWR) 106–7, 190
  Mersey Railway 104–5
  Metropolitan Railway 99, 104
  North Eastern Railway 106
  semaphore signals 77, 190
  semi-automatic 1–2, 100, 105
  United States 108–9
automatic train control (ATC) 202, 359, 361–2
automatic train operation (ATO) 359–62, 377–8
automatic train protection (ATP) 369–82
  European Train Control System (ETCS) 376
  London Underground 362
  West Coast Main Line (WCML) 368
automatic train supervision (ATS) 361–2
automatic warning systems 298–300
  Hudd-Strowger apparatus 298
  main line electrification 297–300
  signal repeating automatic warning system (SRAWS) 299–300
  standard AWS system 298
  transponders 300

Bakerloo Line 99, 193
Baltimore & Ohio Railroad
  diesel-electric traction 231–2
  Howard Street Tunnel 37–8, 81, 88
  locomotives 37–8, 81, 83, 86, 88, 231–2
Beeching Reports (1963/1965) 210, 267, 284–5, 307, 325, 334, 385
Behn-Eschenburg, Hans 124, 143
Board of Trade 7, 53, 98
British Pneumatic Railway Signal Company 106
British Power Signalling Company Limited 195

British Railways
  Advanced Passenger Train (APT) 300,
    364–5, 386, 394, 399–407
  Beeching Reports (1963/1965) 210, 267,
    284–5, 307, 325, 334, 385
  cab-signalling 298–9
  colour-light signals 285
  Control Centre of the Future (CCF) 373, 380
  E1000/E2001 258–9, 280, 323–6, 329
  ECML *see* East Coast Main Line
  High Speed Train (HST) 279–80, 308, 338,
    386, 405–6
  locomotive workshops 307, 325–6
  single-phase AC 55, 79, 258, 274
  traction policy 211, 261–2, 277–82, 284,
    305–6, 308, 322
British Railways (London Midland Region)
  automatic warning systems 299
  electrification lobby 309
  WCML *see* West Coast Main Line
British Railways Modernisation Plan (1955)
  AC power supply 322
  signalling 287–8
  traction policy 211, 270, 284, 322
  third-rail distribution 63, 74, 262
British Thomson-Houston (BTH)
  control systems 67, 69
  germanium rectifiers 258
  locomotives 83, 215, 327
  mercury-arc rectifiers 176, 180–2, 257
  single-phase AC 127
British Transport Committee report (1951)
    261–2, 273
British Westinghouse Electric &
    Manufacturing 56, 58, 61, 78, 81
Brown-Boveri & Co.
  air-blast circuit breakers 259, 325–6
  asynchronous motors 341, 343–5
  BBC system 342–3, 345, 347
  Brown (C.E.L.) 36, 38
  four-quadrant controller 342
  gridless rectifiers 176
  locomotives 215, 248, 310
  mercury-arc rectifier locomotives 248
  single-phase AC 128–9
  SLM flexible drives 326
  steel tank rectifiers 175
  three-phase system 116–17, 119–20, 122,
    126, 341–4
Brush Electrical Engineering Company 23, 28,
    70, 309, 340
Burlington Northern & Sante Fe 352–3, 372–3
Butte, Anaconda & Pacific Railroad 122,
    140–3

cab-signalling 202, 297–9
Cascade Tunnel 91, 119–20, 131, 241
Central Electricity Board 263–5
Central Line 100, 102, 301, 303
Central London Railway 53, 67, 81–2, 99, 303

Central Railroad of New Jersey
    226–7
Central Train Control 187
centralisation
  control 65, 355–60
  power stations 53, 55, 207
  signal boxes 285–9
Centralised Traffic Control (CTC)
  absolute permissive block (APB) 363
  ACTS 363
  computers 359
  ECTS 363
  Fort Worth centre 359, 371–3, 380
  microprocessors 363–4
  multiplexing 360
  pulse-code technology 359
  solid-state electronics 360
  transponders 364
  United Kingdom 205, 358–60, 363
  United States 202–3, 358–9, 363,
    371–3
Channel Tunnel
  'Eurostar' 282, 349–50, 364, 394,
    397–8
  rapid-transit railways 13, 16, 30
Chicago Milwaukee, St Paul & Pacific Rail-
    road (Milwaukee Road) 145, 315
  HVDC 140–3, 171, 241, 255
  Milwaukee electrification 140–1, 171
  quill-drive locomotives 141–2
Chicago South Shore & South Bend Railroad
    179
City & South London Railway
  automatic signalling 96–8, 185
  coal-fired power stations 52
  electrification 32, 52, 54, 56, 65, 77, 96
  lock and block 97
  locomotives 81
  third-rail distribution 65
coal-fired power stations
  central generation 207
  City & South London Railway 52
  main line electrification 74
  Mersey Railway 60, 63
  Pearl Street (NYC) 15
  thermal efficiency xiii 39, 63, 74, 149, 211–12,
    283
colour-light signals 77, 98–109, 193–8, 205,
    285
computers
  before 1985: 360–2
  Centralised Traffic Control (CTC) 359
  continuous cab-signalling 299
  locomotive-mounted 364–7
  logic circuits 360
  MICAS-S 366
  open systems 375
  signalling 299, 360–3
  traction control 366
  United States 288–9, 359

control
  'Astree' system 376
  automatic control 381–3
  centralisation 65, 355–60
  complexity 374–5
  CTC *see* Centralised Traffic Control
  Integrated Electronic Control Centre 373
  integration of machine-ensemble 1–2
  Pinpoint system 380
  Post Office Railway 187–8
  signals *see* signalling
  telephony 203–4, 288
control systems
  dead man's handle 67, 69, 100
  Sprague-Thomson-Houston 67, 69
  Ward Leonard (H.) 36, 46–7, 132
converter locomotives
  rotary system 245–6, 248
  single-phase AC 124–5, 130–2
converters 171

Daft Electric Light Company
  Ampere locomotive 14, 16, 23, 25, 38
  electric railways 20, 23, 36, 81
  third-rail distribution 24
  trolley systems 24
Dawson, Philip 74, 127, 209, 218, 222–3, 265–6
DC power supply
  HVDC *see* high-voltage direct current
  LVDC *see* low-voltage direct current
Dell, Robert 188, 290–4, 359, 378
Dick, Kerr & Company 77–9
diesel-electric traction
  General Electric 93
  General Motors 93, 163, 226, 229–35, 255, 381
  Soviet Union 165
  United Kingdom 167, 277, 279–80, 322, 338
  United States 93, 163, 226–7, 229–35
District Railway
  automatic electric railways 295
  automatic signalling 67–8, 99, 104, 190
  capital costs 59
  'cut and cover' 32
  electrification 65–8, 77
  fourth-rail distribution 67
  LVDC 61, 66, 117
  power stations 33, 55, 62–3, 72
  rolling stock 66–9, 71
  track circuits 67
  Train Describer/Recorder Receivers 67–8, 204
  Yerkes (C.T.) 32–3, 66–7, 117
Docklands Light Railway 377
Durtnall, W.P. 36, 154, 162–3, 243

East Coast Main Line (ECML)
  Advanced Passenger Train 405
  diesel-electric traction 277, 279–80, 322, 338
  electric traction 349
  electrification proposed 275, 308, 325
Edison, Thomas Alva
  electric traction 13, 23, 27, 31, 61, 71
Edison Company
  DC power house 39, 130
  electric traction 36, 81
  power stations 15, 21, 46
Edison-General Electric 28, 30, 71
electric railways
  1890–1920: 95–113
  1900–20: 73–94
  1920–40: 115–47, 185–205, 207–24
  1940–70: 283–303
  1965–95: 385–418
  United States 11–21
electric signalling
  automatic *see* automatic signalling
  fog signals 102, 104, 108
  lights *see* colour-lights
electric telegraph 2–9
electric traction
  fuel efficiency 207–13
  general electrification xiii, 13, 27
  main lines *see* main line electrification
  thermal efficiency 51, 80, 87, 209
  tramways 13, 16, 18–20, 23–33, 51
electrification
  1900–20: 73–94
  1920–40: 207–24
  1965–95: 385–418
  capital costs 57, 59
  energy policy 209
  general electrification xiii, 13, 27
  locomotive working 81–5
  main lines *see* main line electrification
  nationalisation 63, 75, 144, 264
  return on capital invested 210, 220
  thermal-electric traction 149–68
  United Kingdom 51–72, 219–24, 261–82
electro-diesels 314, 343
Electro-Motive Company 159, 226, 230–1
elevated railways 13, 16, 30–1, 64–8, 95
English Electric
  air-cooled rectifiers 259–60, 330
  locomotives 167, 259, 276, 280–1, 308, 310, 329, 331, 335
  Morecambe/Heysham branch 79
  Southern Railway 312–13
  steel tank rectifiers 176, 180, 184, 258
European Rail Transport Management 374
European Train Control System (ETCS)
  automatic train protection (ATP) 376
  complexity 374
  standardisation 375
  three 'levels' 376
'Eurostar' 282, 349–50, 364, 394, 397–8

444 *Index*

Euston
  central control room 356
fail-safe systems
  automatic train protection (ATP) 362
  complexity 375
  signalling 284
  track circuits 98
Farmer, Moses 11, 71
Ferranti, Sebastian Ziani de 23
Ford, Henry 23, 71–2, 130
  motor-generator locomotives 130
fourth-rail distribution
  District Railway 67
  Great Northern & City Railway 69
  Lancashire & Yorkshire Railway 79, 222
  London & North Western Railway (LNWR) 63, 78
  London Underground 33, 66–7, 80, 222–3, 262
  Mersey Railway 53, 73
  Metropolitan Railway 33, 66–7, 80, 223
France
  automatic electric railways 381
  'Electric Rainhill' 127, 247
  electric traction 44–5
  gas-turbine-electric traction 394–5
  high-speed electric railways 282, 334, 349, 364, 389, 392–8, 410–11
  HVAC 255–6, 392, 395
  HVDC 122, 129, 139, 255–6, 258, 392
  industrial frequency 273
  LGV 395, 397, 411
  mercury-arc rectifier locomotives 255–6, 326
  mercury-arc rectifiers 176–7
  Midi Railway 123, 127, 141, 177, 247
  motor-converter locomotives 133
  motor-generator locomotives 132
  single-phase AC 123, 258, 273–4
  SNCF 132, 139, 256, 273–4, 338, 369, 376, 382, 388–9, 392–8
  synchronous motors 348–9
  TGV 282, 349, 364, 388–9, 410–11
frequency division multiplexing (FDM) 288, 360

Ganz 32–3, 66, 116–17, 120, 122–3, 137
Garbe-Lahmeyer principle 248
gas-turbine-electric traction
  France 394–5
  United Kingdom 258, 278, 280, 323, 329
gate turn off thyristors (GTO)
  choppers 352, 377
  Docklands Light Railway 377
  Germany 345, 347, 351
GEC-Alsthom 305, 374, 411, 413
General Electric
  Butte, Anaconda & Pacific Railroad 140

locomotive-mounted rectifiers 176
locomotives 36, 81, 88
mercury-arc rectifier locomotives 245, 249–53
motor-generator locomotives 55, 131
oil-electric traction 92, 225, 228, 236
petrol-electric traction 93, 157, 164, 166
silicon diode rectifiers 255
'Steamotive' project 236–40
steel tank rectifier 175
three-phase system 119
General Motors
  diesel-electric traction 93, 163, 226, 229–35, 255, 381
  Electro-Motive Division (EMD) 92–3, 229–36, 239, 353, 412, 416
  freight locomotives 233
  La Grange plant 228, 231, 236
  silicon diode rectifiers 255
  standardisation 44, 235
General Railway Signal Company 363, 381
germanium rectifiers
  British Thomson-Houston (BTH) 258
  development 333
  E1000/E2001 259
  Morecambe/Heysham branch 79, 256, 258, 322, 339
Germany
  Association for the Study of Electric Railways 38
  automatic electric railways 381
  Class 120: 345–7, 351
  colour-light signals 109
  Dessau-Bitterfeld scheme 126–7
  gate turn off thyristors (GTO) 345, 347, 351
  high-speed electric railways 9, 282, 336–7, 344, 347, 364, 376–7, 387, 407–10
  Hollenthal trials 133, 138, 247–9, 255, 273
  hydroelectric power stations 212
  industrial frequency 247, 273, 341
  InterCity Express (ICE) 9, 282, 344, 347, 349, 352, 364, 388, 394, 409–10
  mercury-arc rectifier locomotives 245, 247–9, 338
  phase-splitting locomotives 135
  single-phase AC 55, 123, 126, 247
  speed signalling 193
  three-phase system 41, 48, 116, 119–20
  Zossen-Marienfeld military railway 38, 48, 119–20, 345, 388
Global Positioning System (GPS) 380
Gramme, Zénobe Théophile
  dynamos 12, 42–3, 51
Great Eastern Railway
  American methods 73, 75, 215
  'Decapod' 30, 75

Great Eastern Railway *cont.*
  electrification 73–7
  Liverpool Street 73–7, 215, 257, 266, 271, 275, 336
Great Indian Peninsula Railway 215–16
Great Northern & City Railway
  automatic signalling 69, 99
  electrification 32, 69–71, 77
  fourth-rail distribution 69
  loading gauge 69, 71
  rolling stock 69–71
Great Northern Railway
  electrification 264
  track circuits 9
  wireless telegraphy 112
Great Northern Railway (USA)
  mercury-arc rectifier locomotives 249
  motor-generator locomotives 48, 119, 130, 132
  single-phase AC 119, 131
  three-phase system 91, 119–20
Great Western Railway (GWR)
  automatic warning systems 298
  colour-light signals 198
  electric telegraph 4–5, 223
  electrification proposals (1938) 266–7
  gas-turbine-electric traction 258, 278, 280, 323, 329
  route-setting/control 198
  track circuits 10, 203
Grime, T. 209–13, 217

Hamilton, H.L. 92–3, 160, 230–1
Heilmann, J.J.
  electric traction 35–7, 45, 120, 150, 166, 236, 247
  Heilmann thermal-electric locomotives 8001/8002 36, 43
  Brown-Gramme DC dynamo 42–3
  DC traction motors 42, 46, 154
  dual-mode tractor 39, 41
  electric transmission 39, 41
  Fusée Electrique (Electric Rocket) 36, 38–45, 116, 149–50, 153
Henschel 120, 341–52
High Speed Train (HST) 279–80, 308, 338, 386, 405–6
high-speed electric railways
  Channel Tunnel 282, 336, 345, 364
  experimental models 37–8
  France 282, 334, 349, 364, 389, 392–8
  Germany 9, 282, 336–7, 344, 347, 364, 376–7, 387, 407–10
  Italy 388
  Japan 334, 349, 364, 388, 390–2, 410, 413–14
  recent developments 407–18
  South Korea 414, 416
  Soviet Union 415
  United States 411–13
high-voltage alternating current (HVAC)
  advantages 321
  clearances 321
  France 255–6, 392, 395
  locomotives 258–9, 280, 323–5
  Morecambe/Heysham branch 273
  multiple-units 323
  staff training 323–5
  tap changing 327, 329, 331–2, 335, 337
  transductors 331
high-voltage direct current (HVDC)
  acceptance of standard 268–77
  Butte, Anaconda & Pacific Railroad 122, 140, 143
  Chicago, Milwaukee and St Paul Railway (St Paul Road) 140–3, 171, 241, 255
  converter substations 171
  France 122, 129, 139, 255–6, 258, 392
  impact of standard 275–7
  London Midland and Scottish Railway (LMS) 94, 180, 222
  main line electrification 315–19
  Manchester/Sheffield/Wath project 83, 271–2, 275, 283, 312, 315–19, 322
  overhead distribution 83
  Soviet Union 339
  trolley systems 122, 139
  United Kingdom 79, 83, 261, 268–77, 315–19
  United States 55, 122, 139–42
  Westinghouse Company 140
Hollenthal trials 133, 138, 247–9, 255, 273
Howard Street Tunnel 37–8, 81, 88
Huber-Stockar, Emil 124, 143, 230
Hungary
  Kando system 137–8, 340
  mercury-arc rectifier locomotives 338
hydroelectric power stations
  Butte, Anaconda & Pacific Railroad 140
  efficiency 220
  Niagara Falls 17, 19, 99, 115
  Switzerland 129, 212
  United States xiii, 17, 19, 140, 212

Immingham Tramway 74, 145
India 159, 215–16
induction motors 122, 133, 334
industrial frequency
  Europe 247, 262, 273, 341
  Hollenthal trials 247, 273
  Kando locomotives 117
  Krupp locomotives 135, 137
  mercury-arc rectifier locomotives 325
  Morecambe/Heysham branch 79, 322, 339
  Seebach-Wettingen locomotive 124
Inner Circle Line 66–7, 95, 117
Institution of Railway Signal Engineers 191, 370

interlocking
  automatic train protection (ATP) 375
  'Entrance-Exit' system (NX) 28, 285, 355
  first-class railways 4, 357
  level crossings 357–8
  powered operation 195–202
  relay interlocking 285, 288
  route-relay interlocking 200
  solid-state electronics 373–4
  Southern Railway 198–9
  'time and approach' 357
  United States 8, 202
internal combustion engine
  cheap electrification 168, 224, 229, 264
  electric railways 92–4
  hydraulic transmissions 162–3
  Soviet Union 164
  streamliner trains 93
  transmission question 157–68
interurban railways
  Brilliner 146–7
  development 27–8, 30
  HVDC 174
  LVDC 144–6
  overhead distribution 144, 174
  Presidents' Conference Car (PCC) 146–7
  trolley systems 144, 174
  United Kingdom 74, 145
  United States 144–7
Italy
  high-speed electric railways 388
  oil-electric traction 164, 166
  phase-splitting locomotives 120
  three-phase system 117–18, 120, 137, 150
  tilting trains 413–14

Japan
  high-speed electric railways 334, 349, 364, 388, 390–2, 410, 413–14
Jubilee Line 33, 379

Kando, Kalman 116, 120, 133, 137–9, 177, 340
Kennedy Report (1920) 261–2, 268
Krauss-Maffei 346, 351, 407

Lancashire & Yorkshire Railway
  AC power supply 55
  automatic signalling 106
  electrification 55, 60, 63, 73–4, 79, 143, 150
  fourth-rail distribution 79, 222
lever frames
  ground frames 200–1
  miniaturisation 101, 196–200, 290, 357
  Westinghouse electro-pneumatic 101
lineside electronic unit (LEU) 382
Liverpool Overhead Railway
  automatic signalling 8, 77, 97–8, 105, 185
  colour-light signals 77, 98
  electric traction 20, 32, 65, 77
  LVDC 53, 56, 61
Liverpool Street
  AC power supply 322–3
  Great Eastern Railway electrification 73–7, 215, 257, 266, 271, 275, 336
  Integrated Electronic Control Centre 373
  Shenfield line 257, 271, 290, 322
lock and block 6–8, 97–100
locomotive-mounted rectifiers
  General Electric 176
  mercury-arc see mercury-arc rectifier locomotives
  solid-state electronics 176, 255–60, 326
  Westinghouse Company 176–7
Lomonosoff, George W. 63, 161, 164, 211, 213, 217–19
London
  rapid-transit railways 13, 16, 20, 73
London & North Eastern Railway (LNER)
  colour-light signals 193
  electric traction 219, 221, 266, 272
  HVDC see Manchester/Sheffield/Wath project
  route-setting/control 200
London & North Western Railway (LNWR)
  electrification 55, 63, 72, 78–9, 322
  signal boxes 198
London & South Western Railway (LSWR)
  AC power supply 55
  automatic signalling 106–7, 190
  electrification 55, 63, 71, 73, 78, 80
  LVDC 63, 74, 78, 80, 127, 143, 222, 275
  power stations 55, 78
  third-rail distribution 78, 80, 143
  track circuits 10
London Bridge 63, 73, 79–80, 127, 199, 303, 355
London Brighton & South Coast Railway (LB&SC)
  electrification 59, 71, 73, 221
  LVDC 74, 221
  overhead distribution 74, 78–9, 221
  rolling stock 79, 275
  single-phase AC 78–9, 123, 127, 221, 275
  third-rail distribution 74, 79, 123
London Electric Railways 72, 80
London Midland and Scottish Railway (LMS)
  automatic warning systems 298
  colour-light signals 193–5
  diesel-electric traction 167, 340
  Hornchurch substation 180, 182
  HVDC 94, 180, 222
  LVDC 180, 222
  mercury-arc rectifiers 180
London Passenger Transport Board 80, 181, 212
London Underground
  AC power supply 55

Index   447

London Underground *cont.*
  Aldersgate signal frame 189, 286
  American methods 68, 85, 99
  automatic train protection (ATP) 362
  Camden Town control room 302
  colour-light signals 99–100, 102, 205
  District Line *see* District Railway
  electrification 28, 32, 64–5, 77
  fourth-rail distribution 33, 66–7, 80, 222–3, 262
  ground frames 200–1
  London Standard 261, 268
  LVDC 31
  mercury-arc rectifiers 180–4, 306
  power stations 33, 55, 170, 208
  route-setting/control 200
  telephony 203–4
  third-rail distribution 32, 63
  Train Describer/Recorder Receivers 67–8, 204, 289–93
  train stops 100, 102, 104, 300
Lots Road power station 33, 55, 170, 208, 257
low-voltage direct current (LVDC)
  AC power supply 55, 64, 121
  British Railways (Southern Region) 31, 63, 74
  City & South London Railway 53, 56
  converter substations 171
  District Railway 61, 66, 117
  Great Northern & City Railway 53
  interurban railways 144–6
  Lancashire & Yorkshire Railway 63, 73, 79
  Liverpool Overhead Railway 53, 56, 61
  locomotives 84, 309–15
  London, Brighton & South Coast Railway (LB&SC) 74, 221
  London & South Western Railway (LSWR) 63, 74, 78, 80, 127, 143, 222, 275
  London Midland and Scottish Railway (LMS) 180, 222
  main line electrification 85–91, 121, 309–15
  Mersey Railway 53, 61, 73
  Metropolitan Railway 61, 80, 222
  motor-converters 55
  New York Central Railroad 31, 80, 85–7, 104, 143, 149–50, 171
  Pennsylvania Railroad 80, 85–7, 89–90, 122–3, 143, 284
  rapid-transit railways 13, 35, 55, 61, 63, 87, 143, 222, 261
  Southern Railway 31, 63, 77–8, 127, 221, 265, 275, 283
  three-phase AC supply 55
  traction characteristics 115
  tramways 13, 31, 51, 121
  Waterloo & City Railway 53, 61, 78

Maglev project 388, 397, 411
main line electrification

AC power supply 321–38
  automatic control 381–3
  coal-fired power stations 74
  DC power supply 305–19
  HVDC 315–19
  LVDC 85–91, 121, 309–15
  signalling 106–9
  standardisation 150
  traffic density 262
  United Kingdom 257, 261–82, 305–19, 321–38
Manchester/Sheffield/Wath project
  closure 275, 319
  coal trains 271, 315
  completion 83, 271
  HVDC 83, 271–2, 275, 283, 312, 315–19, 322
  locomotives 272, 312, 315–19
mercury-arc rectifier locomotives 245–60, 321, 326, 338
mercury-arc rectifiers
  AEG 182
  British Thomson-Houston (BTH) 176, 180–2, 257
  Cooper-Hewitt valves 175
  excitron 177, 330, 332
  glass bulb 174, 176–8, 180, 184, 245, 259, 306, 324
  grid-control 175, 246
  ignitron 177, 250–3, 256, 326, 329–32, 338
  industrial development 174–7
  London Midland and Scottish Railway (LMS) 180
  London Underground 180–4, 306
  Metropolitan Vickers 257, 259, 329
  static 176–80, 245
  steel tank 174–6, 177, 180, 182, 184, 257–8, 326–7
  water cooled 175–7
Mersey Railway 53–60, 73, 104–5
  third-rail distribution 73
  Westinghouse (G.W.) 56, 64, 78
Merz & McLellan 141, 213, 215, 261, 263, 264, 266, 273
Metropolitan Railway
  automatic signalling 99, 104, 190
  British Westinghouse Electric & Manufacturing 81
  electrification 65–7, 71, 77, 80
  fourth-rail distribution 33, 66–7, 80, 223
  Great Northern & City Railway 71
  locomotives 81, 83–4, 91, 309, 312
  LVDC 61, 80, 222
  three-phase system 66
Metropolitan Vickers
  gas-turbine-electric traction 258, 323
  General Railway Signal 288
  locomotives 83–4, 215–16, 268–9, 272, 309, 312, 315, 323–4, 327

Metropolitan Vickers *cont.*
  mercury-arc rectifiers 257, 259, 329
  single-phase AC 79
  steel tank rectifiers 184
Midland Railway
  Morecambe/Heysham branch 79, 127, 150, 322
  overhead distribution 74, 222
  single-phase AC 74, 80, 127
  telephony 203
miniaturisation
  advantages 378
  'Entrance-Exit' system (NX) 285, 288, 355
  lever frames 101, 196–200, 290, 357
  semaphore signals 98–9
Morecambe/Heysham branch
  closure 79
  germanium rectifiers 79, 256, 258, 322, 339
  HVAC 273, 339
  industrial frequency 79, 322
  Midland Railway 79, 127, 150, 258, 322
  single-phase AC 79–80, 127, 150
motor-converter locomotives 53–5, 124, 130–4
motor-generator locomotives 48, 53–5, 130–2
multiple-units
  diesel-mechanical traction 224
  experiments 45
  HVAC 323
  LVDC 81
  Northern Line 81
  Sprague company 31, 36, 66

nationalisation
  electrification 63, 75, 144, 264
  research and development 218
  state ownership 217–19
  United Kingdom (1948) 63, 210, 218, 283
New York Central Railroad
  LVDC 31, 80, 85–7, 104, 143, 149–50, 171
  oil-electric traction 228, 241
  signalling 104
  substations 171
New York City (NYC)
  elevated railways 31, 95
  Grand Central 55, 99, 102, 104, 123, 143, 171
  power stations 15, 18
  rapid-transit railways 13, 30
  steam traction abolished 61, 86–7, 102, 265
New York New Haven and Hartford Railway
  electro-diesels 343
  enforced electrification 87
  mercury-arc rectifier locomotives 245, 250–1, 253
  motor-generator locomotives 131
  single-phase AC 31, 91, 123, 139, 143, 178
  third-rail distribution 139, 143, 178, 343

Norfolk & Western Railway 120, 135–6, 145, 240–1, 315
North British Locomotive Company 55, 151–2, 332, 340
North Eastern Railway
  automatic signalling 106
  capital costs 59
  electrification 74, 83, 91, 141, 143, 214–15, 223
  locomotives 91, 213, 268
  mineral railways 83, 143
  overhead distribution 83, 141
  petrol-electric traction 158
  third-rail distribution 74
  wireless telegraphy 112
Northern Line
  signalling 101, 289
  Train Describer/Recorder Receivers 293–4
nose-suspended motor 25, 28–9, 129, 164, 166

Oerlikon Company 38, 87, 116, 124–30, 143
oil-electric traction
  advocates 73
  early 20th century 149, 157–66
  General Electric 92, 225, 228, 236
  locomotive power stations 215
  New York Central Railroad 228, 241
  transmissions 160–1, 164
  United Kingdom 36, 210, 213, 215, 218–19, 264
overhead distribution
  electrification 74, 79, 83, 144, 174
  London, Brighton & South Coast Railway (LB&SC) 74, 78–9, 221
  Midland Railway 74, 222
  North Eastern Railway 83, 141, 143
  Pennsylvania Railroad 37, 108, 177

Pennsylvania Railroad
  automatic signalling 108–9, 202
  Centralised Traffic Control (CTC) 202
  colour-light signals 109
  DD1 locomotives 89–90
  diesel-electric traction 241
  GG1 locomotives 411
  LVDC 80, 85–7, 89–90, 122–3, 143, 284
  mercury-arc rectifier locomotives 249–53
  motor-generator locomotives 132
  New York City Tunnel Extensions 87
  overhead distribution 37, 108, 177
  power stations 284
  rolling stock 135
  signalling 99
  single-phase AC 108, 123, 135, 235
  static rectifiers 177
  third-rail distribution 87
petrol-electric traction 92–3, 36, 157–66
phase-splitting locomotives 120, 135–6

Piccadilly Line 181–2
pneumatic power
  electro-pneumatic signalling 100–2, 104–5, 188, 195–6
  Pneumatic Camshaft Modified (PCM) 297
  points operation 188
  Post Office Railway 185–6
  Westinghouse electro-pneumatic system 30, 58, 100, 104
polyphase distribution/supply 115, 121, 133, 246
Post Office Railway 185–8, 198–200, 381
power stations
  AC systems 53–5
  centralisation 53, 55, 207
  coal *see* coal-fired power stations
  District Railway 33, 55, 62–3, 72
  hydro *see* hydroelectric power stations
  London & North Western Railway (LNWR) 55, 78
  London & South Western Railway (LSWR) 55, 78
  London Transport 33, 55, 62, 72, 170, 208
  low systems efficiency 16
  Pennsylvania Railroad 284
  three-phase system 207
power supply
  see also AC power supply
  HVAC *see* high-voltage alternating current
  HVDC *see* high-voltage direct current
  LVDC *see* low-voltage direct current
  single-phase *see* single-phase AC
  standard systems 321–2
  standardisation 16, 63, 149, 155
  three-phase *see* three-phase system
  United Kingdom 51–3, 321–38
Presidents' Conference Car (PCC) 146–7
Pringle Report (1927) 261–2, 268

radio communications
  Australia 369
  United Kingdom 111–12, 288, 368–9
  United States 2, 368
  West Coast Main Line (WCML) 368–9
railcars
  diesel-mechanical traction 159
  India 159
  oil-electric traction 92, 236
  petrol-electric traction 158
Ramsay Condensing Locomotive Company 155, 211
rapid-transit railways
  automatic electric railways 185, 381
  Chicago 13, 16, 30
  elevated lines *see* elevated railways
  heavy-duty 13, 16, 20, 25, 28, 30, 51, 56
  London 13, 16, 20, 73
  LVDC 13, 35, 55, 61, 63, 87, 143, 222, 261
  New York City (NYC) 13, 30
  reformed operations 75–6

signalling 8, 95
third-rail distribution 31
United Kingdom 13, 20, 51–72
United States 13, 25, 30
Raven, Vincent 141, 209, 213–15
Raworth, Alfred W. 83, 310
Raworth, John S. 83
Raworth-Bulleid locomotive 268, 278, 310–11
rectifiers
  AC power supply 169
  air-blast 176
  arc *see* mercury-arc rectifiers
  germanium *see* germanium rectifiers
  Graetz Bridge 169
  locomotive *see* locomotive-mounted rectifiers 176
  power supply rectification 169–74
regenerative braking
  mercury-arc rectifier locomotives 247
  substations 178
  three-phase system 116, 135
Replacement Automatic Driver Box (RADB) 362
reversible motor-dynamo 12
reversible working 363–4
route-setting/control 198–200, 355–8, 373
Russia *see* Soviet Union

St Gotthard line 43
Secheron Company 139
SEL-Alcatel 378, 381
semaphore signals
  see also signalling
  automatic signalling 77, 190
  colour codes 8, 190
  miniaturisation 98–9
Shildon/Newport electrification 315, 317
Siemens, Ernst Werner von
  DC traction motor 36
  electric traction 13, 27, 71, 72
Siemens & Halske
  electric motor cars 38
  locomotives 25, 71, 81, 126, 388
  three-phase system 119–20, 388
Siemens Company
  automatic train protection (ATP) 376
  computer-based interlocking 374
  locomotives 248, 407
Siemens-Schuckert
  motor-converter locomotives 133–4
  steel tank rectifiers 175
  water cooled rectifier 175
signal boxes 98–104, 189, 192, 197–201, 285–90, 357–60
  London Underground 101, 103, 189, 192, 197, 201, 289
  One Control Switch (OCS) 200, 285, 287
signalling
  1890–1920: 95–113

signalling *cont.*
  1920–40: 185–205
  1940–70: 283–303
  automatic *see* automatic signalling
  British Railways Modernisation Plan (1955) 287–8
  cab-signalling 202, 297–9
  colour-light *see* colour-light signals
  computers 299, 360–3
  electro-pneumatic systems 100–2, 104–5, 188, 195–6
  fail-safe systems 284
  general trends 300–3
  Great Central Railway (GCR) 106, 190
  large-scale integration 370–7
  lights *see* colour-light signals
  lineside signals eliminated 368, 377
  main line electrification 106–9
  manual-mechanical 95–6, 196, 284–5, 355
  pioneer systems 96–8
  post-war reconstruction 283
  powered 106
  pulse-coded electronics 202–3
  routes *see* route-setting/control
  semaphore *see* semaphore signals
  solid-state electronics 355–83
  speed signalling 191, 193–5, 357
single-phase AC
  AEG 123, 126–7, 139
  Austria 123
  British Railways 55, 79, 258, 274
  British Thomson-Houston (BTH) 127
  Brown-Boveri & Co 128–9
  commutator motors 122–4, 126–7, 129, 137
  converter locomotives 124–5, 130–2
  France 123, 258, 273–4
  Germany 55, 123, 126, 247
  Great Northern Railway (USA) 119, 131
  induction motors 122
  industrial frequency *see* industrial frequency
  London, Brighton & South Coast Railway (LB&SC) 78–9, 123, 127, 221, 275
  Metropolitan Vickers 79
  Midland Railway 74, 80, 127
  Oerlikon Company 124, 127–8, 130, 143
single-phase AC *cont.*
  Pennsylvania Railroad 108, 123, 135, 235
  Scotch Yoke mechanical drive 127–8
  suburban railways 222
  Switzerland 55, 119, 123–8, 143
  traction system development 121–6
  WCML *see* West Coast Main Line (WCML)
  Westinghouse Company 31, 122
solid-state electronics
  air-cooled rectifiers 259–60, 330
  Centralised Traffic Control (CTC) 360
  control systems 339–41
  interlocking 373–4
  locomotive drive systems 347–50
  locomotive-mounted inverters 120
  locomotive-mounted rectifiers 176, 255–60, 326
  microprocessors 352, 354, 363–4
  signalling 355–83
  silicon rectifiers 255, 259, 329–40
South African Railways 143, 266, 348
South Eastern & Chatham Railway (SE&CR) 9, 74, 78, 193
South Eastern Railway 5, 7, 73, 112
Southern Railway
  AC power supply 55
  colour-light signals 193–9
  electric signalling 285
  English Electric 312–13
  London Standard 74, 77, 80, 143, 221–3, 265, 268
  LVDC 31, 63, 77, 127, 221, 265, 270, 275, 283
  third-rail distribution 63, 74, 77, 123, 223, 265, 309
  Train Describer/Recorder Receivers 205, 289
Soviet Union
  electrification 220
  high-speed electric railways 415
  HVDC 339
  internal combustion engine 164–5, 280
  three-phase system 341
Spagnoletti, C.E.P. della Diana 5–6, 96–9
Sprague, Frank Julian
  electric traction 13, 20, 27, 35–8, 64, 71, 81
  elevated railways 56, 64–5
  heavy-duty electrification 16, 28, 36
Sprague company 23–31, 36, 46, 66, 80, 91, 164
standardisation
  colour-light signals 7–8, 191
  European Train Control System (ETCS) 375
  General Motors 44, 235
  LVDC 78, 80
  main line electrification 150
  power supply 16, 63, 149, 155
stationary power sources, Grove batteries 12
steam railcars, mechanical transmission 92
steam traction
  20th century improvements 63, 86, 166, 209–17
  British Railways 261, 270, 277–9, 305
  cab-signalling 202, 297–8
  elevated railways 30
  Great Eastern Railway 30, 75–7
  London Underground 64–5, 67
  Mersey Railway 56
  New York City (NYC) 61, 86–7, 265
  steam motor 92
  thermal efficiency 16, 39, 51, 74, 87, 210, 212, 217

Index 451

steam turbines
　coal-fired power stations 149–50, 284
　'electro-Turbo-Loco' 151–5, 211
　Parsons impulse turbine 153
　steam turbine electric locomotives 149
　'Steamotive' project 236–40
steam-electric traction, Heilmann *see*
　　Heilmann thermal-electric locomotives
Sulzer 166, 280, 309
supply networks 16, 18, 63, 223
swing-nose crossing 417
Switzerland 55, 91, 117–30, 143, 151
　hydroelectric power stations 129, 212
　Swiss Locomotive and Machine Works
　　(SLM) (Winterthur) 38, 161, 326, 351
　tilting trains 413–14

telegraph and train order (T&TO), United
　　States 2, 110
telegraphy
　electric *see* electric telegraph
　wireless *see* wireless telegraphy
telephony
　control 203–4, 288
　omnibus system 112, 204
　railway communications 112–13, 203–4, 288
　selective traffic control (STC) 113, 204
　signal post instruments 112, 204, 288
　United States 8
　wireless telephony 2, 109, 111–12, 288
thermal efficiency
　coal-fired power stations xiii, 39, 63, 74, 149,
　　211–12, 283
　electric traction 51, 80, 87, 209
thermal-electric traction 149–68
　'Electro-Turbo-Loco' 151–5, 211
　Heilmann *see* Heilmann thermal-electric
　　locomotives
　Ramsay-Armstrong Whitworth 155–7, 211
third-rail distribution 24, 31, 63–80, 123, 143,
　　178, 222, 262–5, 322, 343
Thompson, Silvanus Phillips 36, 47–8
Thomson, Elihu, electric railways 20, 23, 71
Thomson-Houston Electric Company
　BTH *see* British Thomson-Houston
three-phase system 115–26, 135, 150–4, 345,
　　388
　Ganz 32–3, 66, 137
　Germany 41, 48
　Great Northern Railway (USA) 91, 119–20
　phase-splitting *see* phase-splitting
　　locomotives
　power stations 207
　regenerative braking 116, 135
　Zossen-Marienfeld military railway 48,
　　119–20, 345, 388
thyristors
　gate turn off thyristors (GTO) 345, 347,
　　351–2, 377
　inverters 347

silicon-rectifier locomotives 255, 336
switching power 347
tilting trains 300, 364–5, 386, 394, 399–407,
　　411–14
time division multiplexing (TDM) 288, 360
track circuits
　automatic train protection (ATP) 362
　District Railway 67
　fail-safe systems 98
　first use 8–10, 98
　Great Western Railway (GWR) 10, 203
　track circuit block (TCB) 285
　United States 8, 10, 98
traction policy
　British Railways 211, 261–2, 277–82, 284,
　　305–6, 308, 322
　economics 135, 209, 277–8, 308
　universal flexible tractor 41, 345, 351
Traffic Control System (TCS), United States
　　202
Train Describer/Recorder Receivers 67–8, 199,
　　204–5, 289–94
train service management (TSM) 380
train stops
　London Underground 100, 102, 104, 300
　Mersey Railway 105
　Pennsylvania Railroad 202
tramways
　electric traction 13, 16, 18–20, 23–33, 51
　HVDC 139
　'system' 28–33
　United States 18–19, 23–32, 51, 77
Trans European Express 389, 392
transmission lines
　efficient transformers 53
　United States xiii
transponders 300, 364–5, 368–9
trolley systems
　Daft Electric Light Company 24
　HVDC 122, 139
　interurban railways 144, 174
　Pennsylvania Railroad 37, 177
　Sprague company 24, 28, 46, 91
　three-phase system 116–17, 120
　van de Poele 24

underground railways
　heavy-duty 13, 16
　London *see* London Underground
Union Electric Signal Company 9
Union Pacific Railroad
　automatic signalling 109
　Harriman Dispatch Center 371
　'Steamotive' 237–8
Union Switch & Signal Company 9, 202, 371,
　　373
United Kingdom
　AC traction 79
　amalgamation/grouping (1922) 55, 63, 78,
　　80, 209, 219, 223

United Kingdom *cont.*
  BR *see* British Railways
  Centralised Traffic Control (CTC) 205, 358–60, 363
  communications and control 203–4
  diesel-electric traction 167, 277, 279–80, 322, 338
  diesel-mechanical traction 157, 159
  electrification 51–72, 219–24, 261–82
  gas-turbine-electric traction 258, 278, 280, 323, 329
  HVDC 79, 83, 261, 268–77
  main line electrification 257, 261–82, 305–19, 321–38
  mercury-arc rectifier locomotives 256, 258–60, 326
  mercury-arc rectifiers 176, 180–4
  nationalisation (1948) 63, 210, 218, 283
  oil-electric traction 36, 210, 213, 215, 218–19, 264
  petrol-electric traction 36, 158–9
  pneumatic power 185–6
  power supply 51–3
  radio communications 111–12, 288, 368–9
  rapid-transit railways 13, 20, 51–72
  speed signalling 193–5
  supply networks 63, 223
  Train Describer/Recorder Receivers 204–5
  wireless telegraphy 112
United States
  AC power supply 55
  American Train Control System (ATCS) 369, 371
  automatic electric railways 293, 296–7
  automatic signalling 108–9
  automatic train control (ATC) 202
  Centralised Traffic Control (CTC) 202–3, 358–9, 363, 371–3
  computers 288–9, 359
  diesel-electric traction 93, 163, 226–7, 229–35
  electric railways 11–21
  Electronically Controlled Pneumatic Brakes (ECP) 367
  elevated railways 28, 30–2, 64–5, 68, 95
  general electrification 13, 27
  high-speed electric railways 411–13
  HVDC 55, 122, 139–42
  hydroelectric power stations xiii, 17, 19, 140, 212
  interurban railways 144–7
  mercury-arc rectifier locomotives 245, 249–54, 326
  mercury-arc rectifiers 176–80
  overhead trolley *see* trolley systems
  petrol-electric traction 93, 157
  phase-splitting locomotives 120, 135–6
  rapid-transit railways 13, 25, 30

reversible working 363–4
'road-switcher' 340
semaphore signals 4, 190–1
signalling practice 202–3
speed signalling 191, 193, 357
supply networks 16, 18
telegraph and train order (T&TO) 2, 110
telephony 8
tilting trains 411
time interval system 3
track circuits 8, 10, 98
Traffic Control System (TCS) 202
tramways 18–19, 23–32, 51, 77
wireless telegraphy 109–11

van de Poele 20–8, 71
Victoria Line
  automatic electric railways 293–6, 356–62
  Train Describer/Recorder Receivers 293
Virginian Railway 120, 132, 135, 249, 252–4
vital (safety) circuits 98, 357, 359, 370

Ward Leonard, H. 36, 46–7, 124–32
Warder, S.B. 258, 262, 268, 273–4, 278, 321, 322, 331, 339
Waterloo & City Railway 32, 78–85
Weir Report (1931) 261–8
West Coast Main Line (WCML)
  Advanced Passenger Train (APT) 400–7
  automatic train protection (ATP) 368
  clearance requirements 321, 335
  mercury-arc rectifier locomotives 256–9
  radio communications 368–9
  signalling 285, 300, 368
Westinghouse, George W.
  electric traction 56, 61, 64
  electro-pneumatic system 30, 58, 100, 104
  Mersey Railway 56, 64, 78
  signalling 9, 56
  three-phase system 115–16, 123, 135
Westinghouse Brake & Signal Company 33, 98–9, 369
Westinghouse Company
  AC equipment 87
  automatic signalling 100, 104, 105
  electric traction 31, 56
  HVDC 140
  locomotive-mounted rectifiers 176–7
  main line electrification 31
  mercury-arc rectifier locomotives 245, 250–1, 253
  silicon diode rectifiers 340
  single-phase AC 31, 122, 143
  steel tank rectifier 175
wireless telephony 2, 109, 111–12, 288

Yerkes, Charles Tyson 32–3, 66–7, 117, 145